12th International Scientific Conference of Civil and Environmental Engineering for PhD. Students and Young Scientists

IOP Conference Series: Materials Science and Engineering
Volume 867

High Tatras, Slovakia
15 - 16 October 2020

ISBN: 978-1-7138-3619-3
ISSN: 1757-8981

Printed from e-media with permission by:

Curran Associates, Inc.
57 Morehouse Lane
Red Hook, NY 12571

Some format issues inherent in the e-media version may also appear in this print version.

Institute of Physics
Dirac House, Temple Back
Bristol BS1 6BE UK

Phone: 44 1 17 929 7481
Fax: 44 1 17 920 0979

techtracking@iop.org

Additional copies of this publication are available from:

Curran Associates, Inc.
57 Morehouse Lane
Red Hook, NY 12571 USA
Phone: 845-758-0400
Fax: 845-758-2633
Email: curran@proceedings.com
Web: www.proceedings.com

TABLE OF CONTENTS

IOP Conf. Series: Materials Science and Engineering **867** (2020) 011001 doi:10.1088/1757-899X/867/1/011001

PREFACE

The 12th International Scientific Conference of Civil and Environmental Engineering for the PhD. students and young scientists below 35 years under auspices of the dean of the Civil Engineering Faculty of the Technical University of Košice assoc. prof. Peter Mésároš was planned to be held on 22nd – 24th April 2020 but due to known circumstances was postponed. The postponed conference is planned to be held on 15th - 16th of October 2020 in Štrbské Pleso, the High Tatras, Slovakia. This traditional meeting of PhD. students, their supervisors and young scientists working in the field of civil engineering and environmental science is organised by the Faculty of Civil Engineering of the Technical University of Košice, Slovakia. The Young Scientist 2020 offers an international platform for the dissemination of original research results; provides a pleasant environment to present new, great ideas of young researchers and discovered advances in the field of civil engineering and related interdisciplinary topics. It is an opportunity for PhD. students and young researchers to discussed scientific topics and research results of topical issues in civil engineering, to share their knowledge and experiences, to collaborate in innovative designs and proposals. In addition, the valuable presentations are usually extended by a social program including the interesting field trip in the High Tatras.

The Conference usually assembles more than 60 delegates, from the Slovak Republic (Technical University of Košice, Slovak University of Technology in Bratislava), the Czech Republic (Czech Technical University in Prague; Brno University of Technology, Mendel University in Brno, VSB-Technical University of Ostrava), from the Croatia (Josip Juraj Strossmayer University of Osijek) and from the Poland (Cracow University of Technology of Cracow) including PhD. students and other representatives from the universities. The scientific contributions involved a wide range of research fields: Building Physics and Services, Sustainable Architecture and Energy Efficiency, Statics, Dynamics and Modeling, Structural Engineering and Bridges, Transport and Geotechnical Engineering, Construction Technology, Organization and Management, Construction Economics, Marketing and Management, Building Information Modeling, Innovations in Construction Design and Technology, Environmental Engineering, Healthy Indoor Environment, Material Engineering and Recycling as well as Sustainable Water Management.
Forty contributions have been selected to be published in the volume of Journal IOP Conference Series: Materials Science and Engineering. The organizing committee appreciates the enormous work of the both the editorial board and reviewers in the preparation of this volume. Many thanks also to all authors for their flexibility and meeting deadlines.

assoc. prof. Martina Zeleňáková
assoc. prof. Peter Mésároš
assis. prof. Vlasta Ondrejka Harbuľáková

IOP Conf. Series: Materials Science and Engineering **867** (2020) 011002 doi:10.1088/1757-899X/867/1/011002

Delegates

Young Scientist 2020, October 15 - 16, 2020, High Tatras, Štrbské Pleso, Slovakia

1. assoc. prof. Ing. Mésároš Peter, PhD.
2. assoc. prof. Ing. Zeleňáková Martina, PhD.
3. prof. Ing. Katunský Dušan, PhD.
4. RNDr. Krajníková Katarína, PhD.
5. assoc. prof. Ing. Kápossztásová Daniela, PhD.
6. assoc. prof. Ing. Václavík Vojtěch, PhD.
7. prof. Dr. Ing. Šlezingr Miloslav
8. Ing. Ondrejka Harbuľáková Vlasta, PhD.
9. Ing. Bačuvčík Michal, PhD.
10. Ing. Vacenovská Veronika, PhD.
11. Ing. Dvorský Tomáš, PhD.
12. Ing. Hrudka Jaroslav, PhD.
13. Ing. Bizoňová Silvia
14. Ing. Bokomlaško Jakub
15. Ing. Čákyová Katarína
16. Ing. Charvát Jakub
17. Ing. Csicsaiová Réka
18. Ing. Doubek Rostislav
19. Ing. Figmig Róbert
20. Ing. Gašpárek Jakub
21. Ing. Harčárová Katarína
22. Ing. Húlek Lukáš
23. Ing. Jaroš Pavol
24. Ing. Jeleč Mário
25. Ing. arch. Kabošová Lenka
26. Ing. Kapolka Lukáš
27. Ing. Kardošová Erika
28. Ing. arch. Karľa Viktor
29. Ing. Klosova Daria
30. Ing. Košičan Jaroslav
31. Ing. Kováč Maroš
32. Ing. Kováčová Zdenka
33. Ing. Kovács Róbert
34. Ing. Kowalik Tomasz
35. Ing. Kvapilova Vendula
36. Ing. Kováč Martin
37. Ing. Lavko Martin
38. Ing. Lukić Sanja
39. Ing. Macalová Kateřina
40. Ing. Marko Ivana
41. Ing. Marton Michal
42. Ing. Mokrenko Daria
43. Ing. Moňoková Andrea
44. Ing. Nagy Patrik
45. Ing. Nechyporchuk Yaroslav
46. Ing. Prokopowicz Piotr
47. Ing. Repel Adam
48. Ing. Rózsa Gergely
49. Ing. Smetanková Jana
50. Ing. Sobstyl Patrycja Marlena
51. Ing. Stuchlík Jakub
52. Ing. Štulerová Lenka
53. Ing. Sturkar Kristina
54. Ing. Šutúš Marek
55. Ing. Svoboda Jakub
56. Ing. arch. Tomková Želmíra
57. Ing. Vaňová Patrícia
58. Ing. arch. Wawrek Igor
59. Ing. Wolfová Martina
60. Ing. Zenah Jalal

IOP Conf. Series: Materials Science and Engineering **867** (2020) 011003 doi:10.1088/1757-899X/867/1/011003

EDITORS

ZELEŇÁKOVÁ Martina	Technical University of Košice
MÉSÁROŠ Peter	Technical University of Košice
ONDREJKA HARBUĽÁKOVÁ Vlasta	Technical University of Košice

TECHNICAL EDITORS

KARĽA Viktor	Technical University of Košice
ŠTULEROVÁ Lenka	Technical University of Košice

SCIENTIFIC COMMITTEE

MÉSÁROŠ Peter	Dean of the Technical University of Košice, Faculty of Civil Engineering, Slovakia
ZELEŇÁKOVÁ Martina	Vice-Dean of the Technical University of Košice, Faculty of Civil Engineering, Slovakia
KATUNSKÝ Dušan	Vice-Dean of the Technical University of Košice, Faculty of Civil Engineering, Slovakia
VÁCLAVÍK Vojtěch	VSB - Technical University of Ostrava, Czech Republic
BAJER Miroslav	Brno University of Technology, Faculty of Civil Engineering, Czech Republic
KOZLOVSKÁ Mária	Technical University of Košice, Faculty of Civil Engineering, Slovakia
LEŚNIAK Agnieszka	Faculty of Civil Engineering, Cracow University of Technology, Poland
VAREVAC Damir	Josip Juraj Strossmayer University of Osijek, Faculty of Civil Engineering in Osijek, Croatia
TOMKO Michal	Technical University of Košice, Faculty of Civil Engineering, Slovakia
VILČEKOVÁ Silvia	Technical University of Košice, Faculty of Civil Engineering, Slovakia
VRANYOVÁ Zuzana	Technical University of Košice, Faculty of Civil Engineering, Slovakia

ORGANISING COMMITTEE

ONDREJKA HARBUĽÁKOVÁ Vlasta
KARĽA Viktor
KLOSOVA Daria
LAVKO Martin
ŠTULEROVÁ Lenka

REVIEW STATEMENT

All papers published in this volume of *IOP Conference Series: Materials Science and Engineering* have been peer reviewed per two reviewers through processes administered by the Editors. Reviews were conducted by expert referees to the professional and scientific standards expected of a proceedings journal published by IOP Publishing.

IOP Conf. Series: Materials Science and Engineering **867** (2020) 011003 doi:10.1088/1757-899X/867/1/011003

LIST OF REVIEWERS

BAJZECEROVÁ Viktória
BAŠKOVÁ Renáta
BÁLINTOVÁ Magdaléna
BŘENEK Aleš
CIGÁNEK Jiří
DEMČÁK Štefan
DRANAMIĆ Hrvoje
DRTIL Miroslav
EŠTOKOVÁ Adriana
HARABINOVÁ Slávka
HRUDKA Jaroslav
JANOTKA Ivan
JUNÁK Jozef
KANTOVÁ Radka
KAPALO Peter
KARĽA Viktor
KATUNSKÝ Dušan
KOHNOVÁ Silvia
KORMANÍKOVÁ Eva
KOTRASOVÁ Kamila
KRAUS Ivan
KVOČÁK Vincent
MANDIČÁK Tomáš
MEŠŤANOVÁ Dana
MÉSÁROŠ Peter
MOHAPL Martin
OLEXA Tomáš
ONDOVÁ Marcela
OROLÍN Peter
PLATKO Peter
POÓROVÁ Zuzana
SABOL Peter
SALAIOVÁ Brigita
SEDLÁKOVÁ Anna
SIČÁKOVÁ Alena
SPIŠÁKOVÁ Marcela
STANKO Štefan
STRUKOVÁ Zuzana
ŠIMIČEK Vojtěch
ŠKULTÉTYOVÁ Ivona
ŠOLTÝS Róbert
ŠTEVULOVÁ Nadežda
ŠUHAJDA Karel
TALIAN Juraj
TAŽÍKOVÁ Alena
TERPÁKOVÁ Eva
TKÁČ Štefan
UBYTZ Andrzej
VÁCLAVÍK Vojtěch
VERTAĽ Marián
VILČEKOVÁ Silvia
VRANAYOVÁ Zuzana
ZOZULÁK Marek
ZELEŇÁKOVÁ Martina

IOP Conf. Series: Materials Science and Engineering **867** (2020) 011004 doi:10.1088/1757-899X/867/1/011004

Peer review statement

All papers published in this volume of *IOP Conference Series: Materials Science and Engineering* have been peer reviewed through processes administered by the proceedings Editors. Reviews were conducted by expert referees to the professional and scientific standards expected of a proceedings journal published by IOP Publishing.

Parasitic Architectural Forms (PAF) *S01.E01 "Prologue"*

Tomáš Baroš[1]

[1] Technical University of Košice, Faculty of Civil Engineering, Institute of Architectural Engineering, e-mail: tomas.baros@tuke.sk

Abstract. The aim of this paper is to outline the main functionality of Parasitic Architectural Forms (PAF), extend on potential utilization of it in the digital era, discuss the possibility of architecture being self-designed. I raised a question if architecture could be self-growing, self-developed and self-healing. This issue is being solved by this concept of PAF. Absence of artificial intelligence in the process of design in the architecture field offers a space for new research. Also, it generates options for yet uncharted views for experimental creation and urges us to ask of what are we, people, and computers currently capable.

1. Introduction

By the end of the 20th century, the use of digital tools and techniques in architecture came to the forefront. Initially developed in the aeronautical, manufacturing, automotive, shipbuilding, aerospace and animation industries, these tools and techniques have spread rapidly throughout architectural design since the late 1980s, transforming the way architects design and implement projects forever [1]. Is it possible to move the possibilities even further? We are in the era of Digital and Post-Digital Architecture, why Artificial Intelligence in Architecture? Self-Designing Architecture?

2. Idea

At the very beginning, there was a question that would like to focus scientifically on and bring the answer. What if architecture could design, grow, develop, repair itself? It would instead start by analysing this question than answering it. There has not invented this question. It has been here since time immemorial. I am just trying to bring it to the field of research. Just traditional lamenting during construction accompanied by the lamented, "I wish it were built by itself". Or did it become the germ of futuristic ideas? Furthermore, perhaps we owe it only to the human laziness of ingenuity, comfort, the effort to simplify life and work. For clearly define what should be the subject of research on this theme, it was necessary to define keywords on this issue. Architecture in this question means, for the dissertation of mine, an architectural construction or material. Artificial intelligence should ensure that such an architectural construction, such a material, is capable of being designed autonomously. Software capable of learning and creating. The fact that the structure should grow independently and develop, and heal, understandably repair, should be ensured by a material based on bioplastics or a composite consisting of conventional material and a biological part. Among other things, it should be genetic information material in the form of generic parameters necessary for the growth as mentioned above and development. As usual, the question creates a chain of more and more new questions. These ideas have been translated into the title of the dissertation topic: Physical Rationale for Creating Elements of Parasitic Architecture. It is the creation of the design of architectural constructions by the physics of the environment, and it will be defining and describing why such a form of designing architectural constructions should arise. The notion of parasitic architecture does not exist in principle, I introduce it with the idea of creating a manifesto in architectural design, as a precedent, a new view of creation. It means, in simplicity, additional, secondary architectural elements, it would be better starting with the creation of smaller-scale elements, rather than just the building itself. Naturally, we will eventually reach these scales and complex self-designed objects.

3. Parasitic architecture

What is parasitic architecture? Based on the observed model from nature, the host-parasite, we can apply a mutual relationship beneficial for both sides, to the relationship existing building-additional architectural construction. The different views on this idea, whether in terms of content or form, create a broad scope for defining this model and what we can regard as parasitic architecture. Formally, for example, superstructures, extensions and installations. In terms of content, interventions in the urban structure, Urban acupuncture, where the whole improves through small interventions. If we move to a smaller scale, we can also consider vegetation facades as a parasite where, through the substrate created on the host, the conditions for the growth of greenery that can be beneficial for the building are created. Also, there are installations in the existing urban structure that can positively influence the host environment, whether by expanding space or providing additional design and technology improvements.

(a) (b) (c) (d) (e)

Figure 1. (a), (b) The Green Exhibition House, Rotterdam, Korteknie Stuhlmacher Architecten; (c) Manifest Destiny, San Francisco, Mark Reigelman; (d) Cabin, Athens, Panos Dragonas Christopoulou Architects; (e) Homes for the Homeless, London, James Furzer.

(a) (b) (c)

Figure 2. (a) Balcony, Zalewski Architecture Group; (b), (c) Hutong Bubble, Beijing, MAD Architects.

(a) (b) (c)

Figure 3. (a) One Central Park, Sydney, Australia – Jean Nouvel – Ateliers Jean Nouvel; (b) Vertical Forest Milan, Milan, Italy – Stefano Boeri – Boeri Studio; (c) Green Cast, Odawara-shi, Kanagawa Pref., Japan – Kengo Kuma – Kengo Kuma Associates.

IOP Conf. Series: Materials Science and Engineering **867** (2020) 012001 doi:10.1088/1757-899X/867/1/012001

4. Concept of Parasitic Architectural Forms

Parasitic Architectural Forms (PAFs) can be multiple in parallel. It is artificial intelligence in architecture, building biomaterial with genetic and geometric information, architectural constructions, innovative tool, autonomous building process, experimental design, system-platform, product. In simple terms, it is an experimental creation of the design of architectural structures by physics of the environment. Lighting and acoustics shaping the form, temperature created scale, and moisture created a structure of the parasite. Space is undoubtedly the main determining feature of architecture. Space is different from other arts. Architecture can be considered as a tool that is actively incorporating or using environmental elements by dynamic or passive means. Architecture exists only as part of the whole environment. It lives in a complicated relationship with him. In a delicate balance between use and enrichment [2].

PAF is, therefore, a system consisting of three interconnected parts, linked to each other. Triumvirate. PAF = A.I. + DNA-BIO-MAT + RE.PROD.USE.

Figure 4. Illustration of the essence of Parasitic Architectural Forms.

4.1. Artificial Intelligence (A.I.)

a software capable of processing incoming information and data using neural networks and thus be able to respond autonomously to evaluate, learn, behave and design. Input information is understood to be the influence of the external environment of factors affecting the host. The result will be parametric information for designing, shaping the resulting shape of architectural structures. Conceptually, the whole process and the functioning of the software is set as a course in three cycles repeated in several sequences until the desired forms of the individual structural-spatial parameters are achieved. In the first cycle (Cycle_NTMY), the software aims at recognising the host and recognising the host environment, i.e. it works with satellite image meteorological data, cadastral data, geographic data, geometric data, various contextual and regulatory information compared to sources of example contents. In the second cycle (Cycle_KREA), the task of the software is to create a 3D model of the host and its virtual reality environment, and this output will be able to undergo simulations of external environmental conditions. The whole process takes place on feedback, communication. The result will be generic information needed to simulate the growth, development and shaping of a 3D parasite model. The processing of such a model of the environment and the host building in question will be ensured solely by autonomous software. We have a massive amount of usable material such as the satellite imagery and Google Maps imagined, and using geometric calculations and projections data on the geometry of apartment houses of standardised construction systems when we are talking, for example, of standardised panel systems as possible, suitable hosts. To bring the whole concept of parasitism to perfection, PAF software (A.I.) will also parasitise on other computational and simulation software in the third cycle (Cycle_VERSUS) and thus undergo the 3D model of the parasite, its critical details and individual parts, verification and assessment. Parameterisation, by its very nature, means using sets of parameters or limits as a means of exploring the possibilities of geometry through multiplication, distortion, transformation, or all three at once. The role of parameterisation, useful for creating daring formal abstractions, is to exceed predictable results based on predetermined input data [2]. Broadly, parametric design can be defined as work that is driven by parameters, where specific sets of rules inform the architectural or design output.

It may be surprising that the digital can be traced back so far in history. In fact, it was argued by some architecture historians to begun in the Renaissance [1].

Figure 5. Illustration of how PAF A.I. could work.

4.2. Material (DNA-BIO-MAT)
is containing a bio-component and capable of gradually evolving from element to structure, subsequently clumping to the final form. The definition and creation of material are subject to biomimicry observation, study and conceptual material design. I assume that the material could be based on a combination of biopolymers, lichenized concrete and bio cement, which would gradually change from a resilient to a fossilized material structure providing the basis for further growth, while at the same time being able to describe the purpose and needs of PAF best. Biomimicry adepts are Cocollithophores, a type of coral that can convert CO2 into its building material in the marine environment. Furthermore, mucus, Micro-algae, lichens, mosses, punches, stromatolites, succulents, toxoplasmosis, fungi and animal tissues. Biomimicry studies and adaptation of phenomena such as sedimentation, derivatization, diffusion, desertification, crystallization, karst phenomena of coral chemical phenomena, fusing and chemical expansion reactions are also essential. It will be one and the same material applicable anywhere but varied. Diversity, as in nature, will ensure the ability to survive and adapt, thereby always resist the influences and changes of the environment. Differences will have occurred at the level of the climate zone, regional and political geography, settlement structure, urban-type, building-host. The fit systems are described by the contribution of their components to the maximization of different, though generically related, structural forms. In biological structures, the DNA molecule suggests a minimum inventory-maximum diversity principle from a biomechanical interaction point of view. This new materialism appears to have a diverse and complex influence on contemporary design. Historically there are two distinct ways of influencing design. The first quickly absorbing material as a stylistic and formal phenomenon, while the second is more profound in exploring the complex interrelations between science, technology and design that have only begun to emerge [3]. Architecture is part of its environment as a whole. An environmental approach can be developed, given the awareness of the potential dynamic reciprocity between many species and the environment that makes them conditional. Like cell behaviour, the external stimulus partly determines, complementary systems must foster each other based on symbiotic and cyclic relationships to create a sustainable but variable environment [2].

4.3. Process (RE.PROD.USE)
in this concept means the technological action, phenomenon of autonomous growth, application and reproduction by 3D printing technology using robotization. The whole process of material growth will depend on its structure and composition. While I am working with the concept at the moment I can not precisely define and describe the course of the process, but it will be a combination of growth in terms of chemical expansion reactions, reproduction and guidance system of growth by food, possibly degraded older layers that will serve as a substrate absorbed by new layers, generations of this material. Growth as a kind of movement can turn an otherwise mundane and drab architecture into an infinitely complex and vivid [2]. If we consider that growth is to Nature, what production is to Design, then the notion of shaping and making come to include more than a straightforward manifestation of shape. In nature, the form is informed by the interaction of matter and energy. It is due to the distribution of matter and its properties that such interaction is made possible in the physical realm [3]. Resilience and flexibility of the whole model during its lifetime will be ensured by testing already in the phase of

simulating the interaction of external influences on the form and its feedback, thus generating an update of DNA information for further material growth as it will be a life cycle and new generations of material.

Figure 6. Visualisations of Parasitic Architectural Forms at scene.

5. Biomimicry as a solution

For to transform all these conceptual hypotheses into verifiable theses, biomimicry is the most appropriate approach, taking biomimicry as imitating models, systems, time-tested patterns and strategies from nature, and applying them as solutions to the problems we commonly encounter. The many manifestations in natural organisms provide a rich sourcebook of ideas for structures that could be more efficient than those found in conventional architecture. The principle for architecture that emerges from observing nature is fewer materials, more design. Nature showing us how minimal materials can be used to maximize the effect [4]. For example, Buckminster Fuller explored the dynamics of the form through his synergistic principles. By reducing weight, he sought a structure that he considered minimal but capable of dynamically transforming in response to environmental conditions [2].

Figure 7. The Eden Project: The Biomes, Cornwall, UK, Grimshaw Architects – mimic structure of pollen seed; Figure 16: The Eastgate Centre, Harare, Zimbabwe, Mick Pearce – mimic cooling mechanism of a termite mound; Figure 17: 30 St Mary Axe Tower, London UK, Foster+Partners – mimic shape and lattice structure of a Venus' flower basket.

6. Concept of methodology

The basic idea is to compare the resulting product, which is, in this case, developed artificial intelligence software and bottom-bio-material on contrasting host environments (island-inland, temperate climate-tropical climate zone, urbanism and Europe-Asia culture) with volume and typology comparable hosts. All research should be carried out in the following steps:

- measure values of Barometric Pressure, Temperature, Relative Humidity, Dew Point, Wind speed, Wind direction for 2020 and 2021 in Košice, Taipei (data in situ) and compare with each other and with measured airport's values BTS(Bratislava), KSC(Košice), SLD(Sliač), TAT (Poprad), ILZ (Zilina) - SHMI and data from airports TPE (Taoyuan), KHH (Kaohsiung), RMQ (Taichung), HUN (Hualien), TTT (Taitung) 2010-2021 at half-hour intervals (data) Weather). Define dependencies and determine parameters for parametric design by multicriteria optimization and linear and nonlinear analysis;
- to find a model and principle of functioning and behaviour of selected observed and reference subjects, to adapt biological and behavioural and subsequently to describe and define mathematical (generative geometry) -physical (behaviour) -chemical (composition + structure + behaviour) model;
- creating software PAF A.I.;

- simulations;
- material creation;
- creating model 1:10 host + parasite laboratory tested experimentally.

7. Enlightenment and challenge
The p-a-f.eu website is also an essential part of the development and works on this topic. Through this analytical-informative website, I try to characterize my ideas, inform about the course of the whole research, find and define a variety of moods and opinions of the professional as well as the lay public as a possible input parameter of the system development. The aim is also to open a forum discussion on the topic. Through the open-call challenge to find possible collaborating collaborators from different industries, with enthusiasm for developing this topic, but also the product itself, thus creating a broad interdisciplinary team. Later in this webspace to offer the finished product as an open-source and help to develop further or eventually transmit the imaginary relay.

8. Conclusion
This paper aims to bring my dissertation topic closer to the professional public, to open a discussion on artificial intelligence in architecture, to explore experimental design methods in architecture, and to adapt futuristic ideas in research and practice. The research and development of this subject is an attempt to partially dehumanize architectural design as an experiment. We have an incredible amount of information and examples in nature, but we do not yet know how to use and process this scarce resource effectively. However, unlike the classical artificial intelligence of the cyber age, today's artificial intelligence does not even try to imitate the logic of the human mind. Conversely, advanced electronic computing can now solve seemingly unsolvable problems, problems that we could not solve otherwise, precisely because computers seem to have developed their logic, their scientific method and their way of thinking, which is quite different like ours. Computers do not think, as we think, because of the simple but drastic structural difference between our mind and theirs: unlike computers, we have never been firmly connected to big data. What we call "big data" today means data too large to be managed, but which computers can manage it without any problems. Buildings designed, calculated and built-in this way tend to look very different from everything we have ever designed. They also tend to meet specifications better because, in this spirit, the calculations are ahead of us [5]. Are digital tools that can solve technical problems? Or we can extrapolate their potential to change the way we design, build and inhabit our world for a more sustainable future [1]?

9. References
[1] Claypool M 2019 *The Digital Architecture: Then, Now and in the Future* Space10 Online: https://space10.com/project/digital-in-architecture/
[2] Simitch A and Warke V 2014 *The Language of Architecture: 26 Principles Every Architect Should Know* Rockport Pub
[3] Oxman N 2010 *Material-based design computation* Doctoral dissertation (Massachusetts Institute of Technology)
[4] Pawlyn M 2019 *Biomimicry in architecture* Routledge
[5] Carpo M 2017 *The second digital turn: design beyond intelligence* MIT press

Acknowledgments
The paper presents partial research results of project VEGA 1/0674/18 “Theoretical and experimental analysis of architectural-structural shapes and fragments of building envelope structures designed for harsh weather conditions”.

Adaptive glass system: idea development and design

S Bizonova[1]

[1] Institute of Architectural Engineering, Faculty of Civil Engineering, Technical University of Kosice, Vysokoskolska 4, 040 23 Kosice, Slovakia

Abstract. The aim of the paper is to create a concept of a glass system independently reacting to the parameters of the environment in which is located. Autonomous insulation unit is responded to changes of environmental parameters, capable of evaluating mode efficiency with and without the use of selective layers. Parameter specification of individual modes, description of operation is needed. As a result, a construction scheme of the glass system is proposed.

1. Introduction – State of Art

Adaptive architecture, bearing the idea of adapting to the environment in which it is located, is the logical result of the efforts to continuously reduce energy consumption and, at the same time, a new tool to maximize the use of structures in relation to energy. Adaptability is the ability to change the state / shape / properties of a structure as the environmental conditions have changed at a given time. It can be basic adaptability, where the structure reacts by changing the effect of weather conditions only on the basis of the influence of physical forces (pressure, tension, other stress). Thus the shape changes are based on its material characteristics (tensile strength, compression, elastic modulus). Advanced adaptability is manifested by a change based on a pulse from a device that specifically evaluates current environmental conditions. The device recognizes different scenarios and adapts the structure according to the formula given to it. A passive adaptation is a prerequisite for a flexible / formable material, while in advanced adaptation, the material of the structure can be rigid and only the structure layout itself allows changes. In connection with window structures, basic adaptive constructions include thermochromic glazing. Advanced adaptive systems include electrochromic glass [1], suspended particle device, liquid crystal and others. These currently belong to the glazing with the highest adaptability [2]. By applying voltage, the glass changes its colour and optical-energetic properties. The proposed glass system belongs into the category of advanced adaptive. Its parameters are assessed by numerical analysis.

2. Window that knows what to do

When choosing a suitable glass system, the first of all is its future location. It is both orientation to the cardinal points and the climate zone (see table 1). A properly selected glass system can thus mitigate the effects of weather extremes on indoor thermal comfort for a given site and the energy requirement to achieve it again. The glass system always has the lowest heat transfer coefficient while respecting the other optical-energy properties. In such highly insulated glass systems, where the heat transfer coefficient values are close to those of the peripheral wall, the heat flux between the exterior and the interior is logically very low. This could be a disadvantage if there was a situation where the heat flux would be required. Ideally, in a hot climate, the glass system that reflects all the infrared radiation (from the sun and surrounding surfaces) as well as all UV radiation is suitable. At the same time, it is able to transmit all infrared radiation from the interior to the exterior and at the same time maintain the transparency and allow the visible component of light to penetrate. In cold climates, it would be able to transmit to the interior radiation of all wavelengths except UV while reflecting all infrared radiation back to the interior.

Table 1. Required parameters of the glass system for a specific climate [3].

Climate	τ_V	g	U [W/(m^2.K)]
Cool climate (prevailing heating loads)	>0.70	>0.60	<2
Temperate climate (prevailing heating loads)	>0.70	>0.50	<2.5
Hot climate (prevailing cooling loads)	>0.60	<0.40	<4

The design of the adaptive glass system must be approached conceptually. In the case of a glass system as a specific structure, it is necessary to consider the properties of the structure in individual modes, as it is only mechanical changes (removal of selected parts), not a change in properties of the same structurally unchanged structure. Window films play an important role in the concept. The state they create by their presence/absence defines the thermal-optical parameters of the glass system mode. Window film substitutes the glass pane and its thermal conductivity and emissivity parameters are comparable to conventional clear glass pane. Its advantage over conventional glass is a thickness of less than 1mm and a negligible weight that limits using or increases the cost of multiple glass systems.

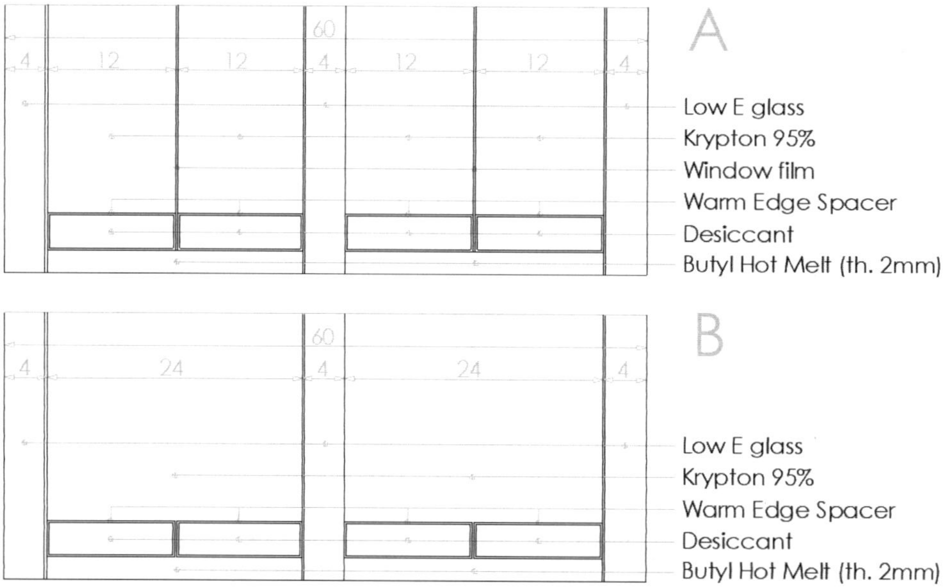

Figure 1. Investigated glass systems.

The glass system consists of three glasses with a low-emissivity layer at positions 2, 3 and 5. Window films without low-emissivity coatings are placed separately in the space between the glasses. The thickness of the gas gaps in the presence of films is 12mm (figure 1A), they are four. In the absence of films state, the gap thicknesses merge to form two 24 millimetre gaps (see table 2, figure 1B). It follows from the theory of heat transfer in closed air gaps that as the thickness of the gas gap increases, the proportion of the convection to the conduction increases (see figure 1). It can therefore be assumed that at a gap-filled thickness of more than 12 mm, the heat transfer is almost exclusively by convection and this ratio does not change significantly with the further increasing thickness. This means in practice that by removing films and doubling the thickness of the gas gap, the heat transfer coefficient of the glass system increases significantly, thus allowing the desired heat transfer from interior to exterior or vice versa. In this way, the glass system is able to adapt to heat conditions. Graphically are the modes of operation shown in figure 3.

Table 2. Description of glass system composition in each mode.

	A	B
Composition of glass system [mm]	4-12-0.1-12-4-12-0.1-12-4	4-24-4-24-4
Overall thickness [mm]	60	60
Glass panes used	Low E glass (4mm) x3	Low E glass (4mm) x3
Low E coating position from exterior	2,3,5	2,3,5
Film used	Polyethylene (0.1mm)x2	-
Gas used and its concentration	Kr 95%, Ar 5%	Kr 95%, Ar 5%

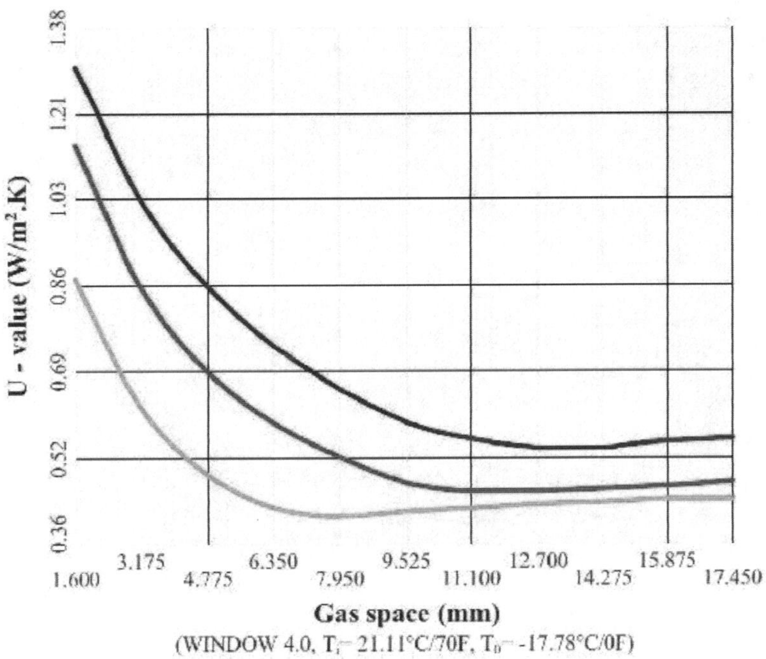

Figure 2. Optimum width for air, Argon and Krypton [4].

Window software was used to obtain numerical results [5]. The model uses the following equation to calculate the heat transfer coefficient U of the glazing [6]:

$$\frac{1}{U} = \frac{1}{h_e} + \frac{1}{h_t} + \frac{1}{h_i} \quad (1)$$

where h_e, h_i are heat transfer coefficients on the outside and inside of the structure; h_t represents total thermal transmittance of the glazing.

The inverse value of the total transmittance of the glazing program calculates:

$$\frac{1}{h_t} = \sum_1^N \frac{1}{h_s} + \sum_1^M d_j r_j \quad (2)$$

where N is number of cavities; d_j is thickness of each layer of the material; r_j represents specific heat resistance of each material (soda-lime glass specific heat resistance is 1.0 m.K/ W); M is number of layers.

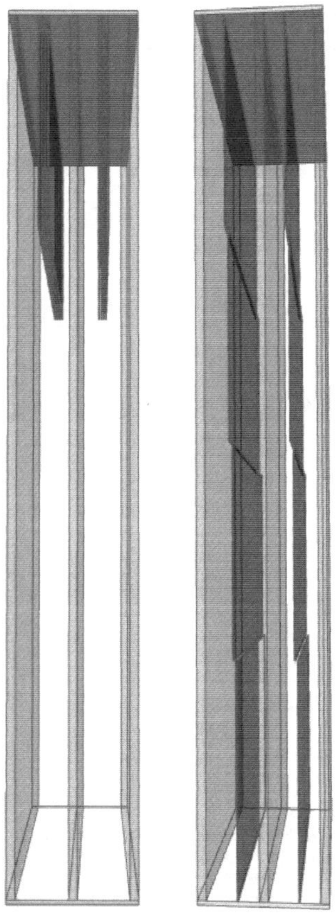

Figure 3. Modes of operation of adaptive glazing concept. Mode 2 is shown on the left; mode 1 is displayed on the right side (see table 2).

The heat transfer of the individual cavities of the glass system is the sum of the radiation transmittance and the *U* value of the gas in the glass system:

$$h_{s,k} = h_{r,k} + h_{g,k} \tag{3}$$

where $h_{s,k}$ is heat transfer through the cavity; $h_{r,k}$ radiation transmittance; $h_{g,}$ is *U* value of gas. The value of thermal transmittance caused by radiation can be calculated by:

$$h_t = \left(\frac{1}{\varepsilon_{1,k}} + \frac{1}{\varepsilon_{2,k}} - 1\right)^{-1} T_m^3 \tag{4}$$

where σ is Stefan-Boltzmann constant; T_m is mean absolute temperature of the gas-filled cavity; $\varepsilon_{1,k}$ and $\varepsilon_{2,k}$ are modified emissivity of the surfaces enclosing the enclosed space between the panes glass at T_m. Model calculates thermal transmittance of gas according to standard EN 673 [5]:

$$h_{g,k} = \frac{N_u \lambda_k}{S_k} \tag{5}$$

where S_k is cavity width; λ_k is coefficient of thermal conductivity of gas; N_u is Nusselt number. The Nusselt number is based on the equation:

$$N_u = A(G_r P_r)^n \tag{6}$$

where *A* is constant; G_r is Grashof number; P_r represents Prandtl number; *n* is exponent. Grashof and Prandtl numbers are based on the equations:

$$G_r = \frac{9{,}81 s^3 \Delta T \rho^2}{(T_m\, \mu^2)} \tag{7}$$

$$P_r = \frac{\mu_c}{\lambda} \tag{8}$$

where ΔT is the temperature difference between the glass surfaces enclosing the gas-filled cavity, ρ is density, μ represents dynamic viscosity, *c is* specific heat capacity, T_m is mean temperature.

3. Results and discussion

Two modes of operation of the glass system were investigated by calculation using software WINDOW [4]. The heat transfer coefficient for Mode B experienced a deterioration of more than 0.334 (107%) compared to state A, table 3. Other thermal-optical parameters visible transmittance τ_V and solar factor g also recorded an increase, but this results from the removal (to a small extent) of the structure that limits heat and light transport.

Table 3. The resulting thermal-optical properties of the glass system in modes A and B.

Thermal-optical properties	A	B
τ_V[-]	0.558	0.687
g [-]	0.432	0.510
U_g[W/(m^2.K)]	0.312	0.646

4. Conclusion

The adaptability of the glass system is particularly important in relation to energy requirements and thermal comfort. Existing adaptive glass systems offer the possibility of changing properties, but at the expense of reducing visual contact with the exterior. The designed glass system capable of adaptation by configuration change offers two modes of operation. One mode with high thermal insulation capability that is able to handle harsh weather and the other mode that can create the conditions for the desired heat gains in winter and heat losses in summer. The properties of both glass system configurations were obtained by numerical analysis. The possibility of verifying the behaviour of the structure under real conditions by means of dynamic simulation is not possible, as the software setting does not assume that the structure under consideration will change over time. It is plan to provide experimental measurements of the glass system properties in a climate chamber to verify and evaluate measured and calculated data.

5. References

[1] Mardaljevic J Waskett R K and Painter B 2016 *Neutral daylight illumination with variable transmission glass: Theory and validation, Lighting Res. Technology* **48** 267–285
[2] Sadek A H and Mahrous R 2018 *Adaptive Glazing Technologies: Bal;ancing the Benefits of Outdoor Views in Healthcare Environments, Solar Energy* **174** 719-727
[3] Piccolo A and Simone F 2015 *Performance Requirements for Electrochromic Smart Window, Journal of Building Engineering* **3** 94-103
[4] Vitro 2016 *Gas Space Convection Effects on U-values in Insulating Glass Units*
[5] Window 7.4.14 2017 *NFRC Simulation manual, Lawrence Berkley laboratory*
[6] EN 673 *Glass in building. Determination of thermal transmittance (U value). Calculation method.* (2011)

Acknowledgments

This research was supported by the Slovak Scientific Grant Agency (VEGA) in collaboration with Slovak Ministry of Education, Science, Research and Sports and Slovak Academy of Sciences (SAS), under Grant number 2/0017/20.

IOP Conf. Series: Materials Science and Engineering **867** (2020) 012003 doi:10.1088/1757-899X/867/1/012003

Stress analysis of asphalt mixture in the indirect tensile strength test

J Bokomlaško[1] and J Mandula[1]

[1] Department of Geotechnics and Traffic Engineering, Technical University of Košice, Vysokoškolská 4, Košice, Slovakia

Abstract. Nowadays asphalt mixture is the most used material in the construction and reconstruction of roads. It is used for its advantages, such as availability of materials, flexibility and no need for dilatation. However, the asphalt mixture has one disadvantage, namely temperature sensitivity. At different temperatures it exhibits different physical-mechanical properties. For a better understanding of the physical-mechanical properties of the asphalt mixture, we can use complex rheological models. Complex rheological models can describe the viscoelastic behaviour of the asphalt mixture with a high prediction. Laboratory tests are required to determine input values for complex rheological models. The article is focused on the analysis of asphalt mixture in the indirect tensile strength test and compare with numerical solution.

1. Introduction

Asphalt mixture is a relatively complex material in terms of determining its physical-mechanical properties. It is a composite that consists of several road materials. These materials which are used in asphalt mixture exhibit different physical-mechanical properties [4]. Therefore, the asphalt mixture is assignment to viscoelastic materials. The filler is aggregate which have low internal damping approaching zero and thus can be classified as a material with high stiffness. Binder is asphalt obtained by crude oil, asphalt lakes or asphalt mines (Gilsonite). Asphalt is a substance that causes the asphalt mixture to be classified as viscoelastic material [1, 2]. Substances in this category are characterized by a change their physical-mechanical properties when the temperature is changed [4, 11]. This results have a change in the physical-mechanical properties of the asphalt mixture as the temperature changes. At minus temperatures it becomes a solid with a high complex modulus [11,12]. At the plus temperatures its physical-mechanical properties changes and the value of the complex module decreases. The asphalt mixture becomes viscoelastic material. Internal damping is higher and closer to value one. Finally, we have additives and admixtures that improve the asphalt mix properties and thus eliminate the negative properties of the asphalt binder.

Because of the composition already mentioned, there is a complication in testing asphalt mixtures. The use of aggregates of uneven shape as fillers results in uneven deposits during compaction. This results in a slight asymmetry in some laboratory tests. The asphalt mixture also changes its physical-mechanical properties mainly due to temperature changes. It is therefore important to carry out a large number of measurements at different temperatures and different types of stress [4]. This article focuses on the static analysis of asphalt mixture in the indirect tensile strength test. On the Marshall sample under laboratory conditions, the decrease of the applied force was observed over time under constant deformation.

2. Materials

Asphalt mixture with label AC16L was used for the test. The mixture was prepared in laboratory conditions with a specified composition (table 1). Aggregate was used from the stone pit Hradova (table 2).

Table 1. Asphalt mixture AC 16L.

Mixture	Max. aggregate fraction	Layer	Type binder	Softening point [°C]	Penetration (25°C, 100g, 5s)/0.1 mm
Asphalt mixture	16	Bounding layer	50/70	49.5	51

Table 2. Aggregate fraction.

Composition of asphalt mixture	Aggregate fraction [mm]				Asphalt binder	Additive
	Stone meal	11 - 8	8 - 4	4 - 0		
Percentage of total quantity [%]	6.6	17.9	19.8	49.9	5.79	0.01

3. Method of laboratory test

Test sample of standard dimensions (figure 1) was subjected to indirect tensile strength stress. The diameter D on the test samples averaged 100.5 mm and the thickness H averaged 68.2 mm.

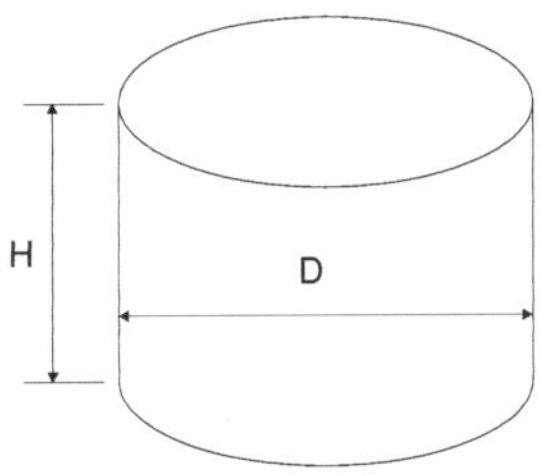

Figure 1. Marshall sample.

The essence of the test is in inducing indirect tension in the test sample. Pressure is applied to the body according to figure 2.

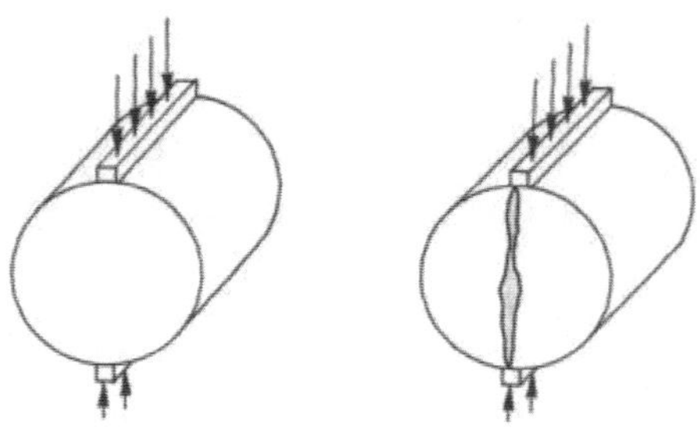

Figure 2. Marshall sample – action force.

For the determination of stress in the middle of the sample was use theoretical relation [7]. The tension intensity is influenced by the dimensions of the loading area. This area is determined by the standard. For the calculation of tensile stress, we can use 1st equation (1) and for the calculation of the compression stress we can use 2nd equation (2):

$$\sigma_{tension} = \frac{2P}{\pi \cdot a \cdot H}\left(\sin 2\alpha - \frac{a}{2R}\right) \tag{1}$$

$$\sigma_{compresion} = \frac{6P}{\pi \cdot a \cdot H}\left(\sin 2\alpha - \frac{a}{2R}\right) \qquad (2)$$

where

P – Force [kN],
a – Loading area [m],
H – Thickness [m],
R – Radius [m],

The geometrical quantities in equation 1st and 2nd and the theoretical stresses along the sample body cross-section are shown in figure 3.

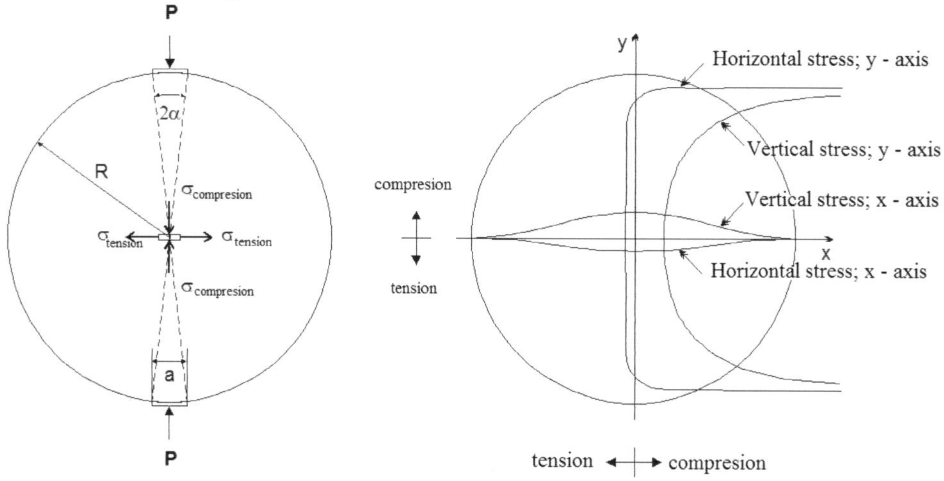

Figure 3. Stresses in the sample body.

4. Procedure of the test

Measurement was realised using hydraulic press device labelled FU160. This hydraulic press has the required sensitivity of the deformation record and allows adjustment of the continuous course of the load intensity. Test sample was measured in one load cycles and consisted of 6 load steps. Figure 4 shows the course of loading of the test sample in a single load cycle. The rate of deformation increase was 0.2 mm/s. Deformation was kept at 0.5 mm for 10 s. Thereafter, the deformation value dropped to 0 mm, which was kept for 20 s.

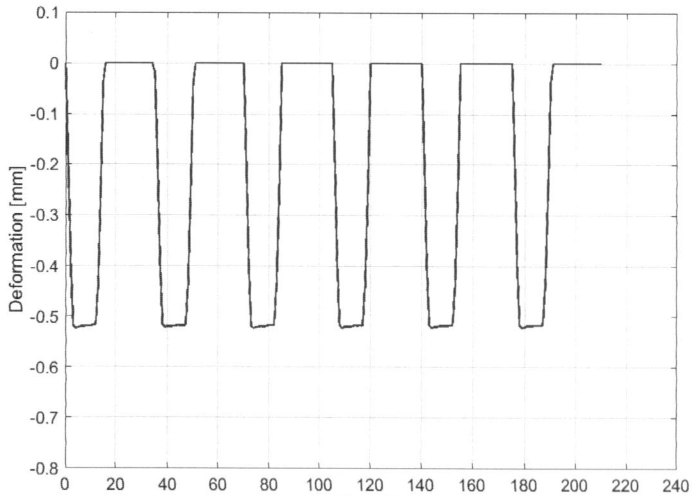

Figure 4. Test run.

5. Results

It is possible to see that force changed (decreasing tendency) during the duration of the selected constant deformation. In the first load step, the highest force was recorded. In the next load steps, force was lower.

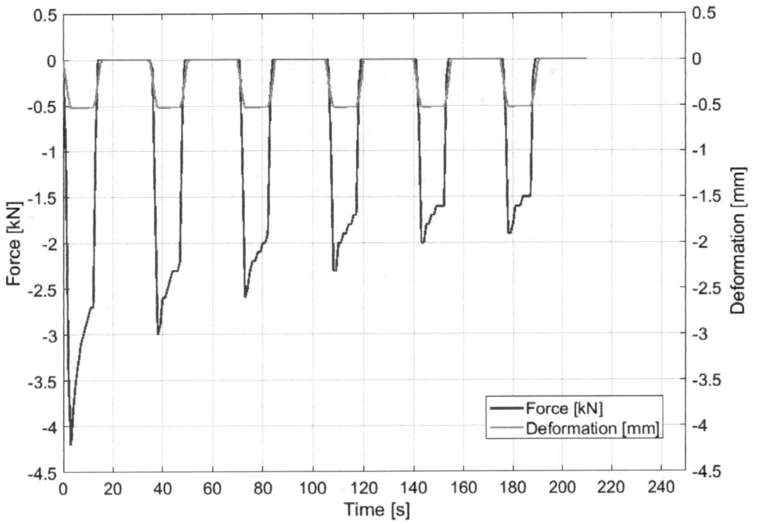

Figure 5. The force changes over time.

The force observed in the test was used to calculate the stress in the center of the sample. Tensile strength stress was calculated with 1st equation and compressive stress was calculated with 2nd equation. Software Abaqus was used to calculate and the results. The results were compared to those obtained by analytical solution. The test and numerical model with basic parameters was made in the program [9, 10]. The values in technical regulation were used as material properties [3]. Figure 6 shows the results from the 1st load cycle. In the plot stress values calculated using 1st and 2nd equation and software Abaqus are compared. At tensile stress there are smaller deviations between the calculated values with equation and the values from Abaqus. On the contrary, at compression stress the value is almost identical.

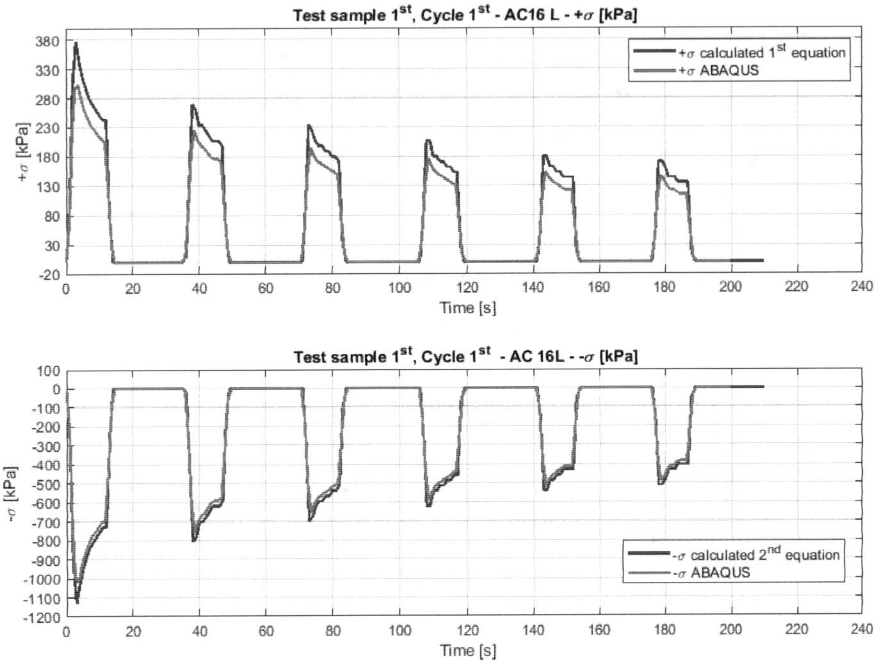

Figure 6. 1st cycle – compare stress.

6. Conclusion

From the results obtained in this study of the asphalt mixture subjected to indirect tensile stress we can conclude that:

1. For the tested sample subjected to indirect tensile loading, the load force decreased at the selected permanent deformation. This points to the fact that the asphalt mixture does not behavior as a linear elastic material and Hook's law cannot be applied [15].
2. The observed non-linear material behaviour leads to rheological models.
3. Comparison of the stress results obtained from laboratory measurements and theoretical models shows the non-linear behaviour of road asphalt mixtures [10,13].

By evaluating given results from numerical and experimental measurements on asphalt mixture stressed by indirect tensile, it leads to the opinion that a rheological model that describes more accurately the behaviour of the material is more suitable for describing the physical-mechanical properties of the asphalt mixture [10,13,14].

7. References

[1] Sobotka Z 1981 *Reologie hmot a konstrukci* (Akademia Praha Československá akademie věd)
[2] Ďoubal S a kolektiv 2012 *Mechanické chování viskoelastických těles* (Praha: Karolinum) 118 – 124 str. (Univerzita Karlova v Praze)
[3] *TP 033 Technické podmienky navrhovania netuhých a polotuhých vozoviek* (účinnosť od: 3. 8. 2009, November 2008)
[4] Mandula J and Bokomlaško J *Deformačné vlastnosti asfaltových zmesí v laboratórnych podmienkach* (Herľany 2016)
[5] Sabol R, Rovnak M, Klein P, Vazquez M and Varga R IEEE Transactions on magnetics *Mechanical Stress Dependence of the Switching Field in Amorphous Microwires* vol. 51 (NO. 1) January 2015
[6] Hýzl P, Dašek O, Varaus M, Stehlík D, Coufalík P, Dašková J, Krčmová I and Nekulová P 2016 Baltic Journal of Road and Bridge Engineering *The Effect of Compaction Degree and Binder Content on Performance Properties of Asphalt Mixtures* ISSN 1822-427X (Vilnius Gediminas Technical University) Vilnius (Lithuania)
[7] Wei C, Louay M and Peyman B 2018 Construction and Building Materials 187 *Use of indirect tension test and viscoelastic continuum damage theory for fatigue characterization of asphalt mixtures* p 38–49 (https://doi.org/10.1016/j.conbuildmat.2018.07.191) 0950-0618/ 2018 (Elsevier Ltd. All rights reserved)
[8] Tomko M, Šoltýs R, Platko P, Bajzecerová V a Demjan I Vedecko-výskumná činnosť ÚIS 2019 *Štúdia vplyvu predĺženia mikrodrôtu na časový prechod doménovej steny* (Prezentácia vedeckých výsledkov projektov ÚIS za rok 2019 Košice Slovensko) Technická univerzita v Košiciach s. 101-106 [print]. - ISBN 978-80-553-3433-2
[9] Šoltýs R 2013 *Nelineárna analýza vynúteného kmitania lanových sústav* (Róbert Šoltýs - 1. vyd. - Košice) :Technická univerzita v Košiciach - 237 s. [print].
[10] Pronk A C 2006 *Theory of the fourpoint dynamic bending test (Part I: general theory)* P-DWW-96- 008 ISBN-90-3693-712-4 (Edition: December 2006)
[11] Michalczyk R 2011 Implementation *of generalized viscoelastic material model in Abaqus code* (Logistyka 6:2883-2890) January 2011
[12] Baoshan H and Shu X 2012 Road Materials and Pavement Design 13(sup1):1-18 *Characterizing viscoelastic properties of asphalt mixtures utilizing loaded wheel tester (LWT)* June 2012 (DOI: 10.1080/14680629.2012.657049)
[13] Estaji M, Coleri E, Harvey T J and Butt A A *Predicting excess vehicle fuel use due to pavement structural response using field test results and finite element modeling* (California Department of Transportation) Caltrans under grant number 65A0542

[14] AKtas B, Aytekin S and Aslan S 2018 *Effects of selected warm mix asphalt additives on viscosity properties of binder* (Volume 1) Issue 1 p 80-84 (Journal DOI: 10.33793/acperpro) Journal EISSN: 2667-5862

[15] Lagos-Varas M, Movilla-Quesada D, Arenas J P, Raposeiras J P, Castro-Fresno D, Calzada-Pérez M A, Vega-Zamanillo A and Maturana J 2019 Construction and Building Materials *Study of the mechanical behavior of asphalt mixtures using fractional rheology to model their viscoelasticity* Volume 200 (10 March 2019) p 124-134 (0950-0618/ 2018 Elsevier Ltd. All rights reserved)

Acknowledgments

This work was supported by the Slovak Research and Development Agency under the contract No. APVV-15-0777.

IOP Conf. Series: Materials Science and Engineering **867** (2020) 012004 doi:10.1088/1757-899X/867/1/012004

The treatment and properties of construction waste for subsequent use in cement composites

J Charvát[1], V Vaclavik[1], T Dvorsky[1], J Svoboda[1], K Macalova[1] and L Gola[1]

[1] VSB - Technical University of Ostrava, Faculty of Mining and Geology, Department of Environmental Engineering, Czech Republic

Abstract. The paper presents the results of construction waste treatment for possible replacement of natural aggregates in cement composites. The individual processes of construction waste treatment for the purpose of obtaining the required fractions 0/4 mm, 4/8 mm and 8/16 mm are described. It deals with the so-called recycled aggregate, which will subsequently be used in cement composites as a substitute for natural aggregate. The properties of individual fractions of recycled aggregate are presented, both in terms of composition, density, water absorption and granulometric, mineralogical and elemental composition.

1. Introduction

Nowadays, the European construction industry generates 820 million tonnes of construction and demolition waste each year, which is about 46% of the total waste generated in Europe. In the USA, construction and demolition waste represents approximately 1/3 of the volume of material in landfills, and in Great Britain, 50% area of a typical landfill is taken by this type of waste [1–3]. According to the average composition of construction and demolition waste, we can find out that about 85% is made of concrete, ceramics and masonry. Although these components constitute the majority, there are often considerable quantities of wood and plasterboard as well [1]. Since it is still more profitable to use natural raw materials instead of recycled materials, a large part of waste is stored in landfills, thereby contaminating and occupying the land resources. Although there is a charge for depositing waste, it is still the most common method of waste management. Some companies illegally process this waste, for example as filler in sacrificial formwork, etc. in order to save money [3,4]. For this and other reasons, the European Commission and other institutions try to control construction and demolition waste, either by reducing its amount or by recycling it [1–3]. Construction and demolition waste can be recycled in several ways. The studies on the use of industrial waste materials in the construction segment may be used as an example [5–9]. The reuse of demolition waste as a secondary raw material has been studied in Asia [10–12] and there are also research projects dealing with the use of construction waste as admixture in concrete to replace Portland cement [13], or to create new building materials [14,15].

2. Materials and methods

Construction rubble from a rebuilt house was chosen as the material for the research of the utilization of construction waste for subsequent use in cement composites. This house is built of bricks, so most of the construction waste was bricks. The construction waste also included wall tiles and mortar, wood, iron, plasterboard, etc.

2.1. Pre-sorting

Construction rubble was free of undesirable admixtures such as pieces of wood, iron, plasterboard, etc. and it was subsequently sorted into the following 3 representative components of construction waste:

- brick waste,
- wall tiles
- mixed construction rubble (bricks, tiles, mortar)

2.2. Crushing

Mixed construction rubble was selected for further processing. Its granulometry was adjusted to the size that could be further processed using a crusher. The crushing itself was carried out in several stages on a single jaw crusher Fritsch pulverisette, type: 01.502. In total, the sample was processed on the crusher in three stages, while each crushing stage was followed by a grain size analysis of the sample.

2.3. Grain size analysis

HAVER EML 300 digital plus screening machine was used to determine the granulometry of the crushed sample and a series of sieves with the mesh size of 0.063, 0.125, 0.25, 0.5, 1, 2, 4, 8, 16, 25, 31.5 and 63 were selected. The sample was sieved into the individual grain-size fractions after each crushing stage and they were weighted using Kern 572-49 laboratory scales. The grain size analysis was used to prepare the grain-size curves, and the trend of the development of the proportion of the individual components of the mixed construction rubble in dependence on the degree of crushing was observed within the range of fractions 8/16 and 16/25.

3. Results and discussion

The grain size analysis results after the individual stages of crushing are presented in figures 1 – 3.

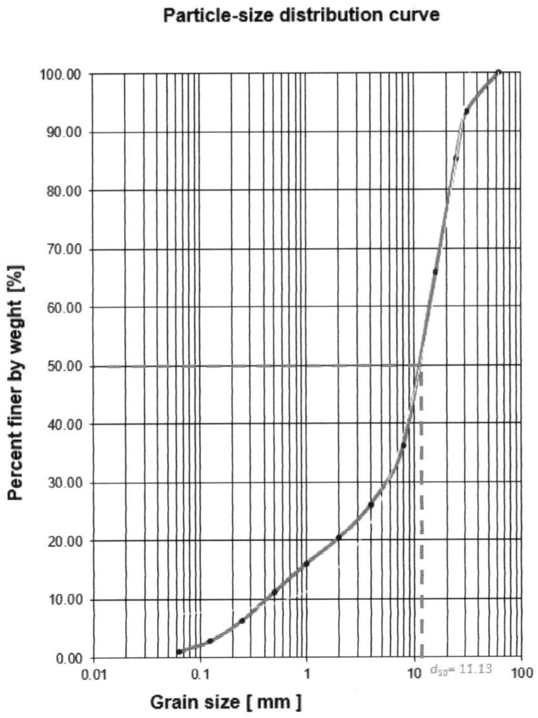

Figure 1. Particle-size distribution curve of the first crushing stage.

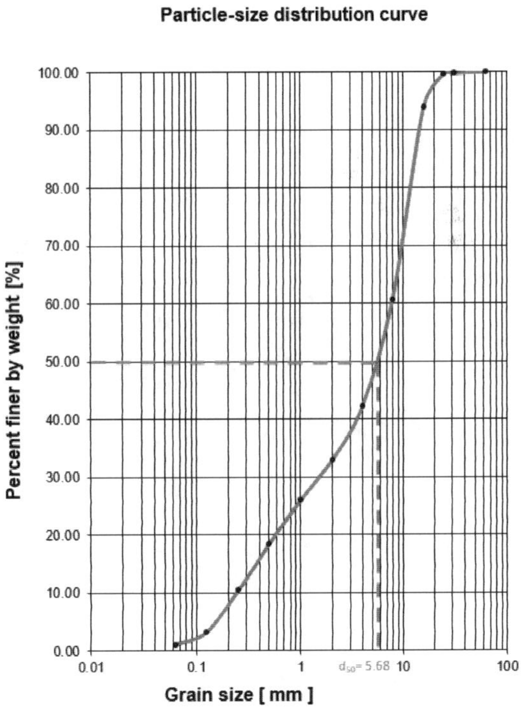

Figure 2. Particle-size distribution curve of the second crushing stage.

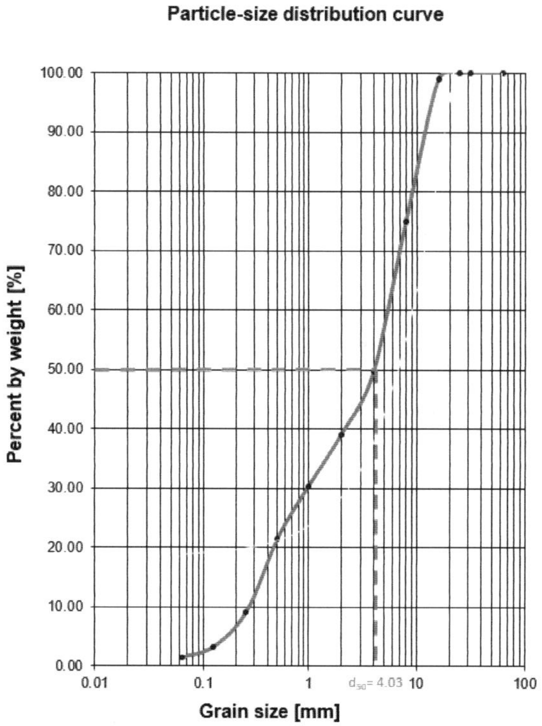

Figure 3. Particle-size distribution curve of the third crushing stage.

3.1. Grain size analysis of the first stage of crushing:
Share of the individual components of input material fraction 8/16:

- Mortar 54.6 %
- Tiles 11.7 %
- Bricks 33.7 %

Share of the individual components of input material fraction 16/25:

- Mortar 10.8 %
- Tiles 40.0 %
- Bricks 49.2 %

Calculation of non-homogeneity number C_u

$$C_u = \frac{d_{60}}{d_{10}} = \frac{14,41}{0,44} = 32,75 \quad (1)$$

where
d_{60} … grain size for 60% siftings
d_{10} … grain size for 10% siftings

During the first crushing stage, the slot size was 1.3 x 6 cm. The subsequent grain size analysis revealed a mean grain size d_{50} = 11.13 mm. The share of the individual components was as follows: fraction 0/4 accounted for 26.3%, fraction 4/8 accounted for 10.0%, fraction 8/16 accounted for 29.7%, fraction 16/25 accounted for 19.4% and fraction over 25 mm accounted for 14.6% of the total weight. On the basis of the calculation of the non-homogeneity number, it can be stated that the mixed construction rubble after the first stage of crushing is non-homogeneous, because it does not meet the criterion of $C_u < 5$ and has the value of 32.75.

3.2. Grain size analysis of the second stage of crushing
Share of the individual components of input material on sieve 8/16:

- Mortar 33.5 %
- Tiles 25.1 %

- Bricks 41.4 %

Share of the individual components of input material on sieve 16/25:

- Mortar 2.7 %
- Tiles 85.6 %
- Bricks 12.7 %

Calculation of non-homogeneity number C_u

$$C_u = \frac{d_{60}}{d_{10}} = \frac{7.85}{0.241} = 31.57 \qquad (2)$$

where
d_{60} … grain size for 60% siftings
d_{10} … grain size for 10% siftings

During the second crushing stage, the slot size was 0.9 x 6 cm. The subsequent grain size analysis revealed a mean grain size d_{50} = 5.67 mm. The share of the individual components was as follows: fraction 0/4 accounted for 42.0%, fraction 4/8 accounted for 18.6%, fraction 8/16 accounted for 33.3%, fraction 16/25 accounted for 5.6% and fraction over 25 mm accounted for 0.5% of the total weight. On the basis of the calculation of the non-homogeneity number, it can be stated that the mixed construction rubble after the second stage of crushing is non-homogeneous, because it does not meet the criterion of $C_u < 5$ and has the value of 31.57.

3.3. Grain size analysis of the third stage of crushing

Share of the individual components of input material on sieve 8/16:

- Mortar 29.1 %
- Tiles 36.5 %
- Bricks 34.4 %

Share of the individual components of input material on sieve 16/25:

- Mortar 0.0 %
- Tiles 93.8%
- Bricks 6.2%

Calculation of non-homogeneity number C_u

$$C_u = \frac{d_{60}}{d_{10}} = \frac{5.63}{0.27} = 20.85 \qquad (3)$$

where
d_{60} … grain size for 60% siftings
d_{10} … grain size for 10% siftings

During the third crushing stage, the slot size was 0.5 x 6 cm. The subsequent grain size analysis revealed a mean grain size d_{50} = 4.03 mm. The share of the individual components was as follows: fraction 0/4 accounted for 49.0%, fraction 4/8 accounted for 25.5%, fraction 8/16 accounted for 24.4%, fraction 16/25 accounted for 0.9% and no material was captured in fraction over 25 mm. On the basis of the calculation of the non-homogeneity number, it can be stated that the mixed construction rubble after the third stage of crushing is non-homogeneous, because it does not meet the criterion of $C_u < 5$ and has the value of 20.85.

The trend monitoring results of the development of the share of the individual components of the mixed construction rubble depending on the stage of crushing within the fractions of 8/16 and 16/25 are presented in figures 4 and 5.

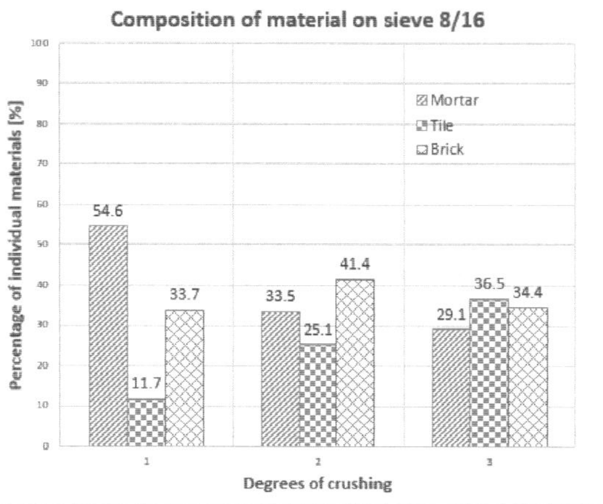

Figure 4. share of individual components, fraction 8/16.

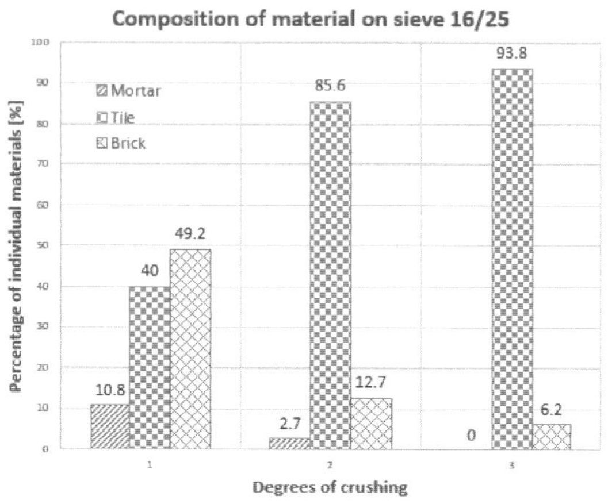

Figure 5. share of individual components, fraction 16/25.

Figure 4 presents the composition of the materials on sieve 8/16. It shows that mortar with the smaller crusher slot has a downward trend in the representation in mixed construction rubble as it passes into finer fractions due to the low strength of this material. The tiles show the opposite trend and smaller slot size means their higher share in the mixed construction rubble. This is due to the shape of the tiles and the characteristics of the crusher design, where the tiles fall through the slot more easily and do not become tinier. The brick material, despite a small fluctuation in the second stage of crushing, retains the share of about 37. Figure 5 presents the composition of the materials on the 16/25 screen. It shows that the mortar has the same trend as in case of the 8/16 screen. The dominant representation of tiles in this fraction follows the trend of the 8/16 fraction, which is due to the design of the crusher. The brick material has a downward trend of representation in the mixture, which implies that it is well crushable by this type of crusher.

4. Conclusion

Based on the results of the tests of the examined material, we can conclude that:

- The crushing of mixed construction rubble using the selected jaw crusher is not suitable, because of the content of ceramic tiles, which fall through the crusher slot due to their shape and are not crushed as result of that.
- The share of the individual components of mixed construction rubble in 8/16 fraction was almost equal after the third crushing stage, however, the grain size curve shows that the percentage of the finer shares of fraction 0/4 has increased to 49% of the sample weight.
- The increase in the amount of fine shares of fraction 0/4 is caused by the low strength of the mortar contained in the mixture, which results in its transition to finer fractions. This corresponds to the decreasing share of mortar in fractions 8/16 and 16/25.

5. References

[1] Gálvez-Martos J-L, Styles D, Schoenberger H and Zeschmar-Lahl B 2018 *Resour. Conserv. Recycl.* **136** 166–78
[2] Chen J, Hua C and Liu C 2019 *J. Clean. Prod.* **212** 190–9
[3] Esin T and Cosgun N 2007 *Build. Environ.* **42** 1667–74
[4] Lu W 2019 *Resour. Conserv. Recycl.* **141** 264–72
[5] Junak J 2016 *Solid State Phenom.* **244** 102–7
[6] Stevulova N, Schwarzova I, Hospodarova V and Junak J 2016 *Chem. Eng. Trans.* **50** 367–72
[7] Stevulova N, Baltakys K, Estokova A and Sverak T 2018 2018 *Adv. Mater. Sci. Eng.* **2018**
[8] Smolakova M, Estokova A, Vaclavik V, Soucek K and Zajicova V 2018 IOP Conference Series: Materials Science and Engineering vol 385
[9] Ondova M, Stevulova N and Estokova A 2012 *Procedia Eng.* **42** 1863–72

[10] Tam V W Y, Soomro M and Evangelista A C J 2018 *Constr. Build. Mater.* **172** 272–92
[11] Duan Z H and Poon C S 2014 *Mater. Des.* **58** 19–29
[12] Bui N K, Satomi T and Takahashi H 2018 *Constr. Build. Mater.* **163** 496–507
[13] Xiao J, Ma Z, Sui T, Akbarnezhad A and Duan Z 2018 *J. Clean. Prod.* **188** 720–31
[14] Lu J-X, Yan X, He P and Poon C S 2019 *J. Clean. Prod.* **234** 1102–12
[15] Sainz-Aja J, Carrascal I, Polanco J A, Thomas C, Sosa I, Casado J and Diego S 2019 *J. Clean. Prod.* **230** 945–55

Acknowledgments

This research was funded by VŠB–TUO, Faculty of Mining and Geology—grants numbers SP2020/30 and SP2019/29; Moravian-Silesian Region—grant number RRC/10/2018.

IOP Conf. Series: Materials Science and Engineering **867** (2020) 012005 doi:10.1088/1757-899X/867/1/012005

Application of mathematical models in design and assessment of sewer network facilities

R Csicsaiová[1], J Hrudka[1], I Marko[1] and Š Stanko[1]

[1] Slovak University of Technology in Bratislava, Faculty of Civil Engineering, Department of Sanitary and Environmental Engineering, Slovakia

Abstract. The application of mathematical models has been expanding in the field of sanitary and environmental engineering. Mathematical models are used in the design and assessment of sewer networks and their facilities. Sewer network models make it possible to create a model of hydraulic and Physico-chemical processes in wastewater flowing through the sewage network. The number of extreme rainfall events is increasing due to climate change. It causes a collapse in the infrastructure of urbanized areas. It is possible to investigate the flow of wastewater under extreme rainfall and to propose measures to eliminate adverse events using mathematical models. Nowadays, it is possible to use modern calculation procedures, which are used to dimension and assess existing facilities. This paper aims to focus on the application of numerical models in the design and assessment of combined sewer overflow chambers. The combined sewer overflow serves to carry away a part of the rain flows from the network to the nearest suitable receiving water body. Their main task is to reduce the uneven load of wastewater treatment plants by rainwater. The combined sewer overflow chambers distribute the inflow into the flow going to the wastewater treatment plant and the lightened flow going to the receiving water. The aim of this paper is to summarize the knowledge of CFD modelling and to get acquainted with the basic principles. In brief, the normal flow describes its simulation using two basic models. Finally, it focuses on the recapitulation of foreign studies and their use in the assessment and design of relief chambers and regulatory objects of the single sewer networks.

1. Introduction

Effective and ecological management of rainwater is an integral part of wastewater disposal solution from urbanized areas. It is necessary to design efficient regulatory facilities of a combined sewer network which would regulate the amount of rainwater flowing into the wastewater treatment plant (WWTP) from the urbanized areas [1].

At present, combined sewer overflow (CSO) chambers and rain separators are used to regulate the stormwater inflow to WWTP. Their effectiveness depends on the appropriate design and regular maintenance. The operation of the CSO chambers also has its drawbacks, but the correct configuration of the shape could at least partially eliminate these negative aspects. The most crucial factor in assessing the impact of extreme precipitation on sewage network facilities in the wastewater flow [2].

The application of mathematical models in sanitation engineering is now commonplace. A lot of research deals with flow modeling in sewer networks. Due to the rapid urbanization of the territory and climate change, it is necessary to assess and reconstruct the sewer networks. The development of computer technology creates the assumption that flow modeling will be used not only for the assessment of the functioning of sewer networks and their objects, but also for the design and dimensioning itself [3].

Many studies deal with the effectiveness of regulatory objects using mathematical models. 1D and 3D models used. Older studies have investigated the interaction of several CSO chambers in one sewer basin. The action of several CSOs in a river basin may create critical conditions in the recipient during tidal rainfall. Alpa and Melching's studies have focused on creating 1D models and simulating hydrodynamic flow [4]. The most significant changes in the flow occur during heavy rainfall events. During such a collision event, the water flow in facilities is unsteady, with rapid changes in velocities and the possibility of creating vortices [5]. In this case, when examining the wastewater flow in regulatory facilities are used 3D models [3].

The aim of this paper is to summarize the knowledge of CFD modeling and to get acquainted with the basic principles. In brief, the normal flow describes its simulation using two basic models. Finally, it focuses on the recapitulation of foreign studies and their use in the assessment and design of relief chambers and regulatory objects of the single sewer networks.

2. Flow modeling

Computational Fluid Dynamics (CFD) deal with Fluid flow modeling. It combines scientific knowledge from physics, mathematics, fluid mechanics, thermomechanics into simulation - the model of physical phenomena. CFD allows simulating fluid movement using numerical models. It used to solve various engineering tasks in the field of aerodynamics, environmental engineering, material engineering, biological engineering, civil engineering, engineering industry, etc.

In the case of water resourcing engineering, CFDs can used to simulate wastewater flow in wastewater treatment tanks, CSO chambers, and sewerage facilities [6]. It is a useful aid in the design and assessment of CSO chambers. In 3D models, we can simulate the behavior of fluid in the CSO chambers. When designing reconstructions of steel structures, the simulations used to determine the most appropriate additional constructions (floating barrage, screens).

CFD examines fluid flow for changes in their physical properties such as speed, density, temperature, pressure, and viscosity. It is necessary to consider every property as a whole to obtain the most accurate results. The individual mathematical models differ depending on the nature of the problem being solved, i.e., heat transfer, changes in the solid phase, or ongoing chemical reactions. [7].

The flow equations which defining fluid flow modeling are based on three basic flow principles:

- the law of conservation of energy,
- the law of conservation of mass
- the low of conservation of momentum.

Navier - Stokes equations, which form a system of partial differential equations, are also derived from these principles. They are formulated from basic equations of fluid mechanics, namely: continuity equation, motion equations, and energy equation [8].

These equations describe not only fluid flow but also heat transfer between different materials.

Mathematical models define the hydrodynamic properties of liquids by four basic equations:

- continuity equation,
- equation of motion,
- energy equation,
- turbulent flow equations.

CFD software solves mathematical models either by finite element method, finite volume method, or finite difference method [9].

These equations present the flow of the laminar liquid. However, this type of flow is rare in nature. The following chapter characterizes the turbulent flow in the CSS facilities and describes the individual solutions of the turbulent flow.

2.1. Turbulent fluid flow

Turbulence as a physical phenomenon affecting the flow pattern occurs in real fluid flow. Therefore, it is necessary to incorporate it into a mathematical model. The flow turbulence determined by the geometry of the object and the buoyancy. Swirls create fluctuations in speed, fluid temperature, and pressure. Turbulent flow generates several vortices with different dimensions. Large eddies gradually interfere to form smaller vortices, which in turn transmit their energy to heat [10].

Simplified solutions have developed to deal with turbulent flow, which can simulate flow relatively accurately. Direct Numerical Simulation (DNS) was created using exact equations.

IOP Conf. Series: Materials Science and Engineering **867** (2020) 012005 doi:10.1088/1757-899X/867/1/012005

It is a direct numerical solution of Navier - Stokes equations. There are two types of models using simplified equations. The first type uses a spatial filter to simulate Large Eddy (Large Eddy Simulation - LES). The second type is based on the solution of averaged equations (RANS). Based on the Boussinesq hypothesis, a model with two equations k - ε or k - ω and a model with one equation k developed. The Prandtl Baldwin - Lomax model, the Cebeci - Smith model, and the mixing length model also belong to this group of models. The Reynolds stress model, with six additional differential equations, is a separate group [9]. Figure 1 shows the distribution of turbulence models.

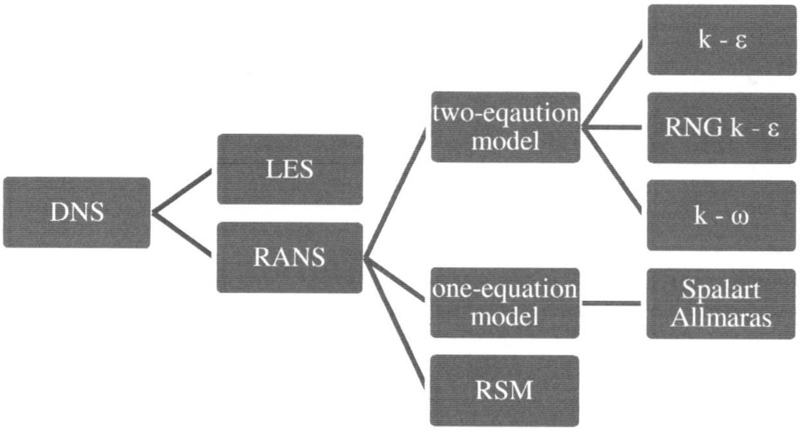

Figure 1. Models of turbulence.

2.1.1. Model Standard k – ε. The k - ε model is a model with two additional equations that describe turbulence using two transport equations. It is a semi-empirical model based on the Boussinesque hypothesis characterized by two variables:

- Turbulent kinetic energy (k),
- The measure of its dispersion (ε).

Unlike previous turbulence models, the k-ε model focuses on mechanisms that affect turbulent kinetic energy. The underlying assumption is that the turbulent viscosity is isotropic, i.e., the ratio between Reynolds stress and mean strain rate is the same in all directions [11].

2.1.2. Model Standard k – ω. The model k - ω is solved similarly to k - ε two differential equations. The equation for the specific energy dissipation rate ω has several disadvantages compared to the equation ε. This model based on the Wilcox k - ω model, which includes modifications for low Reynolds numbers, compressibility, and shear flow propagation. The Wilcox model shows its imperfections by sensitivity in solving the values of k and ω in the free current outside the shear layer. The Standard k - ω model partially eliminates them.

The Standard k - ω model is an empirical model based on the solution of transport equations for the kinetic energy of turbulence k and the specific rate of energy dissipation ω. The scattering rate is essentially the ratio ε / k [12].

3. Use of 3D models in flow modeling in combined sewer overflow chambers

The CSO chambers serve to divert a portion of the rain inflow from the network to the nearest suitable receiving water body. Their main task is to reduce the uneven load of wastewater treatment plants by rainwater by dividing the inflow to the CSO into the flow flowing to the WWTP and the overflow flowing to the receiving water [13].

The basic principle of flow distribution in CSO is the overflow of water over the overflow edge. If the wastewater level reaches the level at the level of the overflow edge, the water will begin to overflow through it and flow to the receiving water [14]. This overflow contains pollutants, including gross suspended solids, suspended solids, and dissolved solids [14], [15]. High demands placed on the hydraulic design of the building. From the hydraulic point of view, the design of the relief chamber is most similar to the weir. However, their complex design does not allow direct measurement of flow and velocities in individual parts of the facilities. Therefore, 3D CFD models currently used in the design and assessment of these objects [15].

Chen et al. used a 3D model of the control object to optimize the design of the Edmonton City Sewer System. The conclusion of the case study concludes that the use of the 3D model is a useful tool for the assessment and design of a sewer system [16].

Another study applied a 3D model to simulate the spread of E. coli in the receiving water body. The simulation was part of the analysis of the sewer network discharging the sink from the system into the marine environment. The study resulted in the application of the 1D model of a combined sewer system (CSS) in combination with the CFD model of a facility in a case study in Italy [17].

Department of Sanitary and Ecological Engineering CTU in Prague, in cooperation with Hobas, developed a new type of CSO, which they called Tube CSO or CSO-T. It is a simple construction of two interconnected fiberglass pipes of different diameters superimposed. To verify the functioning of the CSO, they constructed both a physical and a mathematical model, where they tested different variants of the slope of the sewer and varying flow rates of wastewater [18].

3D CFD model can also be applied in the case of invert traps. The model simulated the flow of water with a large amount of sediment. The study demonstrated the applicability of the VOF model to the stochastic discrete phase model [19].

A study by J. C. Quijano et al. evaluated the receiving water quality for the Chicago Metropolitan Area using the Environmental Fluid Dynamics Code (EFDC) with the Water Quality Analysis Simulation Program (WASP) to perform three-dimensional simulations of hydrodynamics and water quality. Two precipitation events with different properties were evaluated [5].

4. Conclusion

The development of urbanized areas, population growth, and climate change call for a change in the wastewater management concept. Many large cities have prepared sewer network assessments. These assessments often include evaluations of individual network facilities that have an impact on the status of the receiving water. The ecological state of the receiving water is affected by the surface runoff and the discharge from the CSO chambers.

Individual studies examined the interactions of multiple CSOs during heavy rainfall events. Intense precipitation causes flow changes in the CSS facilities. It is possible to simulate the behavior of wastewater flow in the CSO chamber using 3D models. It is possible to modify the individual parts of the constructions with the possibility of improving their operation and reducing the discharged pollution based on the simulations.

Many studies dealt with the development of new structures, and simulations subsequently verified their design. 3D simulations were also implemented in studies of individual river basins while optimizing CSS.

These models help to meet the objectives of the Water Framework Directive (CEC), Directive 2000/60/EC. This paper summarizes the knowledge about the use of 3D models in sanitary engineering and describes the basic principles of flow modeling.

5. References

[1] Niyogi D, Lei M, Kishtawal C, Schmid P and Shepherd M 2017 Urbanization impacts on the summer heavy rainfall climatology over the eastern United States. Earth Interactions 21 (5), DOI: 10.1175/EI-D-15-0045.1

[2] Gao B, Gao L, Xu D, Zhang M, Qu X and Li Y 2019 A novel method for evaluating the potential release of trace metals associated with rainfall leaching/runoff from urban soils. Sci. Total Environ. **664** 37–44

[3] Morales V M, Mier J M and Garcia M H 2017 Innovative modeling framework for combined sewer overflows prediction *Urban Water Journal* **14** (1) 97-111 DOI: 10.1080/1573062X.2015.1057183

[4] Alp E and Melching C S 2009 Evaluation of the duration of storm effects on in-stream water quality *J. Water Resour. Plan. Manag.* ASCE **135** (2) 107–116 http://dx.doi.org/10.1061/(ASCE)0733-9496(2009)135:2(107)

[5] Quijano J C, Zhu Z, Morales V, Landry B J and Garcia M H 2017 Three-dimensional model to capture the fate and transport of combined sewer overflow discharges: A case study in the Chicago Area Waterway System *Science of the Total Environment* **576** 362–373 http://dx.doi.org/10.1016/j.scitotenv.2016.08.191

[6] Hrudka J, Stanko Š and Holubec M 2017 Analysis of flow and sedimentation processes in secondary sedimentation tank *Pollack Periodica* **12** (2) 79–89

[7] Lipeme kouyi G 2011 The use of CFD modelling to optimise measurement of overflow rates in a downstream-controlled dual-overflow structure *Water science and technology* [online] **64**(2) 521-527 [cit. 2020-02-28] ISSN 0273-1223

[8] Jahoda M Počítačová dynamika tekutin- užitečný nástroj pro inženýry [online] [cit. 2020 – 01 - 12] URL: http://uchi.vscht.cz/uploads/pedagogika/bezpecnostni_inzenyrstvi/CFD.shrnuti.pdf

[9] Molnár V 2011 Počítačová dynamika tekutín: interdisciplinárny prístup s aplikáciami CFD (Bratislava: Slovenská technická univerzita v Bratislave) 492 ISBN 978-80-810-6048-9

[10] Wilcox D C 1998 Turbulence modeling for CFD (DCW Industries, Inc. La Canada California)

[11] Zhang Y, Yang X, Yang H, Zhang K, Wang X, Luo Z, Hang J and Zhou S 2020 Numerical investigations of reactive pollutant dispersion and personal exposure in 3D urban-like models *Building and Environment* 169 https://doi.org/10.1016/j.buildenv.2019.106569

[12] ANSYS FLUENT 12.0 2009 Theory Guide. Canonsburg: ANSYS Inc. [online] [cit. 2019 – 01 -10] URL: http://www.afs.enea.it/project/neptunius/docs/fluent/html/th/node1.htm?fbclid=IwAR3rRUZtwDQFMWX6K4ddKfFvd3VUZMqZtOhQrCNcyg1K6BUR3ostJg_AEnQ

[13] Butler D and Davies J W 2004 Urban Drainage. 2nd Edition (London: Spon Press) ISBN 0-203-34190-2

[14] Saul A J 2004 Increasing knowledge of CSO technologies. In: Marsalek J et al. editors. Enhancing urban environment by environmental upgrading and restoration. NATO science series IV: earth and environmental sciences **43** 105-110

[15] Insel S, Dufresne M, Fischer M and Vazquez J 2014 Assessment of the overflow discharge in complex CSO chambers with water level measurements – On-site validation of a CFD-based methodology Flow Measurement and Instrumentation 35 39-43

[16] Chen Z, Han S, Zhou F and Wang K 2013 A CFD Modeling Approach for Municipal Sewer System Design Optimization to Minimize Emissions into Receiving Water Body *Water Resour Manage* **27** 2053–2069 https://doi.org/10.1007/s11269-013-0272-9

[17] De Marchis M, Freni G and Napoli E 2013 Modelling of E. coli distribution in coastal areas subjected to combined sewer overflows *Water Sci. Technol.* **68** (5) 1123–1136 http://dx.doi.org/10.2166/wst.2013.353

[18] Pollert J 2013 Ecological aspect in combined sewer overflows chamber design *Water Practice & Technology* **8** No 3-4

[19] Mohsin M and Kaushal D R 2017 Three-dimensional computational fluid dynamics (volume of fluid) modelling coupled with a stochastic discrete phase model for the performance analysis of an invert trap experimentally validated using field sewer solids *Particuology* **33** 98-111

Acknowledgments

This work was supported by the Scientific Grant Agency of the Ministry of Education, Youth and Sports of the Slovak Republic and the Slovak Academy of Sciences within the project VEGA 1/0574/19, co-funded by the European Regional Development Fund and by the Slovak Research and Development Agency under the contract No. APVV-18-0203.

Share of the volume of work types and financial costs according to the stages of construction of reinforced concrete monolithic buildings

R Doubek[1]

[1] Brno University of Technology, Faculty of Civil Engineering, Institute of Technology, Mechanization and Construction Management, Czech Republic

Abstract. The aim of the paper is to evaluate the technological and economic demands on the construction of multi-storey reinforced concrete monolithic buildings for administrative purposes incorporating the stages of construction, technological stages and selected partial construction processes.

The entry criteria for objects that are included in monitoring and evaluation are their identical location, used technologies, mechanization, built-in materials and their purpose.

The purpose of the paper is to prepare tables for the professional and academic public for a rough idea of the proportional representation of time and financial consumption in the construction of this type of object, for selected stages of construction, technological stages and selected partial construction processes.

1. Introduction

The aim of this paper is to evaluate the technical and economic requirements for the construction of multi-storey monolithic reinforced concrete buildings for administrative purposes, divided into stages of construction, stage processes and selected construction sub-processes.

A subsidiary focus of the paper is the creation of financial indicators within the methodology of construction planning for building construction.

It will be possible to use the indicators listed below to provide a quick and basic informative estimate of the financial costs of reinforced concrete structures in the building industry. They may also see application as a tool providing aid and support in the production of plans for construction work or during pre-production business planning.

These indicators are necessary for rapid and correct decision-making in the field of construction planning. They can be used not only in design activities but also in supplier planning and for study purposes.

The creation of a system summary enables the application of theoretical methodology in a practical and illustrative form that can be used for construction planning activities throughout the preparation of an investment plan.

2. The state-of-the-art

At the end of the last century, large building companies developed their own user-friendly databases of financial expenses for individual types of structures and buildings, divided into construction stages. These databases helped the companies make quick decisions when creating new investment plans. At present, there is no known publicly available work or publication that directly deals with the financial volume of work divided into stages of construction, stage processes and selected subsidiary construction processes for the mentioned use in construction planning.

The currently known investigated issues are dealt with using calculation software products (e.g. Build Power or Kros) which handle finances as part of their data (price) base using technical/economic indicators.

When dealing with the given issues, the investigator must prepare for the acquisition of data and their correct evaluation not only theoretically but also through being a direct participant in realized construction projects. This will enable the correct inclusion of the required delivery and assembly aspects of the relevant investigated parts of a given issue.

3. Methodology for the achievement of results

The fundamental prerequisite for the gaining of useful results is the correct theoretical division and subsequent application of the obtained data. Within the given topic, defined parts (intermediates) that arise chronologically are determined during the construction planning of buildings [1].

These parts are known as stages of construction, which are composed of technical stages. A technical stage is structurally and production-wise a comprehensive part of the building or structure (i.e. a set of structures – not a process). A stage process is an aggregate of subsidiary construction processes whose product is a technical stage. A construction stage process is an aggregate of stage processes whose product is a stage of construction. A stage of construction is thus a part of a building which is comprehensive as far as production is concerned, i.e. composed of several technical stages that are related technically or in terms of purpose [2].

The individual stages of construction (complete parts of the structure) and stage products are clearly defined by characteristics and are divided into sub-units that will be the main indicators in this paper. It is clear from the facts stated above that the indicators of the individual technical stages will then be grouped into stages of construction as follows:

1.) substructure
    ground work (and demolition), foundations, substructure;
2.) superstructure
    superstructure, roofing;
3.) interior works
    partition walls, piping and wiring, application of interior plasters and floor underlays;
4.) finishing works
    construction of floors, completion of surfaces and equipment, completion of piping, wiring and interior works, external modifications.

As a stage process, quality control and handover are not strictly assigned to one specific stage of construction as they can be carried out within all four comprehensive parts of the stages of construction.

Another requirement that should be met in order to achieve relevant results is the determination of the input parameters of the samples from which the data will be obtained. The criteria that the samples (buildings) must have in common are: the location of the building, the purpose of the use of the structure, the supporting structural system, the use of materials (including products and semi-finished products), and building services [3].

The data are taken from specific realized samples (buildings) with the following parameters: the location is in the South Moravian region of the Czech Republic; the purpose is the operation of an administrative building with rental space; the supporting structure is a combined load-bearing system based on deep foundations (pilots) with basements and above-ground floors; the materials are monolithic reinforced concrete with a modular façade and plasterboard structures; and there is a complete array of building services, including measuring and regulation systems along with backup diesel generators. For the relevant determination of the price of the building, it is necessary to place the buildings in a built-up area with a direct connection to utility networks. If the building is not in a built-up area, utility costs will be higher than shown below.

In addition to the input parameters mentioned above, the data used for the determination of indicators also fulfilled the following conditions: the owner and operator of the building are one and the same, and there is a certification programme for the building, designer and contractor. The enclosed space is 63 000 m3 ± 3%.

The operational design of the building is structured as a base building for the subsequent creation of fit-outs for open offices with outdoor landscaping.

The participation of the investigator in the further acquisition of data at the given structures should be direct and active, ideally from the position of being a member of the technical preparation or construction management staff working for the contractor, or a project manager or technical supervisor working for the builder. Thanks to this, the investigator will not only have access to financial data but will also be able to work with them and evaluate them.

4. Results

Table 1. Percentage share in the total costs of each stage of construction.

No.	Stage of construction	Percentage [%]
1.	Substructure	15.05
2.	Superstructure	31.30
3.	Interior work	15.38
4.	Finishing works	38.27
	Total	**100.00**

Table 2. Percentage share in the total costs of the individual technical stages of construction.

No	Stage processes of building construction	Percentage [%]
1.	Ground work and demolition	1.15
2.	Foundations	4.01
3.	Substructure	9.71
4.	Superstructure	29.32
5.	Roofing	1.92
6.	Execution of partitions, piping and wiring	13.75
7.	Execution of interior plasters and floor underlays	1.51
8.	Construction of floors, completion of surfaces and equipment	11.97
9.	Completion of piping, wiring and interior work	22.81
10.	Exterior finishes	3.46
	Total	**99.59**

Table 3. Percentage share in the total costs of utility network stage processes.

No.	Utility network stage processes	Percentage [%]
1.	Ground work	0.20
2.	Structural work	0.06
3.	The construction itself	0.13
4.	Finishing works	0.03
	Total	**0.41**

Table 4. Percentage share in the total costs of subsidiary construction processes and professions.

No.	Selected subsidiary construction processes and professions	Percentage [%]
1.	formwork	3.44
2.	demoulding	1.30
3.	reinforcement	6.85
4.	concreting	6.24
5.	tower cranes	1.01
6.	masonry	0.60
7.	modular facades	18.28
8.	internal fillings of openings	1.83
9.	plasterboard structures	0.90
10.	wiring work	9.41
11.	sanitary installation	3.11
12.	air handling units	6.54
13.	cooling of the building	4.87
14.	heating of the building	2.17
	Total	**66.55**

5. Discussion

Based on the obtained and processed data on the financial volumes of the samples used (structures), the following percentages were determined for the stages of construction and stage processes.

The total financial volume of the construction project is one hundred percent. Each stage of construction and stage process attains a specific number of percent which determines its level of cost as part of the construction process.

It is clear from the results that the chronology (order) of the technical construction processes is not the same as the order of costs from the most expensive to the cheapest, or vice versa.

To illustrate the results from the evaluated data, the total price of building construction is used, e.g. 500 million Czech crowns, alongside the total price in EUR at an assumed exchange rate of 25.50 CZK/EUR. These financial details have been chosen randomly to show the variable and practical use of the investigated issues.

If we maintain the sequence of the technical construction methodology, the cost levels according to the stages of construction and stage processes are as follows:

Table 5. Example cost levels of the stages of construction.

No.	Stage of construction	amount		
		[%]	[CZK]	[EUR]
1.	Substructure	15.05	75,250,000,-	2,950,980.39
2.	Superstructure	31.30	156,500,000,-	6,137,254.90
3.	Interior work	15.38	76,900,000,-	3,015,686.27
4.	Finishing works	38.27	191,350,000,-	7,503,921.57
	Total	**100.00**	**500,000,000,-**	**19,607,843.14**

Table 6. Example cost levels of the stage processes of the building.

No.	Stage productions of the building	amount		
		[%]	[CZK]	[EUR]
1.	Ground work and demolition	1.15	5,750,000,-	225,490.20
2.	Foundations	4.01	20,050,000,-	786,274.51
3.	Substructure	9.71	48,550,000,-	1,903,921.57
4.	Superstructure	29.32	146,600,000,-	5,749,019.61
5.	Roofing	1.92	9,600,000,-	376,470.59
6.	Execution of partitions, piping and wiring	13.75	68,750,000,-	2,696,078.43
7.	Execution of interior plasters and floor underlays	1.51	7,550,000,-	296,078.43
8.	Construction of floor, completion of surfaces and equipment	11.97	59,850,000,-	2,347,058.82
9.	Completion of piping, wiring and interior work	22.81	114,050,000,-	4,472,549.02
10.	Exterior finishes	3.46	17,300,000,-	678,431.37
	Total	**99.59**	**498,050,000,-**	**19,531,372.55**

Table 7. Example cost levels of utility network stage processes.

No.	Utility network stage processes	amount		
		[%]	[CZK]	[EUR]
1.	Ground work	0.20	1,000,000,-	39,215.69
2.	Structural work	0.06	300,000,-	11,764.71
3.	The construction itself	0.13	650,000,-	25,490.20
4.	Finishing works	0.03	150,000,-	5,882.35
	Total	**0.41**	**2,100,000,-**	**82,352.94**

During the investigation of the given issues, several obstacles may occur which limit the collection and evaluation of data.

It is mainly a long-term activity which enables the gradual evaluation of even a small number of samples (structures), these mainly being financially demanding buildings costing in the order of hundreds of millions of Czech crowns.

The obtained results show that the described methodology can be used to estimate the financial demands of buildings.

If more samples were obtained and evaluated not only from a specific town and its surroundings but also from specific parts of the Czech Republic, the European Union and the world, percentage shares in total costs could be determined for regions, states and continents.

6. Conclusion

Based on the obtained results, it can be said that the values for each stage of construction are for the substructure 15.05%, superstructure 31.30%, interior work 15.38% and finishing works 38.27%.

The shares in the total costs of the technical stages of construction of the following are: ground work and demolition 1.15%, foundations 4.01%, substructure 9.71%, superstructure 29.32%, roofing 1.92%, execution of partitions, piping and wiring 13.75%, execution of interior plasters and floor underlays 1.51%, construction of floors, completion of surfaces and equipment11.97%, completion of wiring installations and interior work 22.81%, exterior finishes 3.46%.

For utility networks, the percentage shares in the total costs are divided thus: ground work 0.20%, structural work 0.06%, the structures themselves 0.13% and finishing works 0.03%.

It can be said with regard to construction processes and professions that the percentage share in total costs of monolithic reinforced concrete structures in relation to the whole construction process is 17.83%, which is 89,150,000,- CZK. The involvement of tower cranes is 1.01% (5,050,000,- CZK), masonry 0.6% (3,000,000,- CZK), modular façades 18.28% (91,400,000,- CZK), interior fillings of openings 1.83% (9,150,000,- CZK), plasterboard structures 0.90% (4,000,000,- CZK), building services 16.69% (83,450,000,- CZK) and wiring work 9.41% (47,050,000,- CZK).

7. References

[1] Motyčka V, Henková S and Kantová R 2014 Brno University of Technology' s Faculty of Civil Engineering Considers the Opinions of Construction Companies When Planning Study Programme Content *Conference Proceedings of 3rd Cyprus International Conference on Educational Research* Procedia Social and Behavioral Sciences ELSEVIER LTD (Amsterdam - Netherland) ISSN: 1877-0428 pp 423-427

[2] Jarský Č, Musil F, Gašparík J, Motyčka V, Kovářová B, Svoboda P, Václav P and Vyčítal M 2019 *Příprava a realizace staveb* 2 vol. 2 Brno: CERM p 348 ISBN: 978-80-7204-994-3

[3] Šťastný J, Motyčka V, Brandtner M, Bříza L and Doubek R 2018 Monitoring staveb - sledování nasazení zvedacích prostředků *Czech Journal of Civil Engineering* vol. 1 p 85-92 ISSN: 2336-7148

IOP Conf. Series: Materials Science and Engineering **867** (2020) 012007 doi:10.1088/1757-899X/867/1/012007

Efficiency of the crystallizing waterproofing admixture in lower-quality concrete

R Figmig[1]

[1] Technical university of Kosice, Faculty of Civil Engineering, Institute of Environmental Engineering, Department of Material Engineering, Slovakia

Abstract. The article deals with the topic of a crystallizing waterproof admixture using in concrete. This type of admixture is designated for making a hardened concrete more durable by minimizing its permeability. Considering that concretes of higher strength classes meet the requirements for durability parameters in most cases, in this experiment 3 low-durability (in terms of high w/c ratio) concrete ($D_{max}8$) batches were mixed. From each of mixture, 2 sets of samples (with and without the admixture) were prepared for depth of penetration of water under pressure test, strength test and water absorption test after 28 days of curing. Likewise, samples for microscopic visual assessment of crack-filling process were produced. Cracks onto these samples were intentionally introduced after their curing in time of 7 days and subsequently were further cured under water and regularly observed. According to the results, tested admixture acted more counter-productive than helpfully.

1. Introduction

It is well-known and widely accepted that concrete is, due to its technical parameters as well as for price, one of the most, if not the most, used building material in structure assembling. On the other hand, there are some aspects that must be managed to ensure durability and long-term resistance of concrete structure [1, 2] in accordance with requirements of sustainability in civil engineering [3]. The cracking-problem is inherent of concrete, that mainly resides in the low ability of fresh concrete to resist of tensile stress as well as hardened concrete [4, 5, 6]. Also, permeability and porosity are inseparably connected with concrete microstructure and consequently with durability of concrete [7]. All these parameters influence concrete resistance against deterioration. One of the current trends of concrete technology is designing concrete not only with high mechanical properties, but also with environmental added value [8, 9, 10]. For this purpose, research and development is focused on preparing modern additives and admixtures for cement and concrete with the aim to make the concrete microstructure denser with low and very low permeability and porosity or intentionally porous by entraining microbubbles (mainly for freeze-thaw resistance) [11, 12]. The process of crack forming and developing is very complex and has several reasons such as drying up including plastic shrinkage, heat development and autogenous shrinkage, plastic settling or overloading [13]. This problem can be handled by technological-operational curing processes against interface temperature volatility, windy environment or fast water evaporation from fresh concrete. The actions for decreasing (capillary) porosity and permeability reside in using pozzolanic or latent-hydraulic mineral additives [14]. Another opinion is based on using air-entraining admixture as well as waterproofing admixture based on crystalline technology [15, 16, 17, 18, 19, 20].

In this paper, one type of crystallizing waterproof admixture is the object of experimental investigation. Nowadays, it is widely used admixture in concrete for slab foundation and in underground structure, where pressure water occurs. In technical sheets of this kind of admixture can be found, that it has positive effects on lowering the depth of penetration of water under pressure, but also about their ability to make self-healing product by reacting with cement paste under the wet condition. The chemical composition of this admixture is designed for creating the insoluble crystal forms in capillary system of concrete. For this purpose, moisture in concrete is necessary. Considering that most of the concretes of higher strength class and with low cement/binder ratio are able to meet the requirements of durability parameters on itself without using extra additive/admixture, experimental investigation is focused on the concretes with higher w/c ratios (considering local relations).

2. Materials and methods

For the experiment, 3 various batches with two sets of samples (with and without crystalline admixture) were mixed. Consistency by cone-slump and density of fresh concrete were tested as well as crack-filling, density, compressive strength, splitting tensile strength, depth of penetration of water under pressure test and water absorption test were performed on hardened concrete.

2.1. Concrete mix designs

For preparing fresh concretes, cement CEM II/A-S 42.5 R ("CEM"), tap water ("WAT"), natural aggregate fraction ("AGG") 0/2, 0/4 and 4/8, superplasticizer based on PCE ("PCE") and crystalline admixture ("CA") were used. Overall 3 batches were mixed. Within every batch, reference samples were made and then CA in amount approximately 7.35 $kg.m^{-3}$ were added ("+"), batch remixed and samples with CA produced. Labelling with mix compositions are given in table 1.

Concrete mix compositions were designed with aim to prepare 3 various strength classes by using higher content of cement together with decreasing w/c ratio.

Table 1. Concrete mix designs.

Material (wt %)	**M1**	**M1+**	**M2**	**M2+**	**M3**	**M3+**
CEM	10.00		10.85		11.65	
WAT (total)	8.35		8.00		7.65	
W/C total ratio (excl. PCE and CA)	0.835		0.740		0.655	
AGG 0/2	2.45		2.45		2.40	
AGG 0/4	42.40		42.15		41.90	
AGG 4/8	36.70		36.50		36.30	
PCE (wt % of CEM)	0.78		0.91		0.96	
CA (wt % of mix)	0.000	0.175	0.000	0.175	0.000	0.175

2.2. Fresh concrete properties

Fresh concrete consistency ("S") was measured after 5 minutes from first contact of cement and water (for M1, M2 and M3), respectively immediately after CA addition and concrete remix (for M1+, M2+ and M3+). Consistency was performed as cone slump test in 3 layers, each compacted by 25 rod strikes. Fresh concrete density ("D_{mix}") was performed in cylindric mould (slimness 1:2) with volume of 5.3 litres. D_{mix} was also controlled on hardened concrete after demoulding and taking into account evaporated water.

2.3. Hardened concrete curing and properties

Produced samples were cured in laboratory conditions for 1 day protected against water loss. After demoulding, samples were cured in water bath till the test execution.

On 7th day, crack with width <0,4mm was intentionally introduced into the samples, for later observation of the self-healing effect. These samples were consequently cured in water again till 28th day.

Density in saturated state of hardened concrete ("D_{sat}") was calculated as ratio of weight and measured dimensions of saturated samples, density in dried state were determined on samples that underwent water absorption ("WA") test.

Compressive strength ("f_c") in age of 28 days was performed on water saturated surface dry (SSD) cube (a = 100 mm).

WA test were performed on cubes (a = 100 mm), that were dried in the laboratory oven at 105±5 °C for 7 days, until constant mass was reached. Water absorption was calculated as ratio of absorbed water and dried mass. Samples after WA test also underwent compressive strength test ("$f_{c,WA}$")

Maximum depth of water under pressure were determined on cubes (a = 150 mm), when pressure of water 500±50 kPa (5 bars) were applied for 72 hours and then measured the height ("h_{max}") from the bottom, where pressure water reach the maximum. At the end, splitting tensile test ("$f_{ct,sp}$") were performed.

3. Results and discussion

3.1. Fresh concrete properties

Results of consistency and fresh concrete density are given in table 2. Comparing batches with and without CA, it is obvious that consistency decrease within each batch (for M1 – 50 mm, for M2 – 60 mm and for M3 – 80 mm), where CA is applied despite the fact, that CA was applied in liquid form. The density of fresh concrete also decreased when CA was added (for M1 – 15 kg.m^{-3}, for M2 – 20 kg.m^{-3} and for M3 – 30 kg.m^{-3}), what could be caused by adding extra water from CA. Workability behaviour of fresh concrete with CA has character like slightly aerated concrete. For adjustment consistency it would be necessary to add mix water, what should decrease the durability of concrete or add PCE what would make the 1 m^3 more expansive.

Table 2. Consistency of fresh concrete and density of fresh concrete.

Parameter	M1	M1+	M2	M2+	M3	M3+
S [mm]	80	30	100	40	140	60
D_{mix} [kg.m^{-3}]	2345	2330	2340	2320	2395	2365

3.2. Hardened concrete properties

Results of measured parameters are given in table 3. The trend of density decrease with CA addition is obvious also for D_{sat} (for M1 – 20 kg.m^{-3}, for M2 – 20 kg.m^{-3} and for M3 – 30 kg.m^{-3}) and D_{dry} (for M1 – 15 kg.m^{-3}, for M2 – 20 kg.m^{-3} and for M3 – 35 kg.m^{-3}).

In each pair of samples, increase of WA by 0.1% (abs) can be seen, what could be caused by CA addition due to its liquid form and slightly w/c ratio increase. WA varied as supposed, the lowest values reached the concretes (M1 and M1+: 6.4 and 6.5 wt %) with the lowest w/c ratio, the highest the concretes (M3 and M3+: 7.6 and 7.7 wt %) with the highest w/c ratio. Values of penetration height of pressure water were observed in the same way: 9 and 15 mm for M3 and M3+; 20 and 24 mm for M1 and M1+ as well as splitting strengths 3.35 and 3.10 MPa for M3 and M3+; 2.75 and 2.60 MPa for M1 and M1+.

Table 3. Hardened concrete properties.

Parameter	M1	M1+	M2	M2+	M3	M3+
D_{sat} [kg.m^{-3}]	2370	2350	2360	2340	2415	2385
D_{dry} [kg.m^{-3}]	2200	2185	2205	2185	2275	2240
WA [wt %]	7.6	7.7	7.1	7.2	6.4	6.5
h_{max} [mm]	20	24	19	20	9	15
$f_{ct,sp}$ [N.mm^{-2}]	2.75	2.60	3.00	2.55	3.35	3.10

f_c [N.mm^{-2}]	28.5	24.7	35.4	31.1	44.4	39.0
$f_{c,WA}$ [N.mm^{-2}]	43.6	39.7	52.8	45.8	65.5	58.5

The decrease of compressive strength when CA was added can be also observed. Relative decrease of compressive strength for M1 was -13.3%, for M2 -12.1%, and for M3 -12.2%, what could be assumed to be similar decrease. This trend was also approved on the dried sample. Significant increase in compressive strength after drying the samples was also achieved. Despite the fact, that this issue is not the object of this experimental investigation, it can be noted, that f_c increase in this way: +53.0% for M1, +60.7% for M1+, +49.2% for M2, +47.3% for M2+, +47.5% for M3 and +50.0% for M3+.

In the figure 1 and 2 digital microscopy pictures on 7th day (as reference time) and on 28th day of samples M1 and M1+ are illustrated, where no any insoluble product (excluding soluble efflorescence limestone leaking out through cracks) in cracks was observed (the same was observed for all other samples). Thus, in this experimental investigation with pre-set conditions of testing, the self-healing effect of CA was not confirmed.

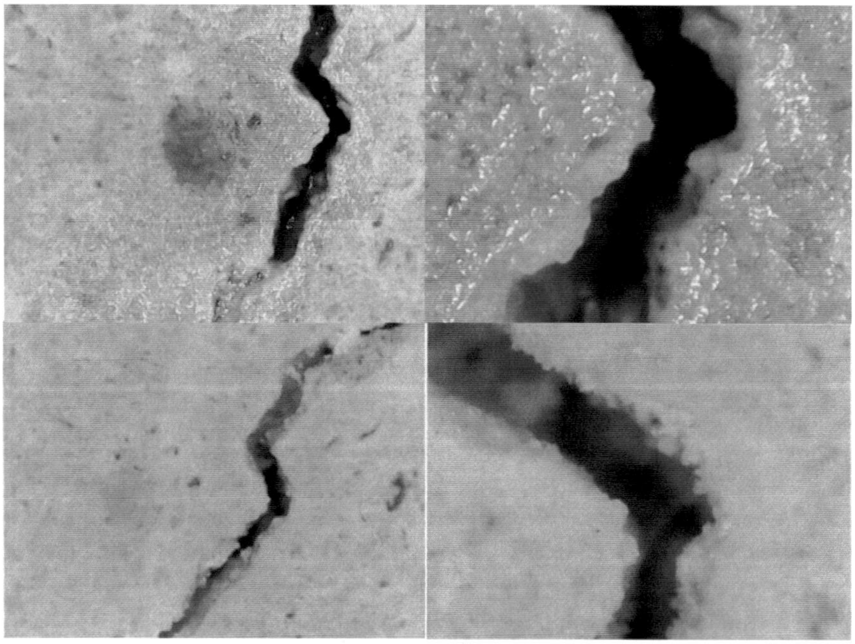

Figure 1. Sample M1 (crack width 0.2 mm) in age of 7 days (up) and 28 days (down). Magnification: 50x (left) and 150x (right).

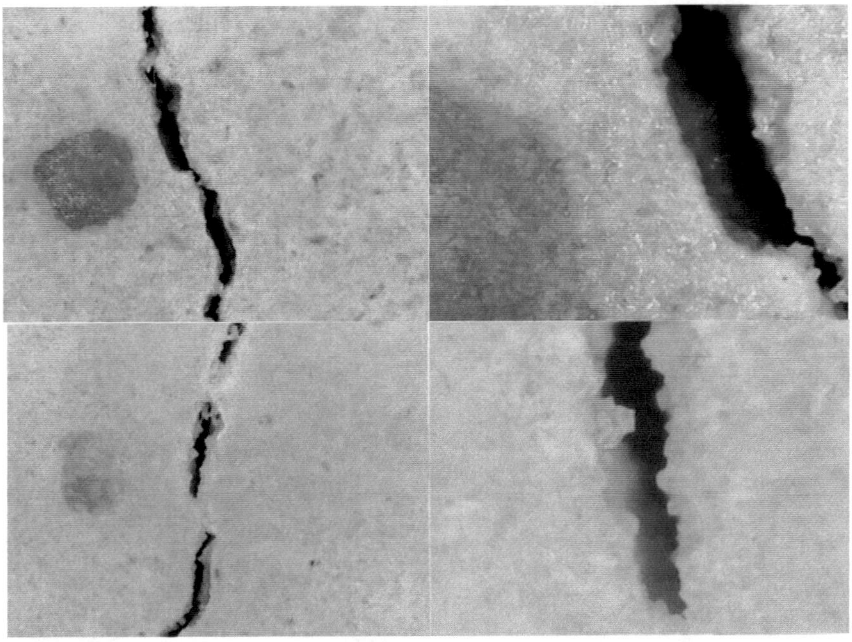

Figure 2. Sample M1+ (crack width 0.2 mm) in age of 7 days (up) and 28 days (down). Magnification: 50x (left) and 150x (right).

4. Conclusion

In this experimental study, efficiency of the crystallizing waterproof admixture in lower-quality concrete was investigated. Based on results of mechanical properties, durability parameters as well as from optical and microscopical observation can be concluding, that in this test with described materials and methods was not confirmed positive effect using this type of admixture. Contrarily, quality decrease in every testing property was achieved. However, that could be caused by increase of w/c ration due to using liquid form of CA. Before using (not only) this type admixture, it should be performed the exhaustive pre-testing, considering the balance between high economical expense of CA on the one hand and technical and environmental contribution for concrete on the other hand.

5. References

[1] Tang SW, Yao Y, Andrade C and Li ZJ 2015 *Cem. Conc. Res.* **78** 143
[2] Gu X, Jin X and Zhou Y 2016 Durability of Concrete Structures *Basic Principles of Concrete Structures* ed Gu X, Jin X and Zhou Y (Berlin: Springer) pp 553-597
[3] Aïtcin PC and Mindess S 2011 *Sustainability of concrete* (Boca Raton: CRC Press)
[4] Wang HL, Dai JG, Sun XY and Zhang XL 2016 *Constr. Build. Mat.* **107** 216
[5] Gérard B and Marchand J 2000 *Cem. Conc. Res.* **30** 37
[6] Picandet V, Khelidj A and Bellegou H 2009 *Cem. Conc. Res.* **39** 537
[7] Mehta P and Monteiro PJM 2017 *Concrete microstructure, properties and materials* (New York: McGraw Hill Education)
[8] Khodabakhshian A, De Brito J, Ghalehnovi M and Shamsabadi EA 2018 *Constr. Build. Mat.* **169** 237
[9] Lieberman R et al 2018 *J. Haz. Mat.* **344** 1043
[10] Kim Y, Hanif A, Usman M, Munir MJ, Kazmi SMS and Kim S 2018 *J. Clean. Prod.* **172**, 3056
[11] Cheung J, Roberts L and Liu J 2018 *Cem. Conc. Res.* **114** 79
[12] Aïtcin PC and Flatt RJ 2015 *Science and technology of concrete admixtures* (Sawston: Woodhead Publishing)
[13] Dyer T 2014 *Concrete Durability* (Boca Raton: CRC Press)
[14] Tokyay M 2016 *Cement and Concrete Mineral Admixture* (Boca Raton: CRC Press)

IOP Conf. Series: Materials Science and Engineering **867** (2020) 012007 doi:10.1088/1757-899X/867/1/012007

[15] Flores MR, Pirritano F, Serna P and Ferrara L 2016 *Constr. Build. Mat.* **114** 447
[16] Cappellesso VG, dos Santos Petry N, Dal Molin DCC and Masuero AB 2016 JBPR01(2016)009
[17] Pazderka J and Hájková E 2016 *Acta Polytechnica* **56** 306
[18] Cuenca E, Tejedor A and Ferrara L 2018 *Constr. Build. Mat.* **179** 619
[19] De Nardi C, Bullo S, Ferrara L, Ronchin L and Vavasori A 2017 Mater. Struct. **50** 191
[20] Hassani M, Vessalas K, Sirivivatnanon V and Baweja D 2017 *ACI Mater. J.* **114** 911

Acknowledgments
This paper has been prepared with a support of the Scientific Grant Agency of the Ministry of Education, Science, Research and Sport of the Slovak Republic and the Slovak Academy of Sciences (VEGA Grant No. 1/0648/17).

IOP Conf. Series: Materials Science and Engineering **867** (2020) 012008 doi:10.1088/1757-899X/867/1/012008

Indoor air quality in classrooms of a newly built school

K Harčárová[1]

[1] Technical university of Kosice, Faculty of Civil Engineering, Department of Environmental Engineering, Slovakia

Abstract. The presented paper deals with the evaluation of the quality of the indoor environment in a new school building with modular construction. The required short-term measurements were carried out in two selected classrooms. Selected physical-chemical factors such as thermal comfort, sound level, lighting conditions, concentration of CO_2, particulate matter ($PM_{0.5}$-PM_{10}) and total volatile organic compounds (TVOC) were monitored. Based on the obtained results, it was shown that the values of operative temperature, relative humidity and air velocity measured in selected rooms meet the requirements given by legislation. The average CO_2 concentration was exceeded in both classes, which may lead to headaches and concentration problems in pupils. Concentrations of $PM_{2.5}$ were below the acceptable limit, while PM_{10} concentrations were exceeded in both classes. The recommended TVOC concentration of 200 μg/m^3, which does not yet cause discomfort and has no adverse health effects, was exceeded in both classrooms.

1. Introduction

Polluted indoor air is currently considered to be one of the most important determinants of environmental quality and public health. The indoor environment in schools is an important factor, as children belong to a particularly vulnerable group of population. Nowadays, it is generally known that children spend on average six hours a day and five days a week in one classroom at school. There is plenty of evidence of the potential harmful health effects of many indoor air pollutants that can be found in school environments that originate from ambient air or arise indoors from various materials, products or activities. The presence of pollutants in schools can also affect children's growth, their learning performance as well as their cultural and social development. Indoor air pollution in schools is complex and variable and can have different origins (exterior, interior) and nature (physical, chemical and biological). The most frequently monitored indoor pollutants include particulate matter ($PM_{0.5}$-PM_{10}), carbon dioxide (CO_2) and total volatile organic compound (TVOC). Kalimeri et al. evaluated the quality of the indoor air in three schools. Based on the results, it was shown that the concentrations of selected volatile organic compounds were within acceptable limits. Average indoor temperature and relative humidity during pupils occupancy ranged from 20-26 °C and from 40.1-52.2% respectively for the non-heating period and from 15-24 °C and from 31.4-50.7% for the heating period [1]. According to Simanic et al. median values of indoor air temperatures were between 19-23.5 °C in all monitored classes during school hours. The 90-percentile levels of CO_2 concentrations were below 1000 ppm in 60 of 61 classes [2]. Benett et al. found that the average indoor level of PM_{10} during the school day (30.1 μg/m^3) was significantly higher than the outdoor level (8.9 μg/m^3). A more detailed analysis of the samples revealed that the indoor PM_{10} was primarily composed of bark (soil) elements likely to be introduced into children's footwear. The presence of $PM_{2.5}$ was due to external infiltration from motor vehicle emissions [3].

The main aim of this study is to evaluate the quality of the indoor environment in the school building. Air monitoring was performed in two representative school classes. In each class, in addition to thermal, acoustic and light comfort, concentrations of CO_2, particulate matter and TVOC were also monitored.

2. Methodology

2.1. School location and classroom characterization

The monitoring of selected physical and chemical factors of the internal environment was carried out at a primary school in the village of Veľký Biel situated in western Slovakia. The building is specific in that it is the first school building with a modular construction in Slovakia. Indoor air quality measurements were realized in winter season in two selected classes (class A and class B), which represent the types of classes in which different conditions and measured values can be expected to occur. During the measurements, there were 12 children and 1 adult in the class A and 15 children and 1 adult in the class B. In both classes, regular air exchange was ensured by natural ventilation. The classrooms were equipped by standard school furniture consisting of wooden school-desks and chairs, an interactive whiteboard, several particleboard cabinets and playing mat.

2.2. Instrumentation

All measuring instruments were located in the back of the classrooms for security reasons and to avoid disturbing the teaching process. The devices were placed approximately at a height of about 1.1 m in the breathing zone of occupants. Thermal comfort parameters, light comfort parameters and CO_2 levels were determined using a TESTO 435-4 multifunction instrument with appropriate probes (Testo, Inc.; Germany). The concentration of CO_2 was measured in classes with natural ventilation. The illuminance (E) was measured within the lighting network at a height of about 1.1 m. A hand-held noise analyzer - Brüel and Kjaer Type 2250 (Brüel & Kjaer; Denmark) was used to measure equivalent sound levels (LAeq). These measurements were carried out for 1 hour. The particulate fraction concentrations of 0.5-10 µm ($PM_{0.5}$-PM_{10}) were determined using a HANDHELD 3016 IAQ (Lighthouse Worldwide Solutions, Inc., USA) measuring instrument, which uses a laser-diode light source and collecting optics for particle detection. The measurement was performed with a real-time direct-reading device for 50 minutes. Total concentrations of volatile organic compounds were determined by a ppbRAE 3000 photoionization detector with UV lamp (RAE Systems, Inc. USA). Two-point calibration, zero and standard reference gas (isobutylene) was performed prior to measurement. All measured concentrations were converted and are expressed in toluene equivalents.

3. Results and discussions

The results of the measurements of the monitored indoor air quality (IAQ) parameters are summarized in table 1. This table also includes statistical processing of measured data.

Table 1. Summary of IAQ results in classroom A and B.

Classroom	Statistics	Indoor air temperature [°C]	Relative humidity [%]	CO_2 [ppm]	$PM_{2.5}$ [µg/m^3]	PM_{10} [µg/m^3]	TVOC [µg/m^3]	LAeq [dB(A)]	E [lx]
A	average	23.6	43.8	1238	16.0	154.6	2256	79.5	1122
	min	22.3	37.4	732	10.6	46.0	1416	71.6	-
	max	25.0	51.5	1709	20.4	272.6	3367	85.1	-
	standard deviation	0.8	4.5	293.5	3.7	78.9	581.4	5.4	-
B	average	22.4	45.2	1415	12.7	88.6	2847	73.6	364
	min	21.9	37.2	705	9.1	33.8	2490	59.1	-
	max	22.7	47.9	1615	14.0	112.4	3150	85.7	-
	standard deviation	0.2	3.6	224	1.0	17.2	174.9	12.3	-

3.1. Thermal comfort

The average air temperatures in the selected classes were 23.6 °C (class A) and 22.4 °C (class B). The average relative humidity values in the selected classes A and B were 43.8% and 45.2%, respectively. Required range of relative humidity (30-70%) according to Decree no. 259/2008 Coll. was fulfilled in all the monitored classes [4]. The average air velocity should not exceed 0.2 m/s. This legislative requirement has been met in both cases. As can be seen in the figure 1, the air temperature during the measurement had a decreasing tendency. The lowest recorded air temperature was 22.3 °C and the highest was 25 °C. Figure 2 shows the course of temperature and relative humidity during the hourly measurement in class B. The lowest recorded air temperature was 21.9 °C and the highest was 22.7 °C.

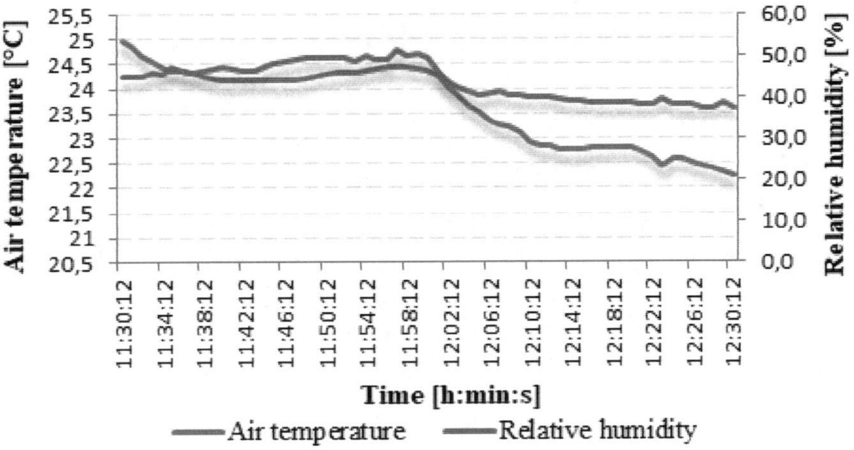

Figure 1. The course of temperature and relative humidity in class A.

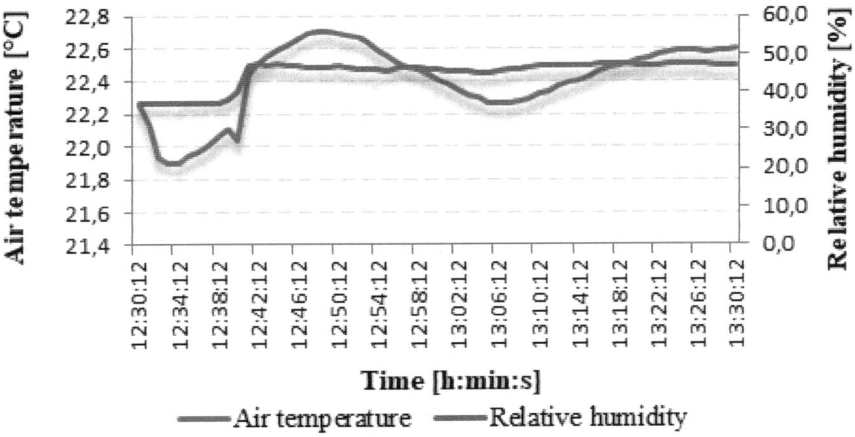

Figure 2. The course of temperature and relative humidity in class B.

3.2. Light and acoustical comfort
The average level of illuminance was 1122 lx in class A and 364 lx in class B. In the second case, the measured value did not reach the minimum allowed value of 500 lx, which is given by Decree no. 259/2008 Coll. [4]. This value was 27.2% lower than the limit value.

The measured values of equivalent sound levels (LAeq) are shown in Table 1. The lower average LAeq value was found in class B, i.e. in a class with less number of pupils. The limit value of 40 dB (A) as given by Government Decree no. 115/2006 Coll., was exceeded in both classes almost twice [5].

3.3. Concentrations of CO_2
The course of CO_2 concentrations in class A in the presence of 12 children and 1 adult is shown in figure 3. The CO_2 concentration at the beginning of the measurement was 1185 ppm and at the end 732 ppm. Figure 4 shows the course of CO_2 concentrations in class B in the presence of 15 children and 1 adult. In this class, the temperature at the beginning of the measurement was 722 ppm and at the end of the

measurement 1559 ppm. In both classes, the average CO_2 concentration values exceeded the recommended value of 1000 ppm given by Pettenkofer [6].

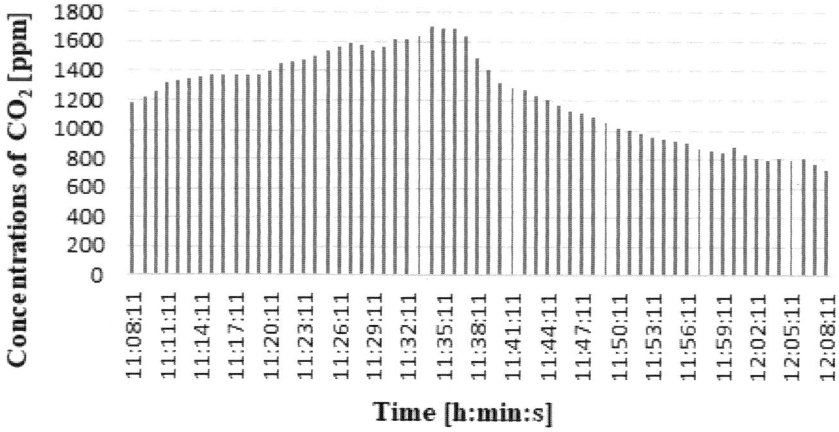

Figure 3. The course of CO_2 concentrations in class A.

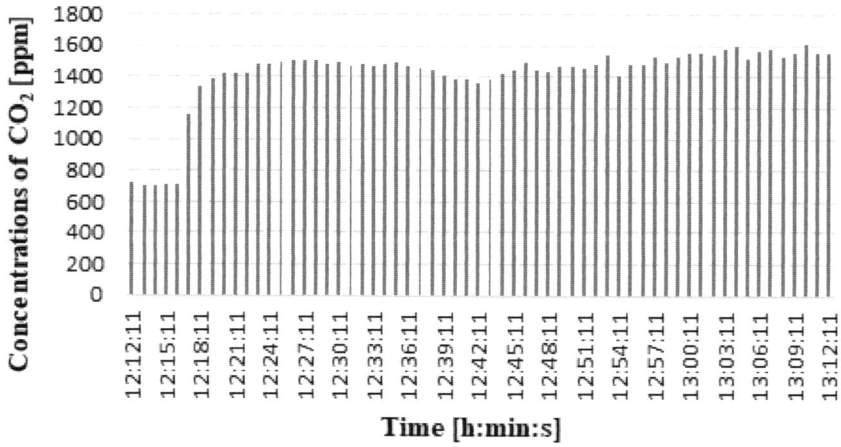

Figure 4. The course of CO_2 concentrations in class B.

3.4. Concentrations of particulate matter

In class A, $PM_{2.5}$ concentrations ranged from 10.63 $\mu g/m^3$ to 20.38 $\mu g/m^3$ and PM_{10} concentrations from 45.97 $\mu g/m^3$ to 272.60 $\mu g/m^3$. In the class B, $PM_{2.5}$ concentrations ranged from 9.07 $\mu g/m^3$ to 13.97 μg m^3 and PM_{10} concentrations from 33.82 $\mu g/m^3$ to 112.36 $\mu g/m^3$. The average concentrations of $PM_{2.5}$ and PM_{10} in the classes were 15.96 $\mu g/m^3$ and 154.63 $\mu g/m^3$ for class A and 12.65 $\mu g/m^3$ and 88.65 $\mu g/m^3$ for second class B. The maximum permissible value of PM_{10} particulate concentration is according to Regulation 210/2016 Coll. 50 $\mu g/m^3$ [7]. This limit value is set for 24-hour human exposure. The measurement of the particulate matter concentrations did not take 24 hours, but to understand the really high levels recorded in the selected classes, the measured values were compared with this limit value. Concentrations of PM_{10} in class A were three times higher than the limit value, which represents a 67.7% exceedance of this limit value. In class B, it was observed that this limit was exceeded by 38%. Figure 5 and 6 show the course of particulate matter concentrations of all measured fractions over 50 minutes of measurement. As can be seen, concentrations fluctuated because measurements were made under natural ventilation conditions.

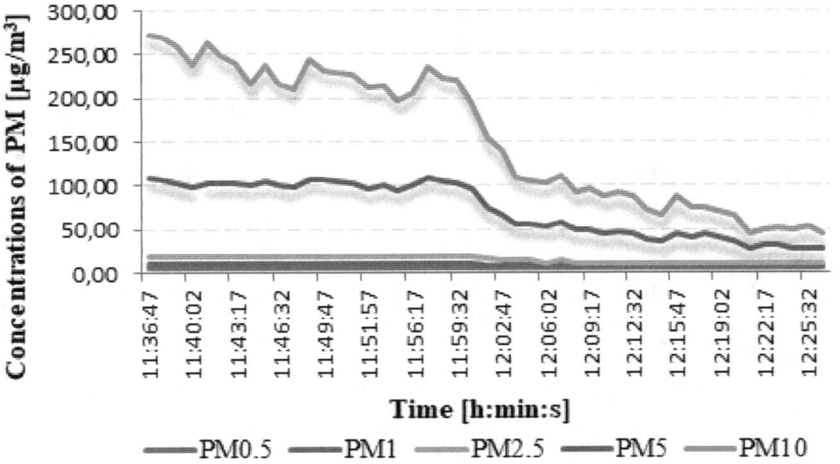

Figure 5. The course of $PM_{0.5}$-PM_{10} concentrations in class A.

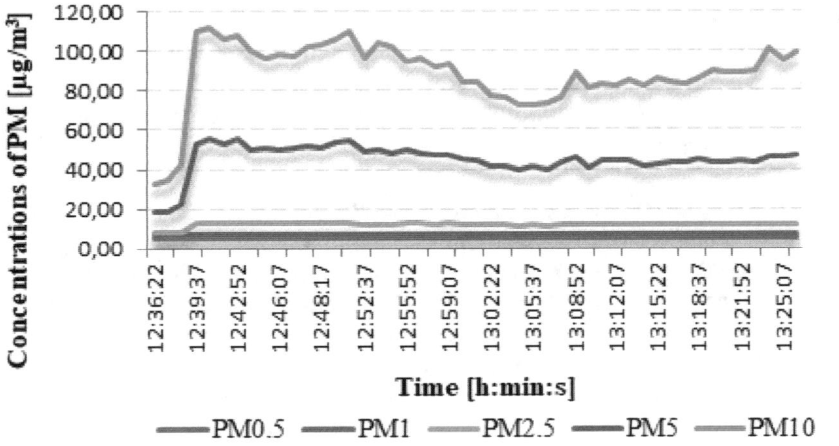

Figure 6. The course of $PM_{0.5}$-PM_{10} concentrations in class B.

3.5. Concentrations of total volatile organic compounds

In class A, concentrations ranged from 1416 to 3367 μg/m^3 with an average value of 2256 μg/m^3. In class B, concentrations ranged from 2490 to 3150 μg/m^3 with an average value of 2847 μg/m^3. The recommended concentration of TVOC, which does not yet cause discomfort or adverse health effects according to Mølhave with a value of 200 μg/m^3, was exceeded several times in both classes [8]. In class A the limit value was exceeded by 91.1% and in class B by 93%. A significant decrease in TVOC concentrations was observed in class A, as the air exchange in the class was ensured by natural ventilation.

4. Conclusion

The average relative humidity values in the selected classes were 43.8% and 45.2%, respectively. Based on the measurement, it can be concluded that the relative humidity has reached the optimum range of 30-70% given by legislation. The permissible air velocity for the activity class should be $0 \leq 0.2$ m/s. This legislative requirement has been met in both cases. The average level of lighting in class B did not reach the required limit set by legislation. The lower average LAeq value was found in the second class, i.e. in a class with less pupils. The limit value of the equivalent sound level was almost doubled in both classes compared to legislative values. The average CO_2 concentration was exceeded in both classes, which can cause discomfort and concentration problems. Concentrations of CO_2 mostly fluctuated because pupils were present in the classroom and natural ventilation was maintained in the classroom. Higher $PM_{2.5}$ and PM_{10} concentrations were observed in the first class. In both cases, the concentrations were fluctuating, as the measurements were carried out under natural ventilation. The recommended TVOC concentration of 200 μg/m^3, which does not cause discomfort or adverse health effects according to Mølhave, was exceeded in both classes by 91.1% and 93%. On this basis, it can be assumed that pupils and teachers may experience health problems due to high concentrations of TVOC.

5. References

[1] Kalimeri K. K, Saraga D. E, Lazaridis V. D, Legkas N. A, Missia D. A, Tolis E. I and Bartzis, J. G 2016 Indoor air quality investigation of the school environment and estimated health risks: two-season measurements in primary schools in Kozani, Greece *Atmospheric Pollution Research* **7(6)** 1128-1142.

[2] Simanic B, Nordquist B, Bagge H and Johansson D 2019 Indoor air temperatures, CO_2 concentrations and ventilation rates: Long-term measurements in newly built low-energy schools in Sweden *Journal of Building Engineering* **25** 100827

[3] Bennett J, Davy P, Trompetter B, Wang Y, Pierse N, Boulic M, Phipps R and Howden-Chapman P 2019 Sources of indoor air pollution at a New Zealand urban primary school; a case study *Atmospheric Pollution Research* **10(2)** 435-444

[4] Ministry of Health, Slovak Republic, Governmental Regulation Number 259/2008 Coll. Detailed Requirements for the internal environment of buildings and on minimum requirements for apartments of lower standard and accommodation facilities 2008

[5] Ministry of Health, Slovak Republic, Governmental Regulation Number 115/2006 Coll. Minimum Health and Safety requirements for the protection of workers against the risks of exposing to noise 2006

[6] Pettenkofer, M V 1858 Über den Luftwechsel in Wohngebäuden. München: Literarischartistische Anstalt der Cotta'schen Buchhandlungen.

[7] Ministry of Health, Slovak Republic, Governmental Regulation Number 210/2016 Coll., Detailed Requirements for Indoor Environment of Buildings and minimum Requirements for Low-standard Flats and Accommodation Facilities 2016

[8] Mølhave L 1990 *Toronto Canada* Volatile organic compounds, indoor air quality and health. Indoor Air *In Proceedings of the 5th International Conference on Indoor Air Quality and Climate.* 15-33

Acknowledgments

This study was financially supported by Grant Agency of Slovak Republic to support of project No. 1/0307/16 and project NFP313010T578.

IOP Conf. Series: Materials Science and Engineering **867** (2020) 012009 doi:10.1088/1757-899X/867/1/012009

Construction of modular green roof panel

M Hejl[1]

[1] Brno University of technology, Faculty of Civil Engineering, Department of technology, mechanization and construction management

Abstract. This article is dedicated to make people familiar with process of creating new construction of modular green roof panel. Article also shows creative process of prefabricated panel and all versions of its construction from the first attempts to final ecological modular panel for green roof. Article also describes problems of every version of modular panel which was created, and solutions which were made for final versions of green roof modular panel. Article also shows a final modular panel with its construction and possibility of its use on the roofs. In the article you can also read about tests which will be done in the future for verification of right construction of green roof panels and its behaviour in the real outside conditions.

1. Introduction

Thanks to climate changes the awareness of people about ecological construction is rising. We already know about Heat Island effect which results to continual overheating of city centres. That is the reason why big cities and agglomerations are investing more money in construction of cooling zones made of greenery, especially green roofs and green facades as we can see for example in Wien (Austria) or in Milano (Italy). But these types of constructions are also bringing on the scene some challenging problems with which we have to deal. The main problem is probably long time which we need for transportation of material to the roof and for greening the roof. These constructions are about symbiotic living of organic (greenery) and inorganic materials. This leads to seasonal construction of these systems. For this problem we have an easy solution, making green roofs on the ground and after that transporting whole system on the roofs. We already have technology and possibility to make prefabricated panels for green roofs but in Czech Republic, Slovakia and Middle European countries, we are not using these possibilities and still using old fashioned types of constructions which was used with almost same processes as on the first green roofs built by the Vikings in the Scandinavian Peninsula. On the other hand, in the USA we can see usage of prefabricated panels. But there is a one big problem about it. These panels are built by HDPE plastic, which means these panels are not ecological and also not economical as it could be. This article main aim is about process of creating prefabricated green roof panel which will be ecological and economical as well. Article also describes all stages of research and unsuccessful types of construction panels. The reason of showing unsuccessful types of ecological panels is to show which materials has been already tested and which ways of research can be concluded as the dead end. At the end of the article is presented final successful panel with its characteristics and variations for extensive green roof and for semi-intensive green roof.

2. Design and realization of prefabricated ecological variants for green roofs

As a first step for the construction of prefabricated panel was realized brief review showing us the types of prefabrications which are used in Czech Republic, Europe and USA. Only prefabrication which is used in the Czech Republic as we could find is the prefabrication based on the using of grass or perennial carpets.

First types of panels were chosen with size 600×600 mm. This size was based on the USA model of prefabricated panels and was considered as the optimal thanks to possibility of better manipulation with panels and also the panels should have a lower weight. On the other hand, second type of the panels was decided to be bigger and heavier but with advantage of using current materials which is used on the roofs without necessity of cutting it or long modifications of it. Size of the second type of panel has been 1200×600 mm. This size was also considered as the biggest for possibility for manipulation on the roof and for savings of material which is used on the roof.

For both types of panels had been chosen similar layer composition for extensive green roofs. Extensive green roof was selected mainly because weight of individual panels for easy manipulation and survivability. In the composition had been used retention mats made by recycled polyester fibres interconnected with fusible fibres. This retention layer had 40 mm thickness. Function of the retention mats was protection of the waterproofing layer and retention layer for the green roof. Technical parameters of the retention mats are described in the table 1.

Table 1. Technical parameters of retention mats.

		Mat thickness 30 mm	Mat thickness 40 mm	
Basic weight	g/m^2	3 000	4 000	± 15 %
Weight of saturated mat	kg/m2	22	32	± 15 %
Thickness 0,5 kPa	Mm	30	40	± 15 %
Compressibility	Mm	20	20	
Dynamic stiffness	MPa/m	8.2	7.5	
Maximum water capacity - slope 0°	l/m^2	20	29	
Thermal conductivity – laboratory humidity	W/m*K	0.038	0.038	
Thermal conductivity – saturated state	W/m*K	0.142	0.142	

First were constructed panels with size 600×600 mm. These panels were designed and constructed in four variants. These types are marked as A, B, C and D.

As a first type A was constructed panel with supporting layer made of retention mat. On the mat was put a layer substrate for extensive green roof. Substrate was secured on the sides with polypropylene membrane. The main reason for this construction was its easy design and advantage of using retention mat for more purposes than just using it as retention mat. This construction thanks to its light weight had very good possibility of manipulation and properties for construction of the roof. As the first step for this panel had to be cut retention mats from its manufacturing sized 1200×600 mm to our size 600×600 mm. Around this load-bearing shape were constructed sides made of polypropylene membrane. These sides were tightened with steel profiles to our required shape. Sides were also tied up by metal wires. Reason for this was strengthening sides and bottom of the panel. These wires and polypropylene membrane were mentioned as removable after completion on the roof. This system would allow us to use same sides from polypropylene membrane again and don't waste it for only one usage. Into this load-bearing shape was poured in substrate for extensive green roof with height of 50 mm. Panel has been planted by sedums. For example, we can name Sedum Album, Sempervivum Arachnoideum or Sedum Reflexum. After that panel was observed for several months.

Second type of panel, type B has been constructed similar as type A. Supporting layer of the panel was made of retention mat as well as we can see at the type A. Size of the panel was 600×600 mm. Difference between these construction was that, sides has been moved by 50 mm from the edges to the panels centre. Reason for moving of the sides inside was reinforcing panels and avoidance to any possibility of losing substrate through the sides out of the panel. These 50 mm sides would be filled by substrate additionally during construction process on the roof. Or as the second option these sides could be removed before installation. Sides were from the same materials as at the first type of panel.

With panel type C it was decided to make panel with different supporting layer than just the retention matt. For creation more support of the substrate layer has been used thermal insulation from XPS with height 100 mm. The reason for usage of XPS was simple. It was decided to use material which will help with physical properties of the roof and at the same time won't be heavier than previous version. This type of construction will help to buildings thermal insulation properties and impact insulation of the roof. If we would use higher type of XPS than we could use these panels also as the green roof with opposite construction of the roof. On this XPS panel was layer of retention matt with height of 30 mm. Sides of this panel were made of polypropylene membrane which was fastened to XPS panel with wires. Substrate layer on this type of panel had 50 mm height.

Panel type D was similar to panel type C. On this type of panel has been chosen to make more simple type on the sides. Material for sides was made of Mirelon. Mirelon has been tightened as tight as it was maximally possible around XPS panel (100 mm) and retention matt (30 mm). After that Mirelon has been tightened in more layers for ensuring of the stability. Mirelon has been also fastened to XPS panel by staples. Height of the vegetation layer was 50 mm.

Next two variants of green roof panel-were constructed in different way. With the experiences from making first four types was chosen construction of the panel with size 1200×600 mm. With this size was possible to avoid technical problems which appeared on the construction of first four panel types. Size of the panel 1200 x 600 mm hasn't been chosen randomly but it is the actual size of the XPS panel and retention matt without adjustments. Also after consultation with the expert for perennial greenery prof. Pejchal from Mendel University in Brno, when he claimed that this size of panel will improve effectiveness of the panels by possibility of planting more plants for square meter than on the smaller types of the panel. These two different panels were signed as type E and F.

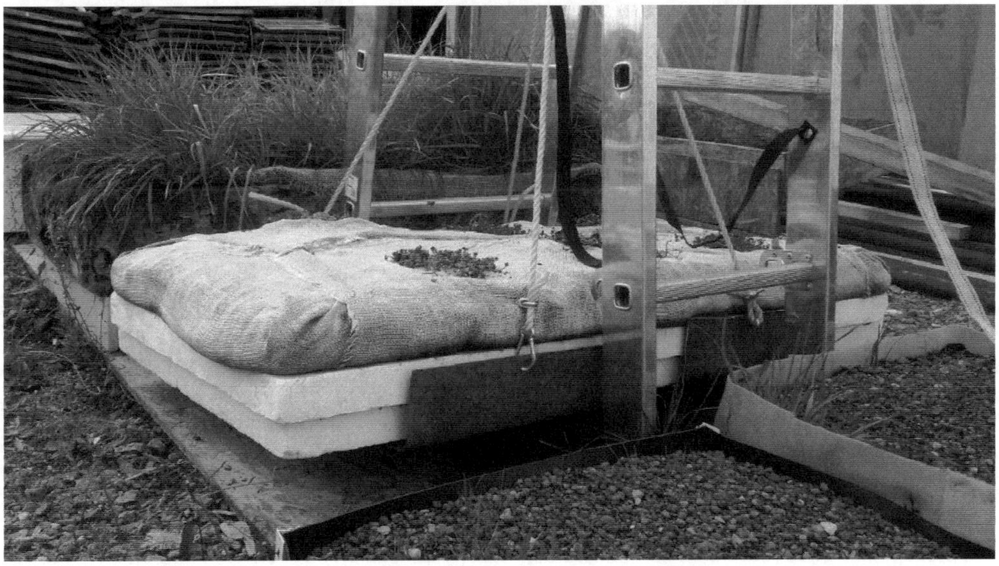

Figure 1. Extensive type E (photo author).

Panel type E was designed as a panel for extensive green roofs with planting perennial herbs, succulents or sedums. As the first thing which appeared to be necessary to do was construction of form in which will be panel constructed. Shape was constructed specifically for making construction of the panel easier and after construction of every panel could be transported to new panel and planted panels doesn't use it anymore. Shape construction was made of OSB boards which were connected on the edges with screws. Boards were higher than future panel by 10 mm. Size of the shape has been bigger by 2 mm on every side than future panel for possibility to remove form after final realization of panels. As the first after construction of form has been putted XPS panel in this shape with height 150 mm and size 1200×600 mm. Sides of this panel has been created by jute net. Jute net has been dimensioned for packaging of retention matt and substrate layer together. This net was after this cut out and on it has been laid retention matt. This construction has been laid on the XPS panel with jute net freely extending beyond the edges of shape. Jute net has been designed to cover all sides of the future panel expect the XPS layer. Retention mat in this panel had height 40 mm. After that was net tightened together to make solid edges for substrate with staples. Only top side stayed free for pouring of substrate. Substrate layer has been 50 mm. After that was fastened to sides top side of jute net around the substrate. After this form was removed and sedums were planted with Sedum Album or Sedum Lydium.

Figure 2. Type F after construction (photo author).

Because type E was the most successful it was decided to make more ambitious version of this panel and make it with same construction but as a panel for semi-intensive vegetation roofs. This type was named type F. Construction of the panel was almost the same as the construction of the extensive type of panel. Substrate layer has been higher than substrate layer for previous types and its highness has been 150 mm. Chosen vegetation for this panel has been mix of meadow grasses. This type had also construction with size 1200×600 mm. Because this panel was higher it was necessary to make a new shape for construction of the panel. Process of the panel construction was the same as for the extensive version type E. After seeding of grasses started the experiment with long-term observation of panels viability.

3. Viability observation for all types of panels

All types of panels were tested in same conditions. All panels were located in research centre ADMAS in outdoor terms. All panels were on higher place for avoidance of contact with ground. The chosen place was chosen-because of presence of natural solar radiation during the day, but at the same time it is protected from the north side strong wind, which often occurs in the given area during summer months.

The surface was made of OSB boards. The individual panels were placed side by side to have as close growth conditions as possible. Individual panels were monitored in regular intervals of one week during three months from June to September. All panels were left without any maintenance to determine their viability. Only the semi-intensive type of panel, due to another type of greening, was regularly watered with clean water during first two weeks.

Type A looked very promising during first weeks but after sometime roots of sedums started to dying. At the same time, constructional deficiencies of the type A began to show up with increasing time. Due to the temperature change, the polypropylene membrane sides have been stretched and substrate gradually fell off the panel. Also the supporting layer from retention matt has proved to be not entirely suitable due to its gradual deflection during transfer.

Panel type B thanks to similar construction as a type A had also the similar problems with dying of plants and sides of the panels lost their tightness during the summer, which led to losing more substrate. Substrate fell of through the sides. In this case, the substrate fell of onto the support construction from retention matt. Onto the sides where it would be poured during implementation on the roof. However, this would be a major issue during the transportation of the panels on the roof.

Panel type C thanks to different construction than the previous two types didn't have problem with supporting layer and stayed compact during whole test and thanks to possibility of better fastened of sides to XPS panel was the amount of substrate falling of the panel lower than on any of the previous type. However, plants didn't survive through summer even on this type of panel.

Panel type D was the construction same as panel type C so its results about the survivability of plants were the same. Sides were made of Mirelon after only one month they started to bend in the middle thanks to pressure of substrate and water in it and that caused a lot of falling substrate out from the panel. In the end of the experiment we could also see some ruptures in the Mirelon material.

First of the bigger types, panel type E, was the most promising panel from the start of experiment. And during first of the three months it didn't have any major problems. Substrate was tightened in jute net and didn't have any possibility to fell off from panel. Sedums during these three months were growing as it was expected before the launch of experiment. Even after a long-time jute net didn't had any problem with its entirety. After three months this panel was also measured on the weights when it was fully saturated. Weight of the panel was 28 kg. So its weight is operable for one worker on the roof even if its fully saturated but we can assume at least two workers who will help with construction of vegetation roof to each other. This variant was considered as the best variant but with necessity of future upgrades.

Figure 3. Semi-intensive type F (photo author).

Semi-intensive type F was the most successful type. After almost the no-time started grow grasses through jute net. Roots of the grasses also tightened all substrate and jute net together so this variant was the most stable for transporting as well. Panel survived whole summer without any complications. Only problem of this type was its weight. Weight of this panel was 35 kg and when it was saturated it weight was around 75 kilograms. Which is too much for working on the construction just with one worker. This type was concluded as a success but inappropriate for more testing.

4. Conclusion

During this research were made six experimental types of panels for prefabricated vegetation roofs. All of these panels were constructed as economical and ecological variants appropriate for installation in ecological construction which green roof is. Panel types A and B were a failure because of its bad construction of supporting layer. Types C and D could be successful but because its small size of panels they didn't storage enough water and minerals for the plants which couldn't survived for the summer. Type F was the complete success, but it was experiment with possibility where we can go to make panel to comfortable transport and construction and from this side, we can say about this type it was also a failure. On the other hand, we found the best construction possibility in panel type E which we can consider as a complete success.

The best type, type E, has been chosen as a best variant from realised panels. This type had to be also modified during the next research and tested again on experimental roof but its first construction as you can see in this article was promising from the start. After making these adjustments the panel was patented as a utility model thanks to junior specific research FAST-J-19-5867.

As we can see in this article, we cannot expect every solution with which we are counting as a possible improvement and on the paper is looking as ideal solution is not every time a good solution. But thanks to experiments and further testing we can always find a way how to make an improvement to unsuccessful research and find the solution which will be successful.

5. References

[1] Korol E, Shushunova N, 2016 *Benefits of A Modular Green Roof Technology* WMCAUS 2016
[2] Williams K J H, Lee K E, Sargent L, Johnson K A, Rayner J, Farrell C, Miller R E, Williams N S G 2019 *Appraising the psychological benefits of green roofs for city residents and workers* Urban Forestry and Urban Greening
[3] *Heat Island effect* United States Environmental Protection Agency [online] Environmental Protection Agency 1200 Pennsylvania Avenue N.W. Washington DC 20460: Environmental Protection Agency 2019 [cit. 2019-09-05] Available: https://www.epa.gov/heat-islands
[4] *Green Roof Modular System* Urbanscape [online] Knauf Insulation d.o.oTrata 32 4220 Škofja Loka Slovenia: Knauf Insulation d.o.o 2019 [cit. 2019-09-05] Available: http://green-urbanscape.com/en/solutions/green-roof-modular-system

Acknowledgement
This paper has been worked out under the project No. FAST-J-19-5867 „Possibility of the prefabrication for the green roofs“ (Možnosti prefabrikace zelených střech).

IOP Conf. Series: Materials Science and Engineering **867** (2020) 012010 doi:10.1088/1757-899X/867/1/012010

Analysis of the efficiency of sewerage with the illegal rainwater inflow

J Hrudka[1], I Marko[1], M Šutúš[1] and Š Stanko[1]

[1] Slovak University of Technology in Bratislava, Faculty of Civil Engineering, Department of Sanitary and Environmental Engineering, Slovakia

Abstract. In our research, we are focused on analysing the impact of illegal rainwater connections on the sewage network. During the normal operation of the sewage network, rainwater cannot be present. These heavy rains flows are extreme during rain events and the sewerage system is significantly overloaded and functionality it is reduced. Our study is based on the real data of the sewerage network and selected area of interest. In the assessment of illegal connections of rainwater, we use rainfall data according to the relevant ombrographic station.

1. Introduction

The basic function of the sewerage infrastructure is to protect the population and the environment from the adverse effects of the produced pollution in municipal waste water. The waste water disposal and treatment process must respect the principles of sustainable development, environmental protection and the application of legal and conceptual requirements for water management [1]. The Public Sewer Development Plan also mentions a reduction in the gap between the quantity and quality of water consumed and the water supplied by the sewage system to the aquatic environment, as well as the trouble-free and safe drainage and treatment of wastewater under any operating condition without formation and discharge. Urcikán and Rusnák [1] also describe the issue of rainwater today: Rain surface runoff is gradually less and less naturally draining. The number of impermeable areas increases and the outflow of rainwater increases. Rainwater draining from asphalt and concrete roads washes away surface pollution caused by human activity. There are also known negative impacts such as lowering of groundwater level, increase of urban wastewater, but also worsening of self-cleaning conditions in watercourses [1] [2].

The Member States of the European Union have agreed on the conditions for the disposal and treatment of municipal waste water pursuant to Council Directive 91/271 / EEC on urban and waste water treatment. The Slovak Republic has also undertaken to fulfill these conditions through conceptual and strategic plans, such as the Plan for the Development of Public Sewerage Systems for the Territory of the Slovak Republic. The goal of fulfilling public sewerage development plans is, on the one hand, the development of municipal infrastructure, the improvement of sanitation, housing comfort and the standard of living of the population and, on the other hand, increased protection and improvement of the natural resources in order to achieve the WFD (Water Framework Directive) objective of achieving good water status [2].

2. Methods

We chose the sewage network of the municipality of Veľký Lapáš as an interest site for our research. The branch under consideration is located in the newly built part of the village, there are localized new residential buildings with adjacent paved areas (pavement and parking). The selected locality is located in the southern part of the village (figure 1).

IOP Conf. Series: Materials Science and Engineering **867** (2020) 012010 doi:10.1088/1757-899X/867/1/012010

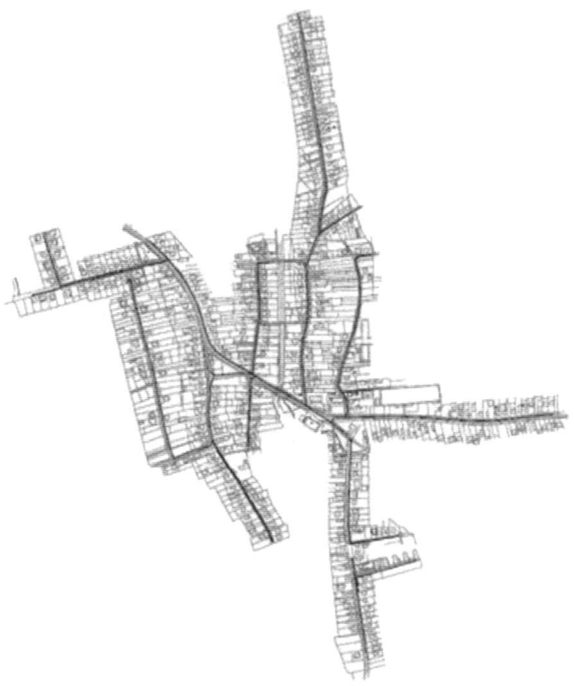

Figure 1. Sewerage network in village.

According to rainfall and duration we divide the rains into normal (normal) and extreme (abnormal). Extreme rainfall events are either short-lived (downpours, flushes) or long-term (regional, landraces). In terms of the design of sewer profiles, short-term, more intense rains are more significant. [2]

For calculations of characteristic runoff from roads, roofs, sidewalks and yards in urban areas of precipitation are defined only by rainwater. We use boundary rains to calculate the design rain flow. The boundary rain represents an idealized rain intensity profile characterized by a statistically determined periodicity p (r-1) or a repetition time T (r). [2] In this case, a rain curve from the NITRA ombrographic station was used. For periodicity p = 0.50, rain duration 15 min. The resulting boundary model rain rate is q15 = 156.94 l / s / ha (figure 2).

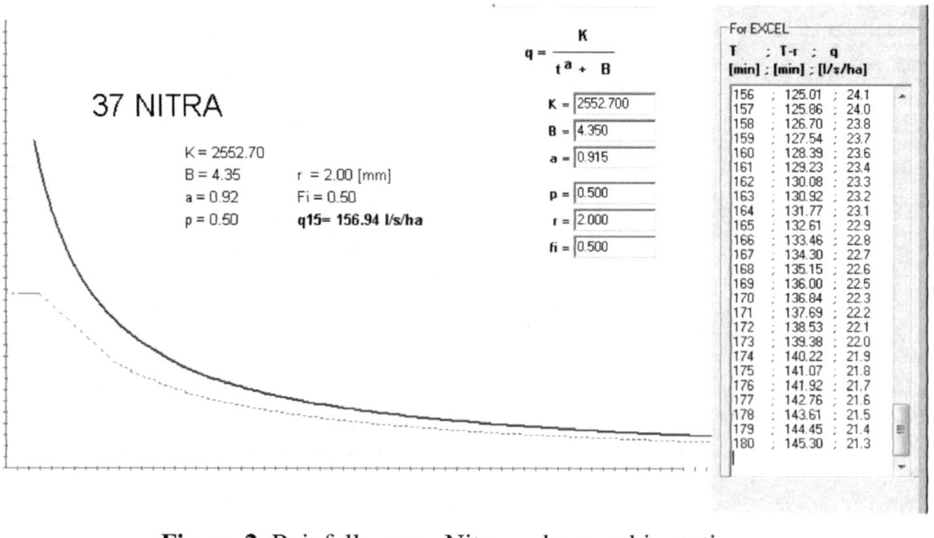

Figure 2. Rainfall curve- Nitra ombrographic station.

3. Results and discussion

To assess the impact of black rainwater connections of buildings was chosen sewer A113, where there are newly built apartment buildings. Figure 5 shows the situation.

Figure 3. Situation on sewer A113.

Water from the roof of apartment buildings and parking spaces next to them shall be taken into account when calculating the impact of black rainwater connections. Table 4 shows the area sizes of the individual objects that we are assessing.

Table 1. List of objects on sewer network - branch A113.

Object	**Area (m²)**	**Area (ha)**
1	209.9	0.0209
P1	300.1	0.0300
2	214.3	0.0214
P2	297.7	0.0297
3	214.3	0.0214
P3	311.0	0.0311
4	214.3	0.0214
P4	264.0	0.0264
5	224.2	0.0224
P5	43.8	0.0043
6	176.8	0.0176

The issue of incorrect operation of the sewer network is devoted to a number of authors who primarily examined the effect of rainwater for use [3] [4] [5]. Connection of rainwater to sewage is prohibited in the Slovak Republic. It is due to the enormous flows from rain events, but also because of sediment flowing with rainwater [6].

4. Conclusion

The assessment of black-connected objects consists of the gradual connection of the objects to the existing sewage system. By gradually adding drains from roofs and parking lots, it was ascertained whether they affect the designed pipe width, which is determined by the calculated flow rate. The aim is to increase the flow of rainwater so much that the already designed pipeline will not be sufficient for the incoming water and it will be necessary to use a larger profile.

Table 2. Analysis of the impact of black connections on sewage.

Objects		Area	Rainwater inlet	Manhole 159 - Branch A	
Flat house	**Parking spot**	**(ha)**	**(l/s)**	**Qv (l/s)**	**DN (mm)**
		0	0	12.69	300
1	P1	0.0509	7.98	28.65	300
1,2,3	P1,P2,P3	0.1545	24.24	61.17	300
1,2,3,4,5	P1,P2,P3,P4	0.2247	35.26	83.19	400
1,2,3,4,5,6	P1,P2,P3,P4,P5	0.2466	112.68	89.87	400

By connecting an area of 0.2247 ha, which corresponds to 5 apartment buildings and 4 parking areas, on the sewer A113 it was found that the designed profile DN 300 is not able to carry the connected amount of rain flow. In this case, the lower shafts on the sewer network may be flooded, which would mean flooding of the third class road - 1642 and houses and cellars in the affected area.

5. References

[1] Stanko Š, Kriš J and Škultétyová I 2011 Type of sewer system - technical and environmental advances and arrangements doi:10.1007/978-94-007-0280-6_11

[2] Urcikán P and Rusnák D 2011 Stokovanie a čistenie odpadových vôd I. Navrhovanie stokových sietí (Bratislava) p 324 ISBN 978-80-227-3435-6

[3] Dillman M, Snyder D and Chandramouli V 2013 *Calibrating a sanitary sewer system for low and high flows* World Environmental and Water Resources Congress 2013: Showcasing the Future - Proceedings of the 2013 Congress pp 129-134

[4] Chow M F and Yusop, Z 2014 *Characterization and source identification of stormwater runoff in tropical urban catchments* Water Science and Technology 69(2) pp 244-252 doi:10.2166/wst.2013.574

[5] Campisano A, Nie L M and Li P J 2013 Retention performance of domestic rain water harvesting tank under climate change conditions doi:10.4028/www.scientific.net/AMM.438-439.451

[6] Sercu B, Van De Werfhorst L C, Murray J L S and Holden P A 2011 Sewage exfiltration as a source of storm drain contamination during dry weather in urban watersheds *Environmental Science and Technology* 45(17) pp 7151-7157 doi:10.1021/es200981k

Acknowledgments

This work was supported by the Scientific Grant Agency of the Ministry of Education, Youth and Sports of the Slovak Republic and the Slovak Academy of Sciences within the project VEGA 1/0574/19, co-funded by the European Regional Development Fund and by the Slovak Research and Development Agency under the contract No. APVV-18-0203.

Old cement-based render as the cause of extremely low carbonation of 100-year bridge concrete

L Húlek[1], M Bačuvčík[1] and J Gašpárek[2]

[1] Building Testing and Research Institute (TSÚS), Bratislava, Slovakia
[2] Faculty of Civil Engineering, Slovak University of Technology, Bratislava, Slovakia

Abstract. In-situ research and laboratory examination provided on concrete core samples from old bridge structures confirmed that a thin layer of ordinary cement-based render coat protects the underlying concrete from carbonation. The average measured value by phenolphthalein test was less than 2 mm after more than 100 years of service life in exposure class XC3 of EN 206. Low carbonation depth is explained by the presence of a thin (2-4 mm) layer of the protective render coat (PRC) situated on the bridge concrete. The place, where the PRC was of good quality was almost impermeable and the carbonation of the concrete underneath was even 0 mm. If the PRC locally spalled, the carbonation depth of the same concrete, at the same structure and environment reached at this place even more than 80 mm. It was observed that low carbonation of concrete significantly depends on the non-permeability of the PRC. A narrow free space filled with the carbonates causes increased non-permeability of the PRC creating thus the built-in limestone-based (anti-carbonation) barrier with the ability to dramatically reduce CO_2 penetration into the beneath concrete over time. Values from in-situ and laboratory research are presented in the article with the explanation of these phenomena.

1. Introduction

In-situ research of a 125 years old Monier type concrete bridge at Krásno nad Kysucou in 2014 observed that a (2-4) mm thick PRC protected the underlying concrete from carbonation [1]. The repeated measurements performed in 2015 by three independent workplaces (STU Bratislava, TU Žilina and TSÚS Bratislava) proved the previous result. The same results were found on the 100-year bridges Sládkovičovo and Rimavská Sobota. The overall diagnostics of the old bridges were divided into four main phases. The first phase was concerned by visual observation of the bridge and site-determinations for consequent testing, the second included the whole non-destructive testing and taking the concrete cores for laboratory mechanical, microstructure and pore structure tests. The third phase consisted of extensive laboratory tests described below. The fourth, now experimentally ongoing phase, deals with the development of modern surface treatment from today's materials.

2. Methods of testing

In-situ tests applied in the current project concerned the following procedures: 1) testing concrete by hardness method using Schmidt hammer, model N according to STN 73 1373, 2) ultrasonic examination of concrete by STN 73 1371 using PUNDIT apparatus searching for the homogeneity of concrete material and hence the mechanical properties, 3) measurements of surface tensile strength of concrete according to STN EN 12 636, 4) measurements of surface permeability of uncovered and covered concrete by the PRC measured by Torrent method [2] (figure 1(a) and 1(b)) and 5) carbonation depth by phenolphthalein test. The taken cylindrical cores with diameter 100 mm and length at least 300 mm were tested for volume density (VD), dynamic modulus of elasticity, Young's (static) modulus of elasticity and compressive strength according to STN 73 1371, STN ISO 6784 and STN EN 123903, respectively. The cylindrical concrete specimens were half broken in the laboratory by longitudinal splitting. Phenolphthalein test was repeated on the first part and the second part was used to study the microstructure and pore structure. Consequently, two sub-samples were chosen for in-deep detail verifications – surface layer: directly exposed to prolonged exposure to CO_2 from the air and – internal layer: covered with an overlay of concrete kept out of direct contact with open air.

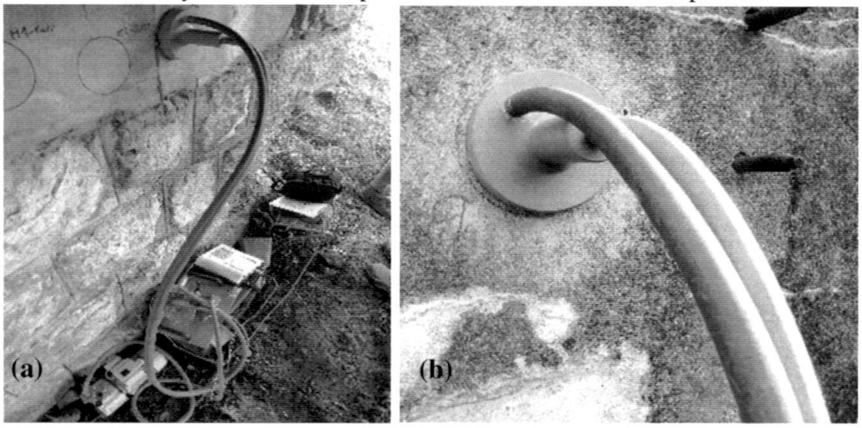

Figure 1. Torrent device for the on-site surface permeability measurement.

Subsequently, the sub-samples were crushed, heated at 60 °C to constant weight and ground for fineness passing 0.063 mm for next analytical procedures. Chemical analysis of the ground concrete matrix (without large aggregate grains) was performed in concordance with STN 72 010. The X-ray diffraction analyses were made on the automatic theta/theta X-ray diffractometer STOE & Cie GmbH (Germany) using the program Bede ZDS W95/98/NT for assessing the records and run in a 2Θ range of 5-65°. CuK_α radiation and Ni-filter were used. Thermal analyses were conducted at the NETZSCH apparatus STA 449 F3 Jupiter (Germany). Samples weighing 100 mg were heated in flowing air within the heating range (20-1100) °C and at a heating rate 10 °C/minute. The microstructure of bulk concrete samples (without large aggregate grains) was studied by scanning electron microscopy on the JEOL 7500F apparatus (Japan). The samples were fixed using a two-component epoxy paste in aluminium discs with a diameter of 25 mm, which are used to transfer the sample to the interior of the microscope. These prepared samples were plated with a mixture of gold-palladium under argon. For plating the spreader device Balzers was used. The image was captured electronically using the program "PC SEM" (version 3.0.1.8). Optical microscopy was carried out on a digital device VHX-5000 with CMOS image sensor 1/1 8 virtual pixels 1600 (H)×1200 (V) from Keyence International (Belgium). The present state of concrete is assessed on the base of the above set of the obtained results.

3. Results and Discussion

In-situ investigation revealed the relationship between the permeability measured on the PRC and the carbonation depth of the underlying concrete. Most of the measured placed appeared to be almost impermeable (measured coefficients of permeability were below $0{,}01\times10^{-16}$ m²) and at these places, the underlying concrete showed negligible carbonation depth [3] (max. 2 mm) even after being exposed to XC3 exposure class of STN EN 206+A1 for more than 100 years. At some cases, negligible carbonation was measured also if the permeability results were a bit higher. It was found that even a small crack (crack width below 0.05 mm) caused higher permeability results and also a higher carbonation depth at this location. The results of the in-situ measurements are presented in tables 1 - 3 and figures 2 and 3 for different more than 100 years old bridges.

IOP Conf. Series: Materials Science and Engineering **867** (2020) 012011 doi:10.1088/1757-899X/867/1/012011

This article is mainly focused on the bridge Sládkovičovo and Rimavská Sobota (only partially to Krásno nad Kysucou) with the observed negligible depth of carbonation. The carbonation was assessed by the own methodology based mainly on TG-DTA analysis supplemented by the XRD technique. These techniques in concrete containing limestone and dolomite stone failed to distinguish the origin of carbonates: the one when calcite is formed by the carbonation from the one when it is derived from the used aggregate. The reliable use of TG-DTA and XRD techniques is guaranteed in concrete made with siliceous aggregate. The mineralogical composition of the aggregates confirmed SiO_2-based character by the XRD analysis [4].

Table 1. Results of in-situ measurement of carbonation depth at the bridge in Sládkovičovo.

Sample		Coefficient of permeability	Average carbonation depth	Notes
Denotation	Surface	kT($x10^{-16}$ m^2)	(mm)	
G1	PRC	0.073	<1	High-quality cement-based render
G2		0.002	<1	
M3		0.039	<1	
M4		0.246	<1	
M9		0.189	<1	
M10		0.062	<1	
M6		0.123	10	Lower quality cement-based render
M2		0.178	15	
M7		4.943	20	
M8		8.464	20	
M1		>10	25	Small crack of width less than 0,05 mm in the render
M5		>10	30	Lower quality render, cavern below the cement-based render
B1	Uncovered concrete	>10	60	Spalled render
B2		>10	80	

Table 2. Results of in-situ measurement of carbonation depth at the Bridge in Krásno nad Kysucou.

Sample		Coefficient of permeability	Average carbonation depth	Notes
Denotation	Surface	kT($x10^{-16}$ m^2)	(mm)	
E1	PRC	Not measured	<4	High-quality cement-based render
E2		Not measured	<4	

Table 3. Results of in-situ measurement of carbonation depth at the Bridge in Rimavská Sobota.

Sample		Coefficient of permeability	Average carbonation depth	Notes
Denotation	Surface	kT($x10^{-16}$ m^2)	(mm)	
C1	PRC	0.001	<2	High-quality cement-based render
C2		0.001	<2	
C3		7.163	15	Small crack of width less than 0.05 mm in the render
D1	Uncovered concrete	>10	50	Spalled render

Carbonation of the underlying concrete is closely related to the impermeability of the PRC layer as stated in tables 1 to 3. Figures 4(a) and 4(b) illustrate differences between the microstructure of the unprotected concrete surface and surface-protected by the PRC.

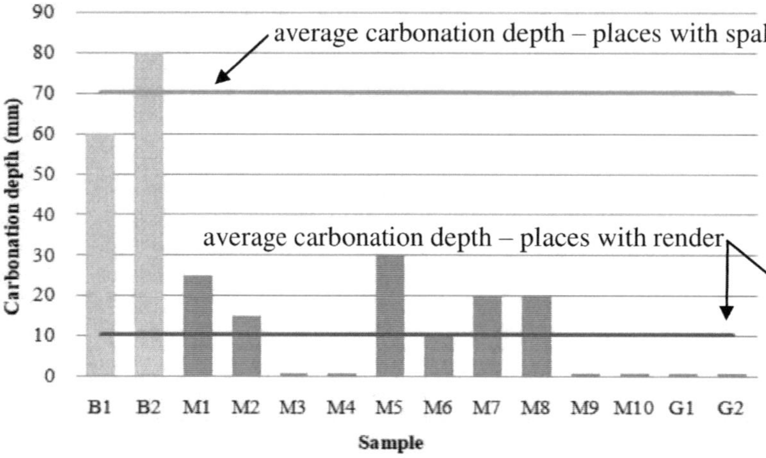

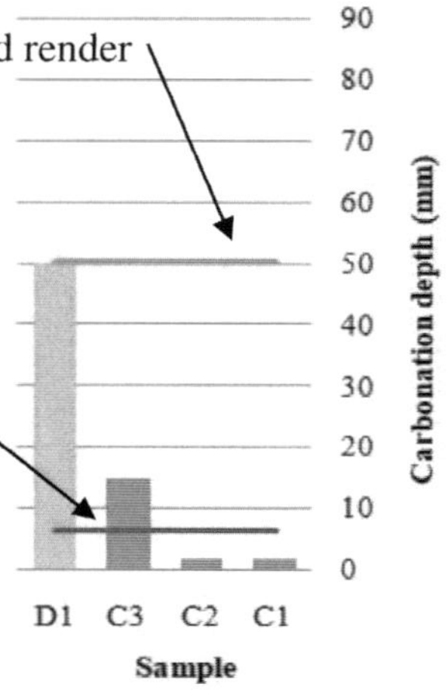

Figure 2. Results of in-situ measurement of carbonation depth at the Bridge in Sládkovičovo.

Figure 3. Results of in-situ measurement of carbonation depth at the bridge in Rimavská Sobota.

Concrete with direct contact to air demonstrates the evident amount of the coarse crystallized carbonation products, while that PRC-protected is densified by fine-grained carbonate alternates. The reduced content of coarse-grained $CaCO_3$ in the original, 100 - year - old PRC is attributed to a narrow space capable for its formation. When yet formed, the spatial deficiency of the PRC does not allow further CO_2 penetration into the concrete beneath. The carbonation depth of the concrete specimen with the uncovered and covered surface is illustrated in Figure 5(a) and 5(b).

Figure 4. The microstructure of uncovered (0-20) mm surface concrete B1 (a) and surface-protected concrete by the PRC situated over M3 (b).

4. Degree and stage of carbonation

The degree of carbonation (CD) as the CaO_{Carb}/CaO_{Total} ratio×100 (% wt.), was calculated from TGDTA results and chemical analysis. CaO_{Carb} means the CaO content bound in $CaCO_3$ originated from carbonation reaction and CaO_{Total} concerns CaO bound in $CaCO_3$, $Ca(OH)_2$, gel-like hydration products (C-S-H, C-A-H) and $CaSO_4 \times 2H_2O$. The extent of carbonation attack is also specified by the degree of modification changes (DMC) given by the $CaCO_{3,\ coarse\text{-}grained}/CaCO_{3,\ fine\text{-}grained}$ ratio. The fine-grained $CaCO_3$ is detected by TG-DTA between (600-750) °C and coarse-grained $CaCO_3$ between (750-

1100)°C. The extent of CO_2 attack is characterized by four stages of carbonation (I<55 %; II: 55-65 %; III. 66-80 %; IV.>80 %) [4]. Table 4 summarizes the carbonation characteristics of uncovered concrete, PRC-protected concrete and PRC themselves.

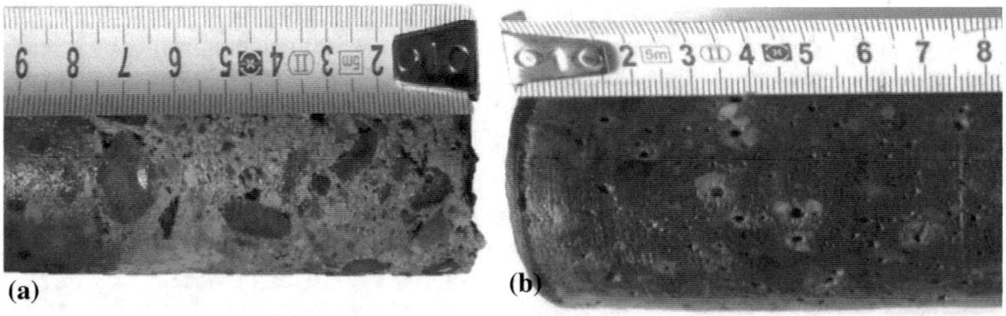

Figure 5. Carbonation depth of uncovered concrete B1 (a) and concrete below the PRC M3(b) characterized by the low permeability of 0.073×10^{-16} m^2.

Table 4. The assessment of carbonation attack by the results of chemical and thermal analysis.

Sample as marked in Tables 1 to 3 and Figure 2	CaO_{total} (% wt.)	CaO_{Carb} (% wt.) occurring in			Assessment of carbonation		
		fine - grained $CaCO_3$	coarse-grained $CaCO_3$	total $CaCO_3$	CD (%)	DMC (-)	Carbonation stage (-)
B1 - surface	9.09	2.40	7.87	10.27	113.0	3.3	IV
B1 - internal	8.51	1.40	3.08	4.49	52.8	2.2	I
M3 - surface	9.86	1.77	3.87	5.64	57.2	2.2	II
M3 - internal	7.06	1.41	2.83	4.24	60.1	2.0	II
PRC over M3	29.11	5.05	10.19	15.24	52.4	2.0	I
D1 - surface	15.66	4.91	9.55	14.46	92.3	1.9	IV
D1 - internal	19.41	1.75	9.48	11.23	57.9	5.4	II
C2 - surface	20.68	2.10	10.79	12.89	62.3	5.1	II
C2 - internal	19.42	1.63	8.41	10.04	51.7	5.2	I
C1 - surface	19.90	1.80	8.78	10.58	53.2	4.9	I
C1 - internal	16.07	1.67	7.36	9.03	56.2	4.4	II
PRC over C2	20.25	4.32	12.14	16.46	81.3	2.8	IV
PRC over C1	20.15	4.42	13.40	17.82	88.4	3.0	IV

Concrete samples (surface is 0-20 mm and internal is approximately 280-300 mm from the surface) from Sládkovičovo bridge show lower CaO_{Total} content compared to those of Rimavská Sobota. This fact indicates differences in cement content in both types of concrete. Concrete samples from Rimavská Sobota contain more carbonation products. The PRC over M3 records more CaO_{Total} than PRCs over C1 and C2. This indicates a higher proportion of cement binder in PRC over M3 compared to PRCs over C1 and C2. Both bridges consistently confirm that in the surface application of PRC, the carbonation stage (SC) of covered concrete moves at levels II. and I, while uncovered concrete is characterized by stage IV. Significantly higher degree of modification changes in concrete from Rimavská Sobota compared to Sládkovičovo bridge indicates the higher conversion of fine-grained $CaCO_3$ to coarsegrained crystals. This may be due to a lower proportion of cement-based binder and therefore to the easier conversion of the carbonates present to the coarse-grained modification. The same assumption applies to the PRCs. The PRC from Sládkovičovo is characterized by the SC I, while from Rimavská

Sobota by the SC IV. Despite this fact, PRCs from Rimavská Sobota bridge eliminate the carbonation attack in 100 years of operation by the same way as that from Sládkovičovo with the SC I. The achieved results confirm the anti-carbonation effect of applied PRCs regardless of the achieved degree of carbonation, degree of modification changes and carbonation stage. It is apparent from this finding that an important property of PRC in eliminating the carbonation attack is the sealing ability by accumulating carbonates in the narrow PRC layer and thereby increasing the impermeability against further penetration of CO_2 into the interior of the underlying concrete.

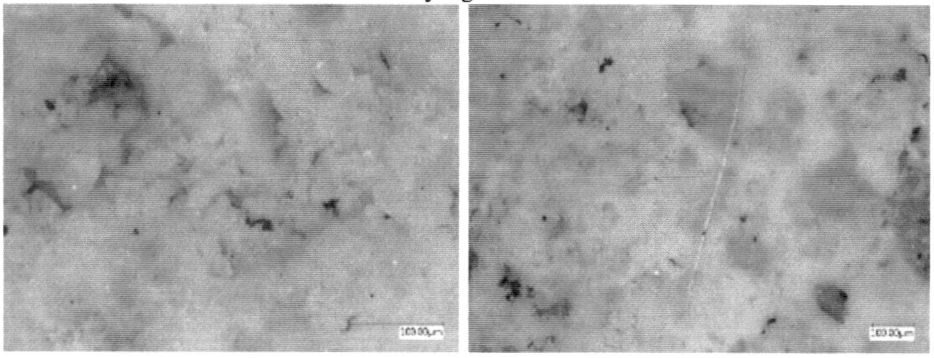

Figure 6. Dense barrier created from the carbonation products in the space-tight PRC M3.

The thin PRC layer is characterized by a dense carbonate substance even up to the amorphous state boundary, which is referred to as "carbonate milk" (a designation that has not been used in the literature yet). The images from the optical microscope at various magnifications show invisibly but later demonstrably accumulated carbonation milk within the PRC (figure 6). Shining needles confirm also the presence of crystal-developed coarser $CaCO_3$ compressed within the PRC thickness. The carbonate milk has such a low permeability with optimally functional PRCs that the spatially sealed barrier of carbonation products is created, preventing thus carbonation of old concrete effectively.

5. Conclusions

Based on the results of the series of the performed tests, the following conclusions are drawn:

- The in-situ measurements of carbonation depth show that a high quality (2-4) mm thin cementbased protective render coat can effectively protect the underlying concrete against carbonation within the 100-year service life of a bridge.
- Carbonation depth of the underlying concrete depends upon the non-permeability of the PRC. The close mutual correlation between the depth and extent of carbonation phenolphthalein, TGDTA) and the coefficient of surface permeability (Torrent method) was found.
- The very low permeability of the PRC is caused by the accumulated carbonates in a narrow spatial area of the PRC; the formed carbonation milk condenses the space up to the impermeable barrier for CO_2.

6. References

[1] Janotka I, Bačuvčík M Ševčík P and Paulík P 2016 *Proc. Int. Conf. Concrete Days (Wisla)*, vol 1(Stowarzyszenie Producentów Cementu, Kraków) p 493

[2] Torrent R and Frenzer, G 1995 *Study on methods to determine and judge characteristic values of the cover concrete on site (in German),* (Bundesamt für Strassenbau, Bern), p. 105

[3] Paulik P, Janotka I, Bačuvčík, M and Gajdošová K. 2019*Proc of the fib Symposium, Concrete – Innovations in Materials, Design and Structures* (W. Derkowski et al.), p.2134.

[4] Janotka I, Bačuvčík M and Paulík P 2018 *Low carbonation of concrete found on 100-year-old Bridges*, Case Studies in Construction Materials p 97

Acknowledgements
The authors wish to express their gratitude to the Slovak Agency for Research and Development (project APVV-17-0204) for financing this research work.

IOP Conf. Series: Materials Science and Engineering **867** (2020) 012012 doi:10.1088/1757-899X/867/1/012012

Coupled heat and moisture transport in building material – water absorption coefficient and capillary water content

P Jaroš[1] and M Vertal'[1]

[1] Technical university of Kosice, Faculty of Civil Engineering, Institute of Architectural Engineering, Slovakia

Abstract. At present, restoration of historic buildings is a difficult process. The most common problem of historic buildings is increased moisture, because waterproofing is missing and thanks that water can be transported in porous material by capillary forces. There is several simulation software available, which can include many factors such as water transport from the ground, sun radiation, impact of driven rain. Also it includes thermal conductivity, specific heat capacity and water vapor diffusion factor, which are depended on content of water. For correct assessment of hygrothermal behaviour of these buildings and find suitable retrofit strategies, hygric and thermal properties are needed. When we want to use simulation software for assessment of hygrothermal behaviour of historic structures, we need to know heat and water transmission parameters of sandstones. In this paper are presented two water transmission parameters and that water absorption coefficient and capillary water content. For the test measurement of these parameters was used clay brick. The water absorption and capillary water content are obtained by experiment of one-dimensional transport water. The results are compared with values of clay brick, which are presented in German material database.

1. Introduction

The moisture is one of the most common issue of existing buildings. It can affect not only service life and buildings energy efficiency but also their air quality in interior of buildings [1]. Water transport is crucial factor in terms of building design. When designing new buildings, civil engineers consider with transport water only as transport water vapor from the interior to the exterior. It is correct, because they can consider with waterproof insulations against ground water or waterproof plasters on the facades of buildings envelopes, but in historical buildings are usually missing and thanks that water can be transported to the porous materials of historical constructions. There is simulation software available, which can include influence of water to materials properties, sun radiation, driven rain, orientations of walls, wind and many others important factors. To use this software, it is necessary to know heat and water transmission parameters of historical materials [2]. In eastern Slovakia, sandstone was used in large quantities for construction of buildings. It was used for its compressive strength, good workability and aesthetic appearance. By the end of the 19th century, 336 quarries for sandstone mining were registered in the territory of eastern Slovakia [3]. Only a few scientists have studied the hygrothermal properties of sandstones. Transmission parameters of sandstones were studied by scientist in Canada [4], Czech Republic [5], and Germany [6] [7]. Capillary absorption coefficient (A_{cap}) and the capillary moisture content (w_{cap}) are important material properties. They can be determined by capillary absorption test, which is one of the most frequently performed material characterization test. There are many international and regional standards prescribing the procedures of the capillary absorption test [12], [13], [14], [15], but there are still some differences between procedures and rules. These differences include questions about sample size, area of bottom surface, sealing materials to providing one-dimensional transport water, temperature impact on A_{cap}, operational factors, weighing method. In this paper, the test measurement is performed by using samples of clay brick with two different sealing materials.

2. Material information and sample preparation

2.1. Characterization of material

For test measurement was used common clay brick. It was chosen, because it has a typical tubular pore structure [8] and it is weak in hygroscopicity but strong in [9], this makes it suitable for capillary absorption test measurement.

2.2. Sample preparation

The size of the sample and the area of the bottom surface that is in contact with the water surface are one of the most disputable factors. The standard ISO 15148 requires, that a bottom surface area of sample should has minimum of 50 cm^2 for each sample [16]. The standard EN 1925 requires cubic sample with dimension 50, 70 mm or cylinder sample with diameter 50,70 mm and with height 50, 70 mm. Although there are several standards that set minimum dimensions of samples, many researchers often use much smaller samples with bottom surface areas ranging from 10 to 30 cm^2 [5], [9], [10], [11]. For test measurement were used cylindric samples with diameter 50 mm and height 60 mm figure 1. Two samples were made from one homogenous batch by core drilling machine and the bottom surface was treated by emery paper.

Figure 1. Left: Preparation of cylindrical sample, Right: Cubic sample with diameter 50 mm.

3. Experimental setup and measurement procedure

3.1. Principle of one-dimensional water transport experiment

By performing a water absorption test also called one dimensional water transport experiment, it is possible to determine the water absorption coefficient and capillary moisture content. The principle of this method is that the measured sample is immersed in the water and mass gain over time is monitored. The bottom surface of sample is in contact with water level to allow water absorption [16]. To providing only one-dimensional water transport, it is necessary to seal lateral sides and upper surface of the samples. The insulation must be perfect, because water can move in gap between sample and insulation material. Before the testing, samples must be dried at defined temperature 105°C [16] or 70°C [17]. After cooling down to temperature of the room in desiccator, oven-dried sample can be used for measurement. The test can be performed either with automatic or manual weighing. Some studies show a good agreement between the results of manual and automatic weighing [10]. For this test was used apparatus with automatic weighing shown in figure 2. This apparatus was constructed according measuring device stated in [20]. It consists of a water container, a container lifter, a weighing machine, a rod for suspending sample, a data logger, which is connected to computer for monitoring temperature, relative humidity and weight gain.

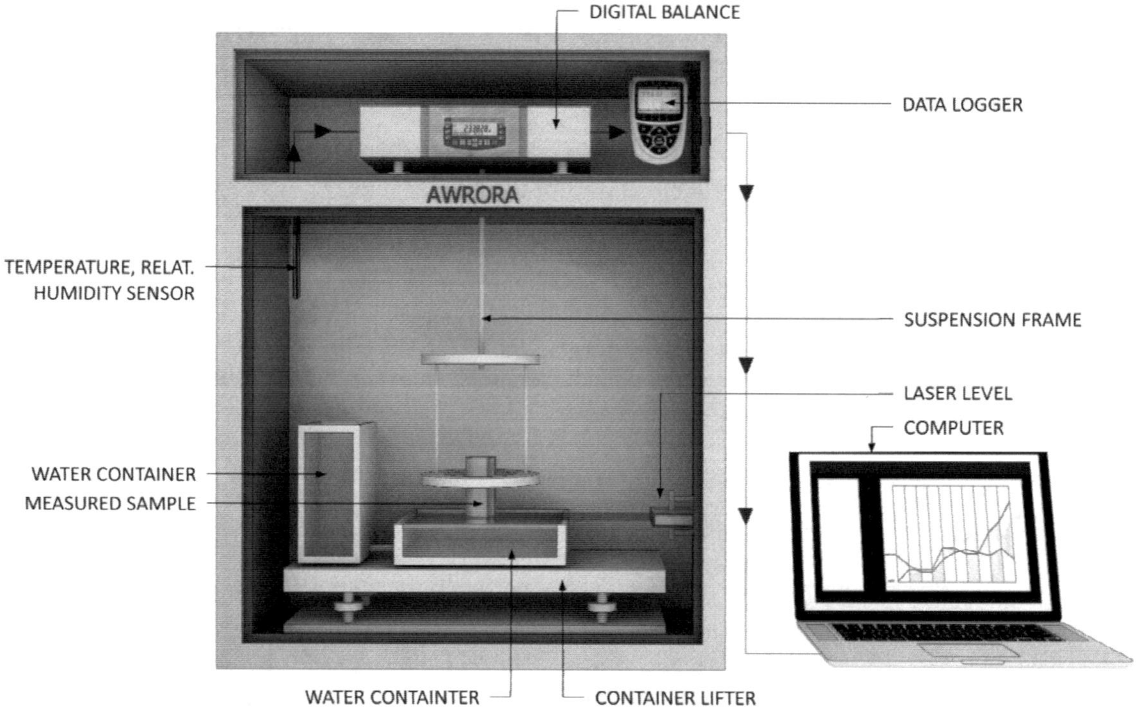

Figure 2. Visualization of apparatus for one-dimensional water transport experiment (AWrora).

3.2. Waterproof insulation

To reduce evaporation of water from the sample, it is necessary to insulate lateral sides of the sample. Insulation material must be vapor-proof and waterproof. Upper surface of the sample, must be vapor-proof but air-permeable, because of air leakage. In general, there are two sealing methods. Samples can be sealed by adhesive, non-adhesive film or by epoxy resin. According standard ASTM C1585 [18], both methods are acceptable. Bomberg *et al.* [10] stated that the self-adhesive tape should not be used, but without any theoretical explanations. Two different insulating materials were used for the test measurement, self-adhesive tape commonly used to insulate the gap between the walls and openings, and a heat shrink tube commonly used to insulate electrical cables. Both materials are shown in figure 3.

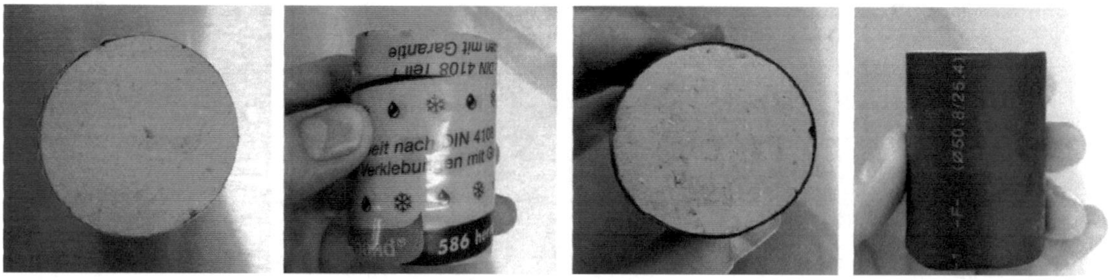

Figure 3. Left: Sample with self-adhesive tape, Right: Sample with heat shrink tube.

Epoxy resin was not used, because it can be penetrated into the sample, what it can lead to reducing the effective cross-section size of the sample [9].

4. Determining of water absorption coefficient and capillary moisture content

One-dimensional water transport experiment can be divided into two stages. The first stage is characterized by a linear mass gain as a function of square root of time. During this stage, water absorption is governed by capillary forces. Water absorption coefficient A_{cap} is defined as the slope of the cumulative water inflow versus square root of time in the first stage of the test figure 4. Second stage is characterized by non-linear mass gain and mass gain is slower. This stage takes a several days or months [19].

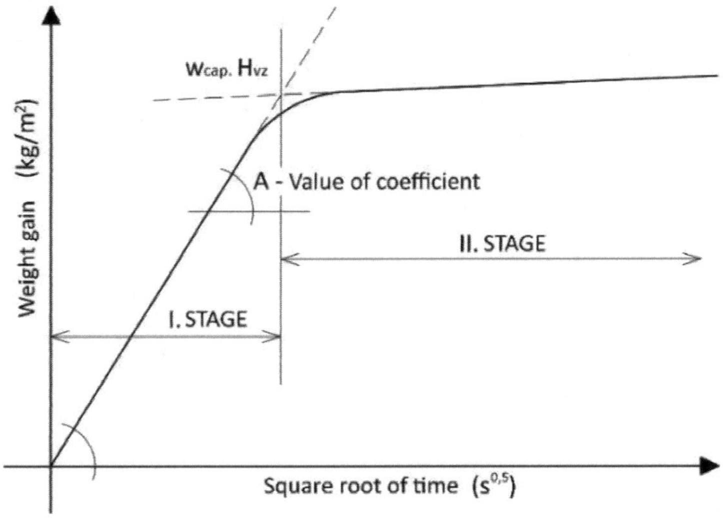

Figure 4. Dependence of the weight gain of water on the square root of time.

Water absorption coefficient A_{cap} can be determined by using following equation (1) [19]:

$$A_{cap} = \frac{m_{wet}-m_{dry}}{t^{1/2}} \tag{1}$$

where A_{cap} is water absorption coefficient *(kg/(m*^{2}*.s*$^{1/2}$*))*, m_{wet} weight of the wet sample *(kg/m*^{2}*)* per unit area *S (m*^{2}*)*, m_{dry} is weight of the dry sample *(kg/m*^{2}*)* per unit area *S (m*^{2}*)*, *t* is time of experiment. Capillary moisture content w_{cap} is defined as the moisture content at the transition from the 1st to the 2nd stage. More specifically, it is the moisture content, which correspond to the cross point with the fitting of the 1st and 2nd stages, as it is illustrated in figure 4. The capillary moisture content w_{cap} can be determined by using following equation (2) [10]:

$$w_{cap} = \frac{m_{wet}-m_{dry}}{S} \cdot \frac{1}{H} \tag{2}$$

where w_{cap} is capillary moisture content *(kg/m*^{3}*)*, m_{wet} is weight of the wet sample *(kg/m*^{2}*)* per unit area *S (m*^{2}*)*, m_{dry} is weight of the dry sample *(kg/m*^{2}*)* per unit area *S (m*^{2}*)*, *S* is the area of the bottom surface *(m*^{2}*)*, *H* is the height of the measured sample *(m)*.

5. Results and conclusion

The resulting water absorption courses in the experiment with two different sealing materials are presented in figure 5 and figure 6. Weight gain recording was performed by automatic weighing at defined times. The weight gain of clay brick with self-adhesive tape is shown in figure 5. The linear course of weight gain is not entirely perfect. It is not because the sample was lifted during the experiment. In figure 6 is shown weight gain of the clay brick with heat shrink tube. The linear course of weight is much better. Such a course is typical of an ordinary clay brick. The measurement results were compared with the different bricks Wienerberger 1, 2, ZD, ZE and Hartman from the German material database called MASEA [21]. They have different values of bulk density. In table 1, the best match can be observed between values of A_{cap} and w_{cap} of brick Wienerberger 1 and values of measured brick with heat shrink tube. But on the other side, there is a bigger difference between values of bulk density. In the case of measurement with self-adhesive tape, the good match can be observed between A_{cap} of brick ZD and measured brick. Also there is a small difference between values of bulk density. The measurement and comparison of the results indicates that there are significant differences between measurements with different sealing material and therefore further measurements are required.

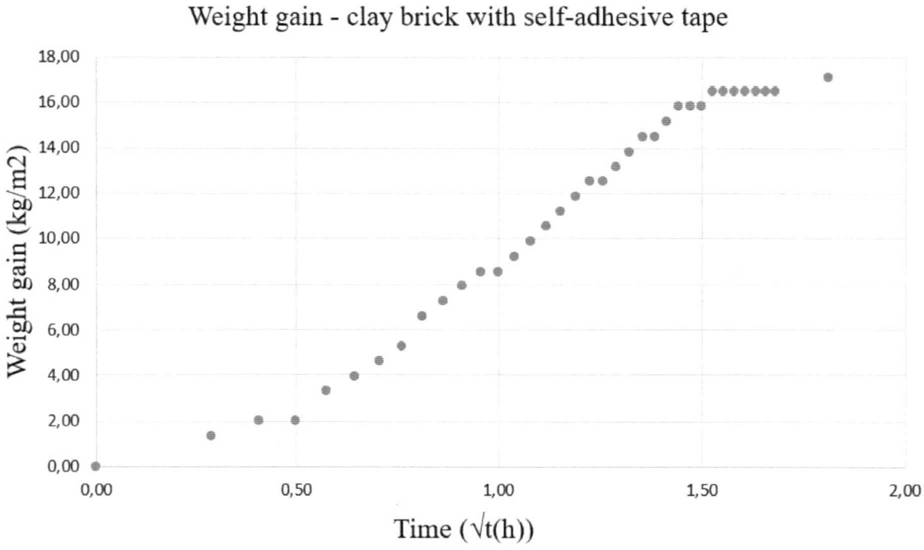

Figure 5. Weight gain during 1D experiment, clay brick with self-adhesive tape.

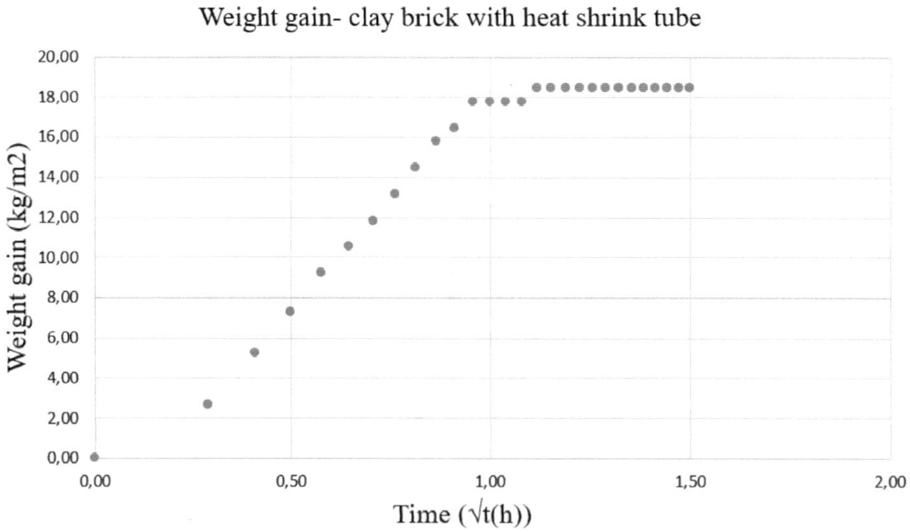

Figure 6. Weight gain during 1D experiment, clay brick with heat shrink tube.

Table 1. Comparison – measured brick and brick from German database MASEA [20].

Material	Bulk density ρ (kg/m^3)	Water absorption coefficient A_{cap} $(kg/(m^2.s^{1/2})$	Capillary water content $w_{cap}(kg/m^3)$
Measured brick – self-adhesive tape	1567	10,96	265,32
Measured brick – heat shrink tube	1579	17,64	286,45
Brick - Wienerberger 1	1744	17,5	287
Brick - Wienerberger 2	1786	11,9	262
Brick – ZD	1611	10,9	216
Brick – ZE	1642	12,9	254
Brick – Hartmann	1655	14,4	228

6. References

[1] Huijbregts Z, Schellen H, van Schijndel J, Ankersmit B 2015 Modelling of heat and moisture induced strain to assess the impact of present and historical indoor climate conditions on mechanical degradation of a wooden cabinet *Journal of Cultural Heritage* **16** 419-427

[2] Roels S, Carmeliet J, Hens H, Adan O, Brocken H, Cerny R et al. 2004 Interlaboratory Comparison of Hygric Properties of Porous Building Materials *Journal of Thermal Envelope and Building Science* **27** 307-325

[3] Čabalová D 2013 *Krása kameňa v živote človeka* (Bratislava: Veda) 374

[4] Mukhopadhyaya P, Kumaran M K, Lackey J, Normadin N, Tariku F and van Reenen D 2007 Hygrothermal properties of exterior claddings, sheathing boards, membranes and insulation materials for building envelope design

[5] Kočí V, Maděra J, Fořt J, Žumár J, Pavlíková M, Pavlík Z et al. 2014 Service Life Assessment of Historical Building Envelopes Constructed Using Different Types of Sandstone: A Computational Analysis Based on Experimental Input Data *The Scientific World Journal* 1-12

[6] Zhao J, Plagge R 2015 Characterization of hygrothermal properties of sandstones—Impact of anisotropy on their thermal and moisture behaviors *Energy and Buildings* **107** 479-494

[7] Krus, M 1996 Moisture transport and storage coefficients of porous mineral building materials: Theoretical Principles and New Test Methods (Fraunhofer IRB Verlag: Stuttgart)

[8] Roels S, Carmeliet J, Hens H, Adan O, Brocken H, Cerny R et al. 2004 Interlaboratory Comparison of Hygric Properties of Porous Building Materials *Journal of Thermal Envelope and Building Science* **27** 307-325

[9] Feng C, Janssen H 2018 Hygric properties of porous building materials (III): Impact factors and data processing methods of the capillary absorption test *Building and Environment* **134** 21-34

[10] Bomberg M, Pazera M, Plagge R 2005 Analysis of Selected Water Absorption Coefficient Measurements *Journal of Thermal Envelope and Building Science* **28** 227-243

[11] Hanžič L, Kosec L, Anžel I 2010 Capillary absorption in concrete and the Lucas–Washburn equation. *Cement and Concrete Composites* **32** 84-91

[12] BS EN 1925, Natural Stone Test Methods. Determination of Water Absorption Coefficient by Capillarity (1999) 1999

[13] UNI 10859, Cultural Heritage – Natural and Artificial Stones – Determination of Water Absorption by Capillarity (2000) 2000.

[14] BS EN 1015-1018, Methods of Test for Mortar for Masonry. Determination of Water Absorption Coefficient Due to Capillary Action of Hardened Mortar (2002) 2002.

[15] UNI EN 15801, Conservation of Cultural Property – Test Methods – Determination of Water Absorption by Capillarity (2010) 2010

[16] ISO 15148:2002(E), Hygrothermal Performance of Building Materials and Products – determination of Water Absorption Coefficient by Partial Immersion (2002)

[17] STN EN 1925: 2002. Skúšky prírodného kameňa. Stanovenie súčiniteľa nasiakavosti kapilaritou

[18] ASTM C1585-13: 2013. Standard Test Method for Measurement of Rate of Absorption of Water by Hydraulic-cement Concretes

[19] Vertaľ M and Ďurica P (2013) *Prenosové parametre vody vo vybraných stavebných materiáloch* (Košice: Technická univerzita v Košiciach, Stavebná Fakulta) 98 0Plagge R Scheffler G and Grunewald J 2005 Automatic Measurement of Water Uptake Coefficient of Building Materials *Bauphysik* **27** 315–323
[20] (online): https://www.masea-ensan.de/

Acknowledgments
This work was supported by the VEGA: 1/0674/18 Theoretical and experimental analysis of architectural-structural shapes and fragments of building envelope structures designed for harsh conditions.

IOP Conf. Series: Materials Science and Engineering **867** (2020) 012013 doi:10.1088/1757-899X/867/1/012013

Shear strength of in-plane loaded cross laminated timber beam elements – experimental and theoretical investigation

M Jelec[1], K Strukar[2] and V Rajčić[3]

[1] Postdoctoral researcher, Josip Juraj Strossmayer University of Osijek, Faculty of Civil Engineering and Architecture Osijek, Croatia, e-mail: mjelec@gfos.hr
[2] PhD student, Josip Juraj Strossmayer University of Osijek, Faculty of Civil Engineering and Architecture Osijek, Croatia
[3] Full Professor, University of Zagreb, Faculty of Civil Engineering, Croatia

Abstract. Cross laminated timber (CLT), as a versatile engineered timber product, has in recent years become well-known and of global interests. The orthogonal laminar structure allows for application as beams, walls or panels for both out-of-plane and in-plane loading conditions. The work presented here concerns experimental investigations of in-plane loaded CLT beams, as very relevant from a practical engineering point of view due to reinforcing effect of transversal layers with respect to stress perpendicular to the beam axis. Experimental results are used for comparison and validation of design proposals from the literature, with the main focus on the shear failure mode III in crossing areas between orthogonally bonded longitudinal and transversal lamination. By using the assumed shear strength values from the literature, analytically predicted shear carrying capacities show good agreement with the experimentally obtained. Although further experimental validation is needed, obtained results indicate that suggested design proposals should be considered in ongoing revision work of Eurocode 5.

1. Introduction

Within the last decade, engineered wood product as cross laminated timber (CLT), has become a very well-known and of global interest. The orthogonal laminar structure allows its application for out-of-plane and in-plane loading conditions, for floor panels, walls, beams etc. Beams made of CLT offer several advantages over solid or glued laminated beams. The transversal layers of orthogonal structure have a reinforcing effect with respect to the stress perpendicular to the beam axis which results in however complex stress state where several failure modes need to be considered in design [1].

For verification of beam shear strength, according to figure 1., three different failure modes (FM) in general need to be considered [2]: gross shear failure (FM I), net shear failure (FM II) and shear failure in the crossing areas between adjacent longitudinal and transversal laminations (FM III).

Experimental verifications of CLT beam shear strength are for example reported by Bejtka [3], Flaig [4], Danielsson et al [5] and Jelec et al [6], where in most cases bending and shear mode III were identified as the critical failure modes. An analytical model for relevant shear failure modes of CLT beams has been presented in [2, 7]. Proposed model has been included in a current working draft of cross laminated timber for future revised Eurocode 5 [8]. This model does, however, suffer from some drawbacks as has been discussed in [9, 10], where alternative improvements were presented in [11]. Comprehensive FE-analyses and comparisons with analytical models have been shown in [12]. Based on these results, an improved model and design proposals were presented in [6, 13].

The aim of this paper is to present experimental results of in-plane loaded CLT beams for comparison and validation of design proposals [6, 13], with the main focus on shear failure mode III.

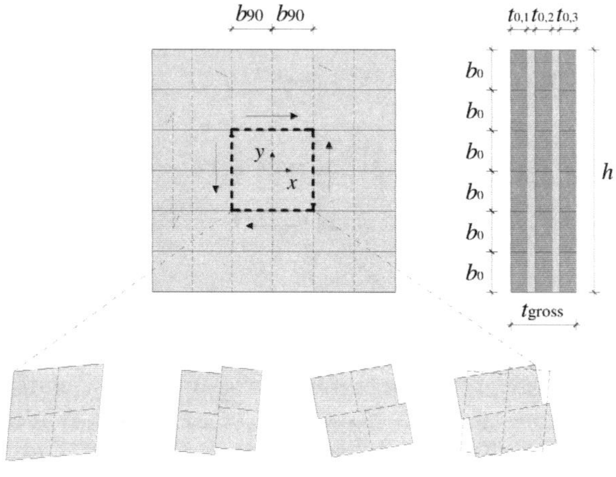

Figure 1. Illustration of geometry and shear failure modes of CLT beams.

2. Analytical model and design proposal

2.1. General assumptions

A brief review of two design proposals [6, 13] is presented below. Both proposals are generally based on conventional beam theory taking into account the orthogonally layered composition of prismatic CLT beams without edge bonding and composed of longitudinal and transversal laminations of width b_0 and b_{90}, respectively. Both proposals are based on assuming equal width b_0 of all longitudinal laminations in the beam height direction and are valid for 3 and 5-layer CLT elements, with the restriction of the 5-layer element lay-up $t_{0,2}/t_{0,1} = t_{0,2}/t_{0,3} \leq 2{,}0$.

2.2. Shear stress analysis

Based on the model and design proposals (DP 1 & 2), the shear stress acting in crossing areas between longitudinal and transversal laminations can be decomposed into (1) unidirectional shear stress parallel to the beam axis τ_{zx} and (2) torsional shear stresses $\tau_{tor,z}$. Design relevant stresses are calculated for the critical crossing area, which relates to the outermost crossing area in the beam width direction and close to the centre-line with respect to the beam height direction, as fallows

$$\tau_{zx}=\frac{6V}{b_0^2}\frac{t_{0,1}}{t_{net,0}}\frac{1}{m^3} \quad \text{for DP 1 \& 2} \tag{1}$$

$$\tau_{tor,z}=\frac{3V}{b_0^2}\left(\alpha\beta-\frac{t_{0,1}}{t_{net,0}}\frac{1}{m^3}\right)k_b \quad \text{with} \quad k_b=\frac{2b_{max}b_0}{b_0^2+b_{90}^2} \quad \text{for DP 1} \tag{2}$$

where m is the number of longitudinal laminations in beam height direction and α and β are weighting factors for shear stress distribution in beam height and width direction, as fallows

$$\alpha=\frac{3m^2-4}{2m^3} \tag{3}$$

$$\beta=\begin{cases}\dfrac{t_{0,1}}{t_{\text{net},0}} & \text{for 3-layer CLT beams}\\[2ex] \dfrac{1}{8}\left(1+4\dfrac{t_{0,1}}{t_{\text{net},0}}\right) & \text{for 5-layer CLT beams}\end{cases} \tag{4}$$

A further simplification is introduced in DP 2 by replacing two weighting factors (α and β) with a single factor γ, as fallows

$$\tau_{\text{tor,z}}=\frac{3V}{b_0^2}\frac{t_{0,1}}{t_{\text{net},0}}\left(\frac{1}{m}-\frac{1}{m^3}\right)k_b\gamma \quad \text{with} \quad k_b=\frac{2b_{\max}b_0}{b_0^2+b_{90}^2} \quad \text{for} \quad \text{DP 2} \tag{5}$$

where

$$\gamma=\begin{cases}1.5 & \text{for 3-layer CLT beams}\\ 0.2\dfrac{t_{0,2}}{t_{0,1}}+1.1 & \text{for 5-layer CLT beams}\end{cases} \tag{6}$$

2.3. *Shear strength verification*

For verification of shear failure mode III, due to stress components in two directions, τ_{zx} and $\tau_{\text{tor,z}}$, a stress interaction criteria may be needed. According to the original model [2], a linear stress interaction criterion is proposed

$$\frac{\tau_{\text{tor,z}}}{f_{\text{v,tor}}}+\frac{\tau_{zx}}{f_R}\leq 1.0 \tag{7}$$

where $f_{\text{v,tor}}$ is a torsional shear strength and f_R is the rolling shear strength. Test results found in literature [1, 7] indicate a mean value of the torsional strength of about $f_{\text{v,tor}}$ = 3,5 MPa and a mean value of the rolling shear strength of about f_R = 1,5 MPa. The corresponding characteristic values are $f_{\text{v,tor}}$ = 2,75 MPa and f_R = 1,1 MPa.

3. Experimental study

3.1. *Test series and materials*

Experimental tests of in-plane loaded CLT beams have been carried out at Lund University and at the Faculty of Civil Engineering at University of Zagreb and a detailed description of the tests and results are presented in [5, 6]. One test series from each testing place is presented below where each consisted of four nominally equal tests resulting a total of 8 individual tests. The main geometry and lay-up parameters are presented in table 1 and shown in figure 2.

Table 1. Geometry parameters of tested CLT beams.

Test series	n [-]	L [mm]	h [mm]	t_{gross} [mm]	b_0 [mm]	b_{90} [mm]	lay-up [mm]
CLT 5s	4	2400	600	160	150	146	40-20-40-20-40
CLT 3s	4	2400	600	100	150	198	40-20-40

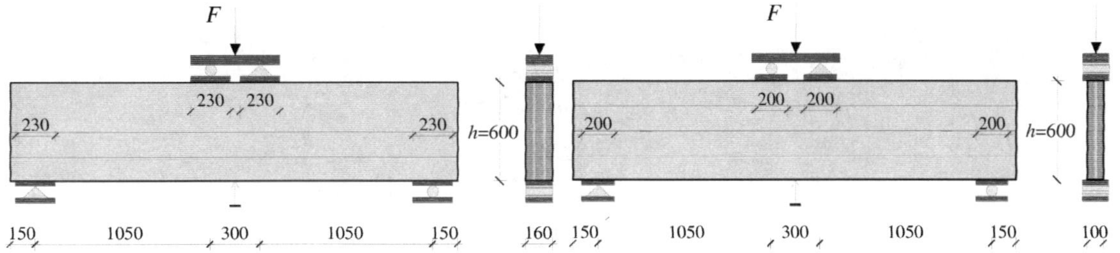

Figure 2. Test setups for CLT 5s (left) and CLT 3s (right), dimensions in mm.

Beams from CLT 5s were produced by Cross Timber Systems LTD according to ETA-15/0906 and beams from CLT 3s by Hasslacher Norica Timber according to ETA-12/0281. The wood species are stated as being European Spruce or equivalent softwood. The mean density of the CLT 5s was 452 kg/m^3 with the moisture content at the time of testing about 10-11%, while the mean density of the CLT 3s was 458 kg/m^3 with the moisture content at the time of testing about 11-12%. At the time of testing, there were no visible gaps between the laminations belonging to the same layer, which is in line with both ETA-s. All beams were cut from larger CLT panels without consideration of the position of the longitudinal lamination in relation to the beam elements edges, with the width of the upper- and lower-most longitudinal laminations in the range $5 < b_0 < 172$ mm with $b_{0,max} = 172$ mm for CLT 5s, and $54 < b_0 < 149$ mm for CLT 3s with $b_{0,max} = 198$ mm, respectively. The laminations widths b_0, as shown in table 1, are thus approximated mean values.

3.2. Test results

The results of all test specimens in terms of maximum loads and graphs of applied loads vs. global deflection are shown in figure 3. All tests were run by deformation controlled testing machine, with the test duration of approximately 15-20 min which allowed careful observations of critical locations where the failure was expected.

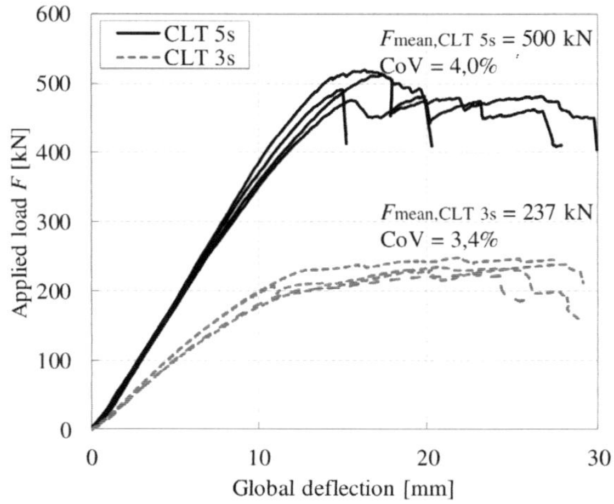

Figure 3. Applied load F vs. global deflection.

In CLT 5s the load bearing capacity in terms of maximum load is related to bending failure of lowermost longitudinal laminations. Before reaching the maximum loads, a gradual decrease in stiffness can however be noted from figure 3. Also, significant sliding between longitudinal laminations in the exterior layer was observed (figure 4), indicating thus partial failure in crossing areas (FM III).

Figure 4. Photos of fractured specimens for CLT 5s (left) and CLT 3s (right).

In CLT 3s the failure modes are related to a combination of shear failures in crossing areas (FM III) and failures due to bending. The initial failure started at 85% of the maximum loads approximately, as shown in figure 3, after which decreasing in stiffness and obvious sliding between laminations was observed, indicating shear damage/failure in crossing areas, as shown in figure 4.

4. Comparisons and discussion

The results for both test series are given in figures 5 and 6 in terms of shear stress components and stress interaction criteria at maximum load for shear failure mode III. The stress values are calculated according to both design proposals from Section 2, based on assumption of equal widths of the longitudinal and transversal laminations, i.e. $b_0 = b_{90} = 150$ mm. The stress interaction criteria according to equation 7 is calculated based on the assumed value of torsional shear strength of $f_{v,tor} = 3{,}5$ MPa and the rolling shear strength of $f_R = 1{,}5$ MPa.

For test series CLT 5s, stress interaction ratios according to both design proposals (DP 1 & 2) are all above 1,0, as shown in figure 6, indicating that shear failure mode III would be expected. These results are in-line with observations during testing where obvious sliding between longitudinal laminations and cracking sounds were noticed before reaching maximum load. Based on assumed shear strength values, $f_{v,tor}$ and f_R, both design proposals are conservative and underestimates the shear capacity compared to test results. Design proposal 2 seems, however, slightly conservative compared to design proposal 1. For three specimens of test series CLT 3s, stress interaction ratios according to both design proposals (DP 1 & 2) are below 1,0, as shown in figure 6, indicating that shear failure mode III would not be expected. This is however not in-line with observations during testing where obvious sliding and cracking sounds were noticed before reaching maximum load. Based on assumed shear strength values, $f_{v,tor}$ and f_R, both design proposals slightly overestimates the shear capacity compared to test results. This overestimation is, however, relatively small compared to the mean value of test results, where the main reason could be on assumed shear strength values. For test series 3, a slightly lower values of shear strength components, $f_{v,tor}$ and f_R, would be thus expected.

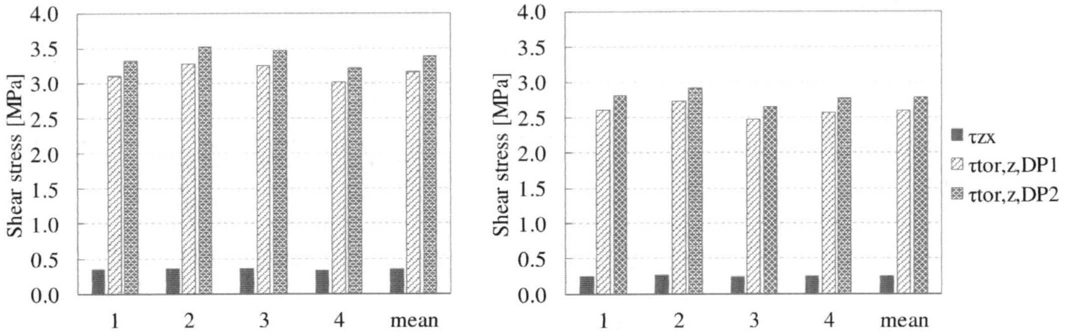

Figure 5. Calculated stress component values at maximum load for CLT 5s (left) and CLT 3s (right).

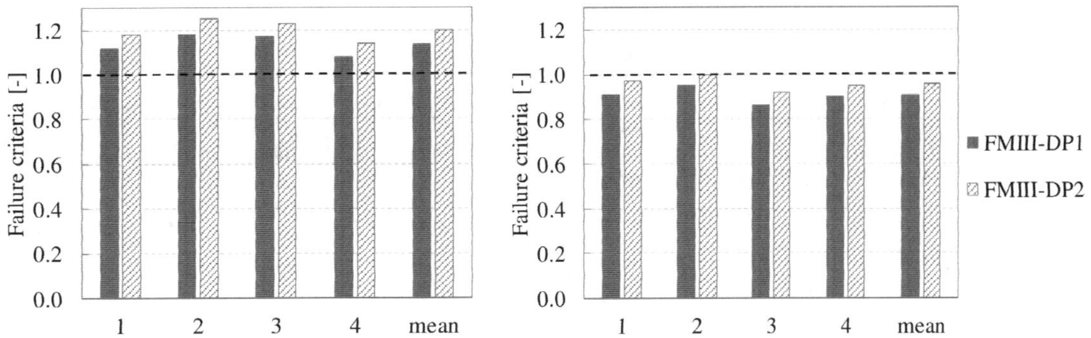

Figure 6. Calculated failure interaction criterion for CLT 5s (left) and CLT 3s (right).

5. Conclusions

Two design proposals, as presented in [6, 13], are reviewed and beam strength predictions are compared to the experimental result. Based on the assumed values of torsional $f_{v,tor}$ and rolling shear strength f_R, as proposed in the literature, in general good agreement between test results and design proposals was obtained. Both design proposals seem suitable for use in practical design situation for CLT beams composed of 3 or 5 layers with the surface layers oriented in the beam length (x) direction. Further studies of CLT beams with inverted layer orientation or studies of 7- layer elements should preferably be carried out.

The proposals are based on assuming equal longitudinal lamination width b_0 for all m laminations in beam height direction. Dimensions and placement of laminations are however in general not know in actual design situation since the CLT beams in general are cut from larger CLT elements, irrespective of their location in relation to the edges of element. The reduced lamination widths close to the beam edges have a small influence on the beam strength, as it is shown in [6]. The lamination widths b_0 and b_{90} to be used in calculation can then be based on values as specified by producer or slightly lower values for conservative design. Following the recommendation in the draft version of new Eurocode 5 [8], in cases where the lamination width is not known, it should be assumed as $b_0 = b_{90} = 80$ mm.

6. References

[1] Jeleč M, Varevac D and Rajčić V 2018 Cross laminated timber (CLT) – a state of the art report *Građevinar* **70** 75-95

[2] Blaß H J and Flaig M 2012 *Stabförmige Bauteile aus Brettsperrholz* (Karlsruhe: KIT)

[3] Bejtka I 2011 *Cross (CLT) and Diagonal (DLT) Laminated Timber as Innovative Material for Beam Elements* (Karlsruhe: KIT)

[4] Flaig M 2013 *Biegeträger aus Brettsperrholz bei Beanspruchung in Plattenebene* PhD Thesis (Karlsruhe: KIT)

[5] Danielsson H, Jeleč M and Serrano E 2017 *Strength and sitffness of cross laminated timber at inplane beam loading* Report TVSM-7164 (Lund: Div of Structural Mechanics, Lund University)

[6] Jeleč M, Danielsson H, Rajčić V and Serrano E 2019 Experimental and numerical investigations of cross-laminated timber elements at in-plane beam loading conditions *Construction and building materials* **206** 329-346

[7] Flaig M and Blaß H J 2013 Shear strength and shear stiffness of CLT-beams loaded in plane *Proc. Int. Conf. CIB-W18/46-2-3* (Vancouver, Canada)

[8] European Committee for Standardization (CEN/TC 250/SC5) *Working draft of design of cross laminated timber in a revised Eurocode 5-1-1*, Version 2018-04-13, part of N892

[9] Jeleč M, Strukar K and Rajčić V 2017 Structural analysis of in-plane loaded CLT beams *Electronic Journal of the Faculty of Civil Engineering Osijek e-GFOS* **8** 20-30

[10] Danielsson H, Serrano E, Jeleč M and Rajčić V 2017 In-plane loaded CLT beams - Tests and analysis of element lay-up. *Proc. Int. Conf. INTER/50-12-2* (Kyoto, Japan)

[11] Danielsson H and Serrano E 2018 Cross laminated timber at in-plane beam loading – Prediction of shear stresses in crossing areas *Engineering Structures* **171** 921-927

[12] Danielsson H, Jeleč M, Serrano E and Rajčić V 2019 Cross laminated timber at in-plane beam loading – Comparison of model predictions and FE-analyses *Engineering Structures* **179** 246254

[13] Jeleč M, Danielsson H, Serrano E and Rajčić V 2018 Cross laminated timber at in-plane beam loading - New analytical model predictions and relation to EC5 *Proc. Int. Conf. INTER/51-125* (Tallin, Estonia)

IOP Conf. Series: Materials Science and Engineering **867** (2020) 012014 doi:10.1088/1757-899X/867/1/012014

Analysis of wind-adaptive architecture

L Kabošová[1]

[1] Technical University of Kosice, Faculty of Civil Engineering, Department of Architectural Engineering, Slovakia

Abstract. As an architectural response to the changing climate, a wind-driven design approach is developed. By adopting the digital design tools, the wind can be integrated into the design process, from the large scale of urban planning, through the form-finding of buildings, as well as designing the texture and roughness of building surfaces. In its more extreme forms, it is even more notable, how the wind flow is shaped and influenced by architecture. Former docks in Stockholm serve as a case study site to test architectural interventions that alter the wind flow on the site. This wind-driven design approach enables the evaluation of the wind performance of various design options, which informs design decisions. Grasshopper, the graphical algorithmic plug-in for Rhinoceros, creates one working environment, where the geometry is parametrically designed, and subsequently analysed in the virtual wind tunnel through the Swift extension for Grasshopper. The Computational Fluid Dynamics analysis results retroactively help decide which of the design options performs the best in the given wind conditions.

1. Introduction

The paper presents a methodology for the wind-driven designing, with the wind factor incorporated into the early stage of architectural form-finding. The recent decades showed that aiming only on the quality of the interior environment when designing buildings will not suffice in the future changing climate [1, 2, 3]. Climate-conscious design as a form of adaptation to climate changes along with substantial and sustained reductions in greenhouse gas emissions can limit the consequences of ongoing climate change [4]. Moreover, to create a comfortable outdoor environment for pedestrians, the factors such as the wind effects, or outdoor thermal comfort should be fully considered throughout the whole process of designing the built environment [5].

1.1. Built environment shaping the wind flow

The rapid pace of improving hardware and software, including CFD (Computational Fluid Dynamics), as well as parametric design techniques, motivates the research of reciprocal interactions of architecture and the wind [6], enables exploring the aerodynamic behaviour of building shapes [7, 8], natural ventilation of urban districts and buildings [9], prediction of pollutant dispersion from high-density urban areas [10], or utilizing the wind for renewable energy not only in high-rise buildings [11] but also in low-rise residential areas [12]. Architectural design, from urban configurations, through building shapes, to the local form of the façade, impacts the character of the wind flow, as well as the creation of turbulence, and acceleration or deceleration of the wind speed, which further influences the wind pressure on buildings.

1.2. Designing the shape-morphing facades

The progressive computational design enables the creation, as well as single or multi-criteria optimization of urban and architectural designs, as early as in the conceptual design stage. More specifically, techniques such as parametric designing contribute to creating an environment-conscious, performance-oriented architecture, morphing its shape while interacting with the fluctuating external or internal conditions such as solar radiation, changing temperatures and temperature extremes, precipitation, the motion of the building occupants or the dynamic wind force [13, 14].

2. Design method

The methodology for designing architecture in the era of changing climate is introduced in a case study in Stockholm. The former industrial docks are typical for the turbulent wind conditions, created by the densely built cylindrical concrete silos, which are up to 35 meters high and up to 30 meters wide. The wind-driven design method consists of the following steps: i) firstly, the wind situation of the case study site is examined without architectural intervention, ii) secondly, the influence of an aerodynamic shape, designed between the concrete silos, is investigated, iii) and thirdly, the influence of the wind-adaptive building skin on the wind flow, as well as the wind pressure acting on building surfaces, is investigated.

2.1. CFD simulations of the case study site

The first step in the climate-conscious, wind-driven design is the analysis of wind conditions (figures 1 and 2). The CFD simulations predict where the turbulence (T), as well as acceleration (A), will occur. The wind pressure on silos (positive + and negative -) is depicted in the perspective view in figure 2. Swift for Grasshopper is used for the CFD analysis. The visual algorithmic tool Grasshopper is a plugin for the 3D modeling software Rhinoceros. Swift for Grasshopper runs on the well-known open-source CFD software OpenFOAM®, with the working environment programmed for the parametric use in Grasshopper. The calculations are performed using the RANS (Raynolds-Averaged Navier Stokes) turbulence model. It is set that the simulations should either run for 500 iterations, or converge when the calculated initial residuals of pressure $p < 4e^{-2}$, along with all velocity components $v_{x,y,z} < 1e^{-2}$, and $k\varepsilon < 1e^{-2}$. The convergence criteria are relatively mild as the solution is sought for a large-scale, urban area. The tests are performed with extreme wind speed, 24 m/s, which is used in Stockholm for the structural design of buildings [15]. The surface roughness is set as 'skimming – densely built up, 1H separated large obstacles, buildings', implying that the roughness factor $c_r(z)$ is equal to 1. In the FVM (Finite Volume Method) analysis, the size of one calculation volume in the wind tunnel is 3×3×3 meters, whereas the size of one calculation volume in the refinement region around the tested geometry is 2×2×2 meters. The calculations stopped after reaching 500 iterations, which took more than 36 hours.

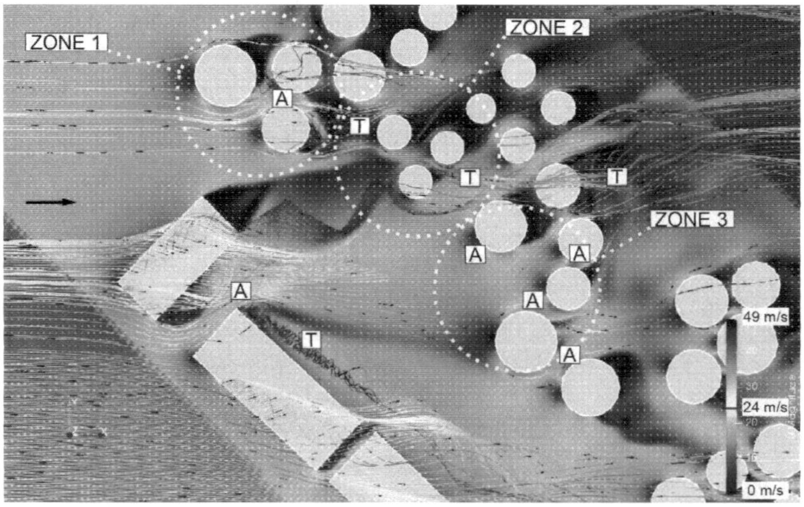

Figure 1. CFD analysis of a part of the former industrial site with silos and few other buildings on the western side (top view at 1.8 m from the ground).

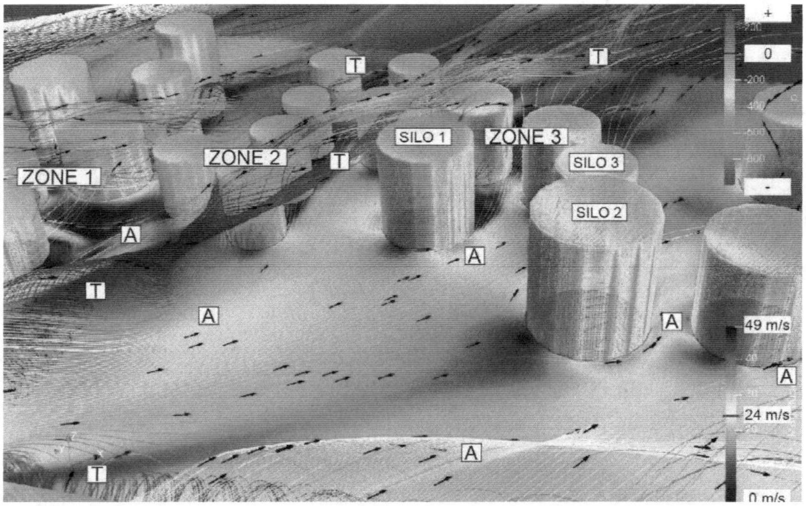

Figure 2. CFD analysis of a part of the former industrial site (perspective view). T – turbulence, A – acceleration of the wind flow.

2.2. *Design interventions*

Design interventions in the former industrial park include a protecting membrane around swimming pools in Zone 1, silos wrapped in the electricity-producing membranes in Zone 2, and an auditorium with a membrane building skin in Zone 3. Three shapes winding around the silos, which are designed parametrically in Grasshopper, deflect the wind in Zone 1, accelerate the wind in Zone 2 [16], and act as a minimum resistance to the wind flow in Zone 3 (figure 3). This paper will focus on one of the shapes, the auditorium in Zone 3. The parametric design of architectural shapes enables the fast creation of several design options.

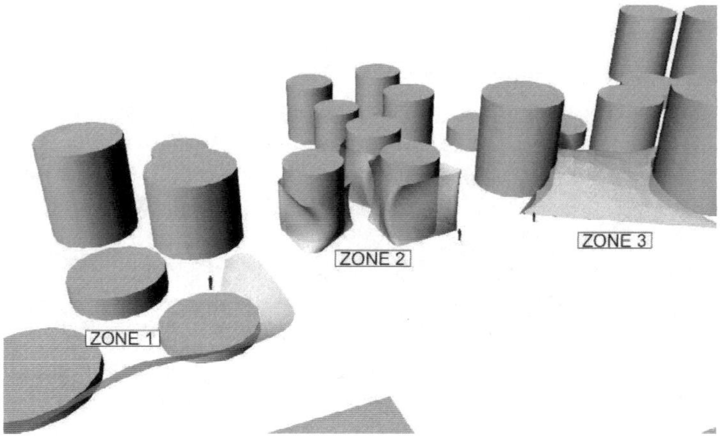

Figure 3. Architectural interventions around concrete cylindrical silos affecting the wind flow.

2.3. *Wind-adaptive building skin*

Tensegrity-membrane building skin, adapting to the wind in real-time, was designed and tested as a prototype in the previous research [17]. Thanks to the tensegrity geometry, as well as the material properties of the structure, every element of the adaptive façade bends in the acting wind (figure 4). Depending on the intensity and the direction of the wind flow, adaptive elements passively (with the embedded material and geometrical properties) respond to the dynamically changing wind conditions. Although lightweight, such building skin can withstand high wind speeds by distributing the wind loads within the adaptive structure. One of the benefits of the passive (without computer control) response is the reversible shape adaptation under load. The developed adaptive skin is applied to the designed auditorium shape in Zone 3. For the investigation of the performance in the wind, the wind-induced shape response of the tensegrity-membrane skin is simplified, neglecting the tensegrity sub-structure. For the CFD evaluation in the early design stage, however, such simplification is possible as the goal is to quickly compare the wind performance of different shape options in the wind. The material characteristics of the membrane, used as a building skin, are derived from [18] and described in table 1. In the Kangaroo 2 [19] simulations, the skin of the proposed aerodynamic shape consists of the membrane and is loaded by the wind force. The wind speed 24 m/s is represented by pressure p_w = 360 Pa, derived from the Bernoulli's equation:

$$p_w = 1/2(\rho v^2) \quad (1)$$

where ρ is the density of air (the value recommended by the EN standard [15] for calculations related to the wind loads is 1.25 kg/m^3) and v is the wind velocity in m/s.

Table 1. Material properties of the membrane in the wind-adaptive element.

Material and thickness h (mm)	Stiffness k (N/m)	Young's Modulus E (Pa)	Pre-tension 0=max.; 1=min. (-)
tensile fabric 0.51	84 150	16.5e^7	0.90

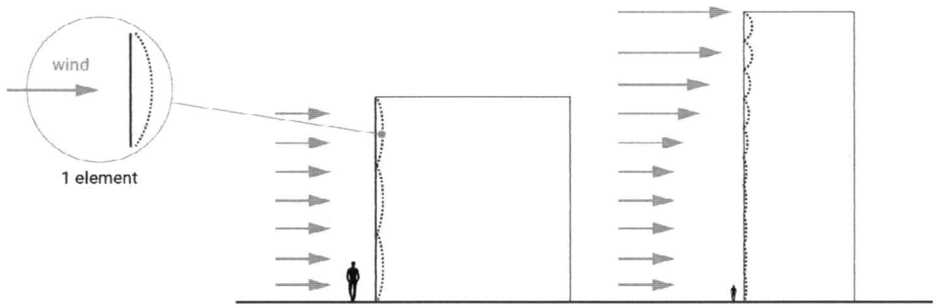

Figure 4. A scheme of the adaptive skin application with a zoom on one adaptive element.

2.4. CFD analysis of the designed shape in Zone 3

The wind analysis of the proposed aerodynamic shape, with and without the intended adaptive skin, is performed in Swift for Grasshopper (figure 5). The same boundary conditions as in the original site analysis, as well as the same initial wind speed 24 m/s is used in the Zone 3 simulations. The wind tunnel (the CFD domain), however, is smaller; the computational volumes are smaller, meaning the wind tunnel will consist of more computational volumes to capture the flow in greater detail. The cell size of the wind tunnel is 3×3×3 meters, whereas the cell size in the refinement region around the investigated geometry is 1×1×1 meter. The convergence criteria are the same as in the site analysis. The calculations are set to run until reaching convergence. By the given boundary conditions, the simulations converged for smooth shape variant and dimpled shape variant with adaptive building skin in 650 iterations and 352 iterations, respectively.

3. Results and conclusion

The designed aerodynamic shape, winding around silos in Zone 3, influences the wind character in its vicinity. The aerodynamic shape directs the wind flow around the silos and contributes to the reduction of the positive and negative surface wind pressure on the three closest silos. The results of the three CFD simulations focus on surface pressure and suction reduction achieved by the designed architecture. The pressure used in Swift is kinematic pressure with units in m^2/s^2. The analysed case without any architectural intervention shows the surface pressure on the silos 1, 2, and 3 in Zone 3, ranging from 1091.64 to +315.50 m^2/s^2 (-1364.55 to +394.38 Pa) (figure 2). Now, with the designed aerodynamic shape, the surface pressure on the three silos ranges from -1048.26 to +411.89 m^2/s^2 (-1310.33 to +514.86 Pa). If the adaptive skin is applied to the aerodynamic shape creating a rough, dimpled surface, there is an even greater reduction of the wind suction; the pressure ranges from -864.96 to +410.50 m^2/s^2 (-1081.20 to +513.13 Pa) (figure 5). The results show that the utilization of computational tools in the early design phase of architectural designing enables a fast exploration of different design proposals. The paper demonstrates that wind-driven architectural design can positively change the wind microclimate around buildings by deflecting the wind from the pedestrian level (1.8 m above the ground). Moreover, architectural interventions can contribute to reducing the negative effects of the wind, such as wind pressure and wind suction, leading towards lighter structures in architecture. To conclude, the application of the wind-adaptive building skin instead of regular, smooth surfaces could contribute to i) much lighter building skin structures even for the more extreme winds, ii) reduction of wind pressure on surfaces, especially the reduction of wind suction, and iii) creating an entirely new reading of the wind through real-time architectural response.

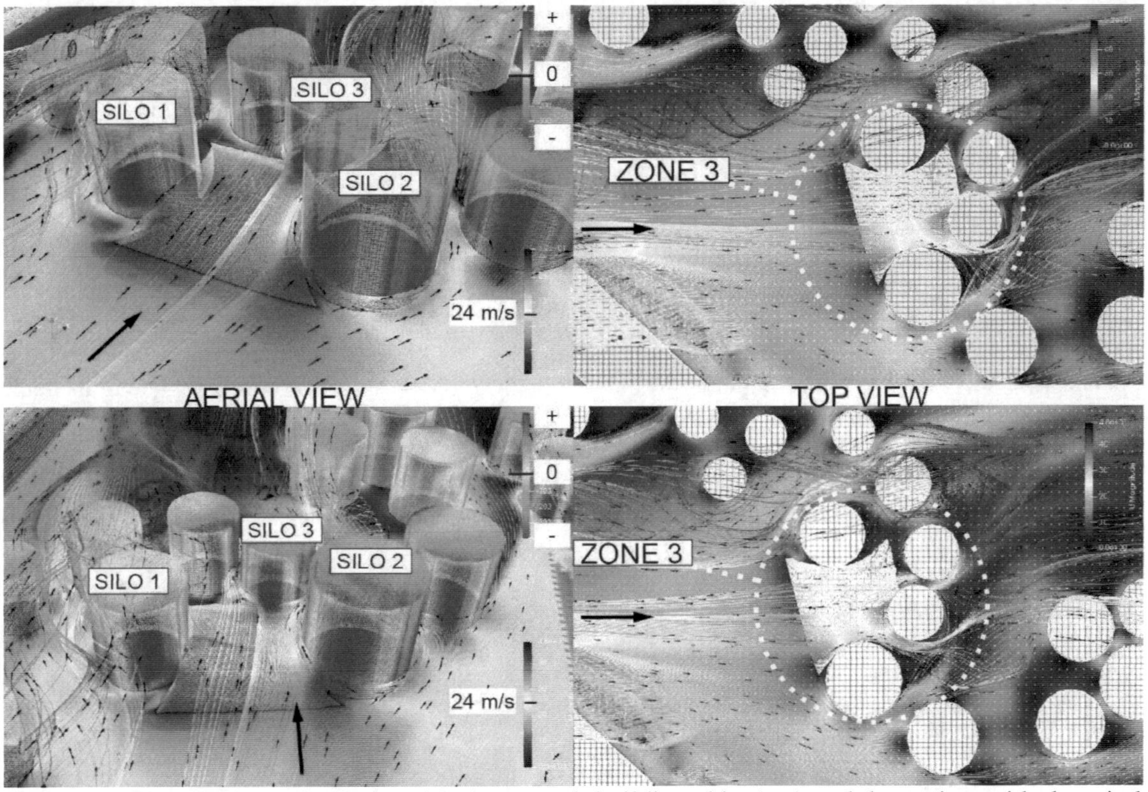

Figure 5. The aerodynamic shape with the smooth building skin (top) and the variant with the wind adaptive building skin (bottom) (top view at 1.8 from the ground).

4. References

[1] Snow M and Prasad D 2011 Climate Change Adaptation for Building Designers: An Introduction, *Environment Design Guide* pp. 1–11

[2] Pellitteri G, Lattuca R, Concialdi S, Conti G and De Amicis R 2009 Architectural shape generating through environmental forces *Joining Languages, Cultures and Visions: Proc. 13th Int. CAAD Futures Conf.* (Montréal) pp 875–886

[3] Kuismanen K 2008 *Climate-conscious architecture: Design and wind testing method for climates in change* University of Oulu

[4] Pachauri, R K, Meyer L A 2014 *Climate Change 2014: Synthesis Report. Contribution of Working Groups I, II and III to the Fifth Assessment Report of the Intergovernmental Panel on Climate Change* IPCC Geneva Switzerland

[5] Jin H, Liu Z, Jin Y, Kang J and Liu J 2017 The Effects of Residential Area Building Layout on Outdoor Wind Environment at the Pedestrian Level in Severe Cold Regions of China *Sustain.* vol **9** no 12 pp 1-18
[6] Kormaníková L, Achten H, Kopřiva M and Kmeť S 2018 Parametric wind design *Front. Archit. Res.* vol **7** no 3 pp 383–394
[7] Mooneghi M A and Kargarmoakhar R 2016 Aerodynamic Mitigation and Shape Optimization of Buildings: Review *J. Build. Eng.* vol **6** pp 225–235
[8] Kabošová L, Kmeť S and Katunský D 2019 Digitally Designed Airport Terminal Using Wind Performance Analysis *Buildings* vol **9** no 3 pp 102–116
[9] Kim H J and Kim J S 2018 Design methodology for street-oriented block housing considering daylight and natural ventilation *Sustain.* vol **10** no 9 pp 1-22
[10] Yuan C *et al.* 2019 Multilayer urban canopy modelling and mapping for traffic pollutant dispersion at high density urban areas *Sci. Total Environ.* vol **647** pp 255–267
[11] Ishugah T F, Li Y, Wang R Z and Kiplagat J 2014 Advances in wind energy resource exploitation in urban environment: a review *Renew. Sust. Energy Rev.* vol **37**, pp 613–626
[12] Zhou H, Lu Y, Liu X, Chang R and Wang B 2017 Harvesting wind energy in low-rise residential buildings: Design and optimization of building forms *J. Clean. Prod.* vol **167** pp 306–316
[13] Turrin M, Buelow P Von, Kilian A and Stouffs R 2014 Parametric modeling and optimization for adaptive architecture *Intelligent Computing in Engineering EG-ICE International Workshop* 1–8
[14] Biloria N and Sumini V 2009 Performative Building Skin Systems: A Morphogenomic Approach towards Developing Real-Time Adaptive Building Skin Systems *Int. J. Archit. Comput.* vol **7** no 4 pp 643–675
[15] Eurocode 1 2010 Actions on structures, Part 1-4: General actions. Wind actions, EN 1991-14:2005 Brussels
[16] Kormanikova L, Chronis A, Kmet S and Katunský D 2018 Wind-formed Architectural Shapes *Proc. 36th eCAADe Conf.* vol **2** pp 377–384
[17] Kabošová L, Foged I W, Kmeť S and Katunský D 2019 Hybrid design method for wind-adaptive architecture *Int. J. Archit. Comput.* vol **17** no 4 pp 1–16
[18] Yang S and Sultan C 2016 Modeling of tensegrity-membrane systems *Int. J. Solids Struct.* vol **82** pp 125–143
[19] Preisinger C 2013 Linking Structure and Parametric Geometry Archit. Des. vol **83** no 2 pp 110–113

Acknowledgments
This research was supported by the Grant Agency of the Slovak Republic. The project numbers are VEGA 1/0674/18 and VEGA 1/0129/20.

IOP Conf. Series: Materials Science and Engineering **867** (2020) 012015 doi:10.1088/1757-899X/867/1/012015

Comparison of results of an adaptive membrane structure with a numerical model

L Kapolka[1], L Stulerova[1] and S Kmet[1]

[1] Technical university of Kosice, Faculty of Civil Engineering, Institute of Structural Engineering, Slovakia

Abstract. This paper discusses results of the adaptive membrane device compared to the model created in FEM software. The numerical model was created according to factory parameters of this device. The adaptive membrane device has to be properly set up, than the device is loaded by load piston with saddle and the adaptivity of this device consisted of two actuators used as anchor rods and another two anchor rods are fixed. The actuator is realised as a hydraulic piston. Based on the results obtained by these measurements and the load results in the numerical model, the comparison was evaluated in the conclusion of this paper.

1. Introduction

Nowadays, membrane structures are a very modern and popular way of roofing large spans. Since there is a demand to increase the economy and technological solution of building structures, these requirements can be met by introducing of adaptive systems. Under adaptive systems, it is possible to imagine a set of elements that allow structures to change their geometric and stress configurations.

This paper describes an experimental device of an adaptive membrane structure, whose design results mainly from the requirement to increase the economy of structures. Due to the fact, that tensile stressed systems are subjected to time-dependent influences (rheology of tensile members) and they are strongly non-linear, they are often designed as oversized compared to conventional steel systems. It would be appropriate to prevent of these effects in some way.

As mentioned above, with the adaptive systems it is possible to vary the stress configuration of the structure in real time, and thus that the cross-sections of the tensile members could be more economically designed. If a lower stress state of the structural system could be achieved in the long term by an adaptive system, the rheological impact could be greatly avoided, but under critical load the adaptive system would be able to provide the required stiffness of the structural system [1], [2].

The adaptability of the experimental device described in this work is based on a system of hydraulic pistons (actuators). It is possible to change the stress state of the whole construction system by extensioning or shortening of the pistons. Conclusion of this paper is the comparison of the results obtained from the experimental device with the results gained form a geometrically detailed numerical model, that was created based on the manufacturing documentation of the device.

2. Description of the experimental device

Based on theoretical assumptions and computational models created in the preparatory phase of the experiment, the shape of adaptive hyperbolic paraboloid was chosen as the most suitable variant. The experimental device [3], [4] shown in figure 1, consists of technical fabric with the 2x2 m floor plan. Total height of technical fabric is 1 m. Membrane surface is supplemented by 4 edge stainless steel cables with 8 mm diameter, which are inserted into welded edge sleeves of technical fabric and they are anchored to the steel frame using ball hinges with 2 actuators and 2 anchor rods.

IOP Conf. Series: Materials Science and Engineering **867** (2020) 012015 doi:10.1088/1757-899X/867/1/012015

Figure 1. The experimental device of the adaptive membrane structure.

The technical fabric is made by Serge Ferrari, type Précontraint 502. It is the polyester fabric coated with PVC coating. The woven of membrane was set in the directions of the individual vertices and it is positioned within the machine such that the x-axis direction, which has an elastic modulus E_x = 1,057 MPa, is anchored between the upper anchor rods (actuator AP2 and anchor rod T2). The modulus of elasticity in the y-axis direction is E_y = 612 MPa. Edge cables are spiral, single-strand with open construction 1x19 (1+6+12) and nominal strength 1 570 MPa. Edge cables are connected to a membrane clamping plates. Actuators have a movement range of 100 mm, but this range is limited mainly by the maximum load capacity of the element of 50 kN.

The steel frame construction is fully welded with one mounting joint at the top and bottom of the frame. It is a welded spatial structure consisting of closed RHS (rectangular hollow section) and SHS (square hollow section) profiles made of S235 steel. The basic profiles are columns SHS160×160×8, beams around the perimeter are RHS140x140x8 and both bases of steel frame box are reinforced with the cross-sections SHS140×140×8. Axial dimensions of the frame are 3.34×3.34×2.50 m.

The membrane is loaded by a cylindrical loading with 700 mm of diameter saddle, which has the same shape as the initial shape of the membrane surface. The piston is anchored in the centre of the upper base of the test frame. The capacity of the load piston is 15 kN, but this value is limited by the maximum movement range of the load piston, that is 100 mm.

3. Numerical model

The numerical model of the adaptive membrane structure of the experimental device was created in the Dlubal RFEM software using the add-on module – Form-Finding, which determines the equilibrium positions of membrane and cable elements.

The model [5], shown in figure 2, includes almost all geometrical and material characteristics according to the experimental device. The equilibrium position is determined based on selected boundary conditions such as the prestressing of the membrane surface ($n_x = n_y$ = 4 kN / m width) and the required geometry (the sag of the edge cables s = 0.20 m). The size of the finite element surface mesh was 50 mm. As the equilibrium shape search process was used the Update Reference Strategy (URS) [6]. This process takes over the originally specified finite element geometry and the surface tension and iteratively shifts the position of network nodes until there is a surface tension in equilibrium with boundary conditions.

After form-finding process, the model was loaded by a free circular load with diameter of 700 mm. The actuators elongation or shortening was simulated by load case using absolute elongation/shortening of anchor rods.

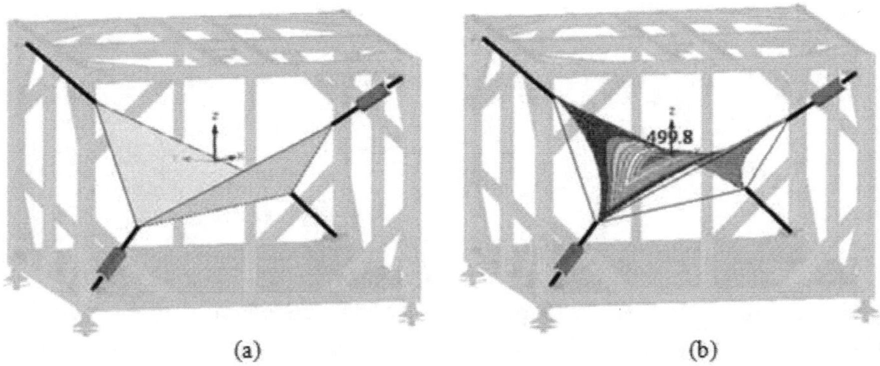

Figure 2. The numerical model of the experimental device:
(a) before form-finding process, (b) after form-finding process.

4. Methodology of measurement

The test set-up was performed in three series. Each serie has three measurements to cover the minimum statistical set. In the first serie of measurements, the membrane surface was loaded by the load piston with pressure force value 1 kN. Each serie of measurements has begun by adjusting of the initial equilibrium state of the membrane surface and stabilizing the tensile force values in the anchor rods and actuators at about 24 kN.

Subsequently, these values were recorded, including the position of the load piston. For the first measurement, load piston has been in contact with the membrane surface, but showed only a minimal signal from the load cell (the force in the load piston was equal to value 0 kN).

Each measurement in the serie started from the initial set-up, followed by loading with the load piston with a specific value for each serie of measurements and one minute system time stabilization was done. This time stabilization was performed after each change in each measurement.

After loading by load piston and recording the measured values, the actuators were shortened in the range of 0 to 3 mm, in steps of 1 mm, then the actuators were returned to the zero position and the measurement continued by extending the actuators from 0 to 2 mm. Values were recorded. The movement of actuators in all measurements was synchronous. In each successive serie the force increment in the load piston was 2 kN.

5. Results

On the basis of the measurements obtained from the experimental device test, the mean values were established. The deformation of membrane surface and tensile forces of actuators and anchor rods were observed. The deformation of the membrane surface was equal to extension of the load piston. These values were compared with the values obtained from the numerical model and the relative measurement error was evaluated for each monitored variable. The results of the comparison of the deformations of the numerical model and the mean values obtained by the experimental device can be seen in the figure 3 and figure 4.

Since the measured values from actuators and anchor rods with the same location have small relative differences (actuator AP2 and anchor rod T2 are located at the upper corners of the test frame, and actuator AP1 and anchor rod T1 are located at the lower corners of the test frame), only the comparison of the elements with the maximum relative difference will be shown. The comparison of the results of the axial forces in the anchoring elements or actuators of the numerical model with the average values obtained by the test assembly can be seen in the figure 5 and figure 6.

IOP Conf. Series: Materials Science and Engineering **867** (2020) 012015 doi:10.1088/1757-899X/867/1/012015

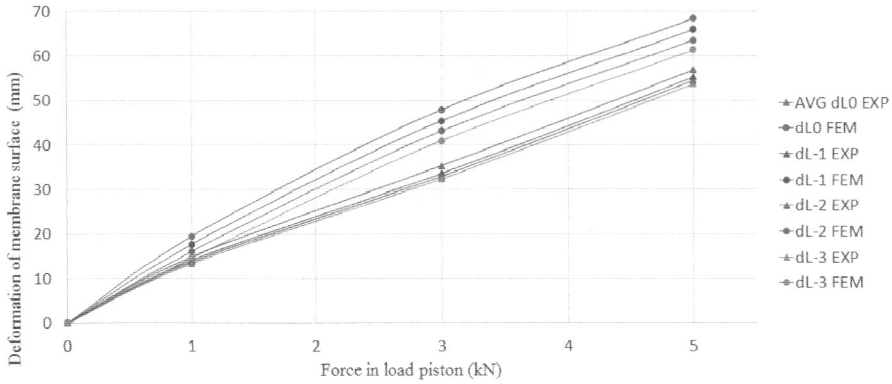

Figure 3. Comparison of the dependence of the deformation of the membrane surface at individual values of forces in the loading piston, for individual steps of shortening of actuators.

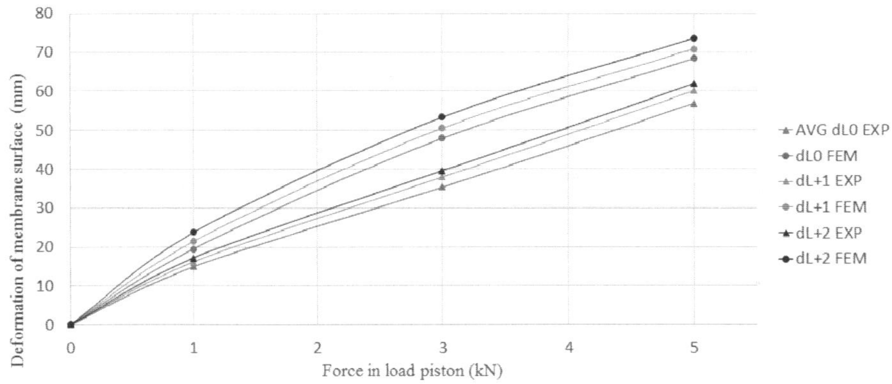

Figure 4. Comparison of the dependence of the deformation of the membrane surface at individual values of forces in the loading piston, for individual steps of extensioning of actuators.

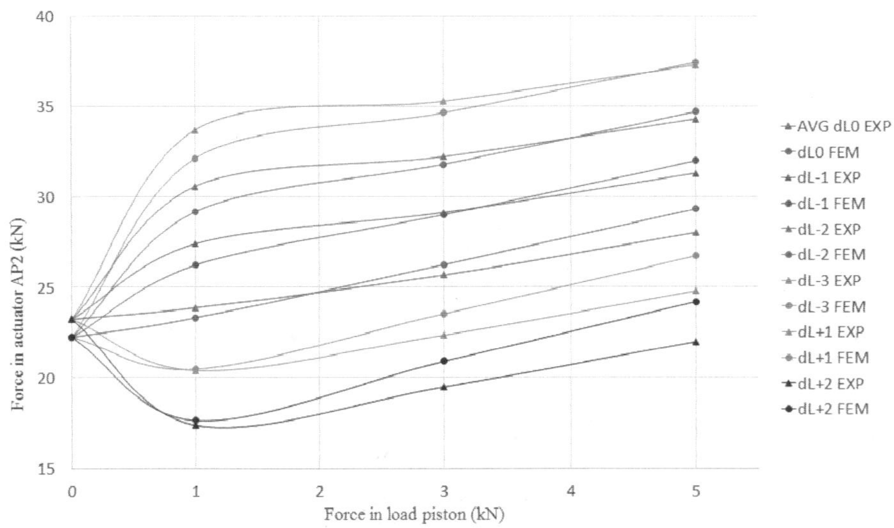

Figure 5. Comparison of the dependence of the axial forces of the actuator AP2 at individual values of forces in the loading piston, for individual steps of extensioning and shortening of actuators.

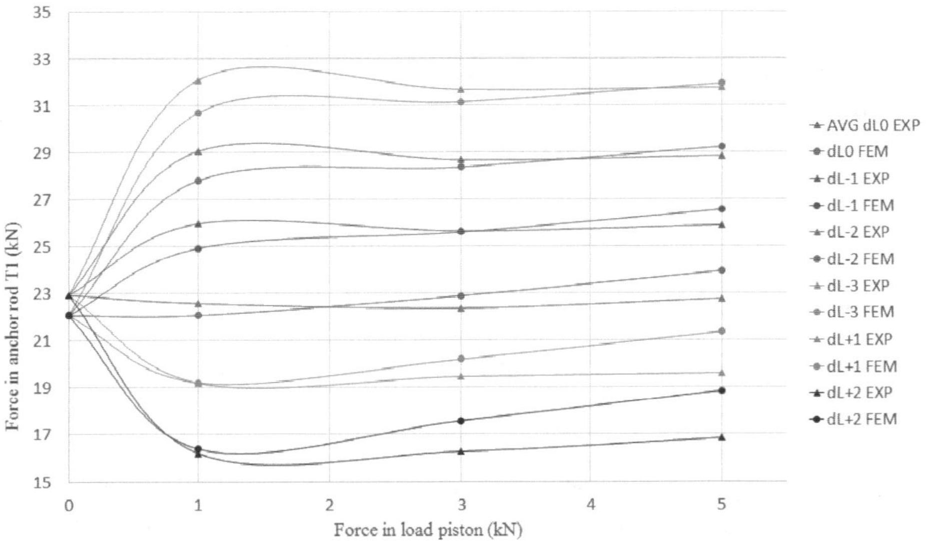

Figure 6. Comparison of the dependence of the axial forces of the actuator AP2 at individual values of forces in the loading piston, for individual steps of extensioning and shortening of actuators.

6. Discussion

Based on the processed results and comparison of deformations between the numerical model and the experimental device test, it can be stated that the minimum relative difference in deformation $\Delta\delta_{R,\,min}$ = 9.53% was at the load of 1 kN. This occurred when the actuator was shortened by 3 mm and it represents difference of deformation $\Delta\delta$ = 1.27 mm between test set and numerical model. The deformation of the membrane surface of the experimental device test was minor compared to the numerical model. The maximum relative difference in deformation was $\Delta\delta_{R,\,MAX}$ = 37.95% at the load of 1 kN and occurred when the actuator was extended by 2 mm, which represents the absolute deformation difference $\Delta\delta$ = 13.75 mm between the experimental device test and the numerical model. The deformation of the membrane surface of the experimental device test was minor compared to the numerical model.

The comparison of the axial forces in the upper group of elements (AP2 and T2), shown that the minimum relative difference in axial force was $\Delta N_{R,\,AP2,\,min}$ = 0.39% at the load of 1 kN and occurred when the actuator was extended by 1 mm, which represents absolute difference between forces from both force sources $\Delta N_{AP2,\,min}$ = 0.08 kN. The axial force of the experimental device test was minor compared to the numerical model. The maximum relative difference in axial force was $\Delta N_{R,\,AP2,\,MAX}$ = 10.10% at the load of 5 kN and arised when the actuator was extended by 2 mm, which represents the absolute difference between axial force $\Delta N_{AP2,\,MAX}$ = 2.22 kN. The axial force of the experimental device test was minor compared to the numerical model.

The comparison of the axial forces in the lower group of elements (AP1 and T1), shown that the minimum relative difference in the axial force was $\Delta N_{R,\,T1,\,min}$ = 0.05% at the load of 1 kN and arised when the actuator was extended by 1 mm, which represents absolute difference between forces from both force sources $\Delta N_{T1,\,min}$ = 0,01 kN. The axial force of the experimental device test was minor compared to the numerical model. The maximum relative difference in axial force was $\Delta N_{R,\,T1,\,MAX}$ = 11.96% at the load of 5 kN and arised when the actuator was extended by 2 mm, which represents the absolute difference between axial force $\Delta N_{T1,\,MAX}$ = 2.01 kN. The axial force of the experimental device test was minor compared to the numerical model.

From the graphs of the axial forces it can be seen that the differences between the numerical model and the experimental device test, increase at 1 kN with increasing stiffness of the structural system, but this varies with increasing load at the same pretensioning stage, when these differences are decreasing during the increasing stiffness of the structural system.

IOP Conf. Series: Materials Science and Engineering **867** (2020) 012015 doi:10.1088/1757-899X/867/1/012015

7. Conclusion

Taking into account the high non-linearity of the whole structural system, the values of relative differences of the axial forces in the range of about 12% can be considered as satisfactory, but the relative differences of large displacement in range of about 40% remains as question. The answer to this question could be the fact that the material characteristics of the technical textiles were taken from the manufacturer and thus these were not be experimentally obtained.

Another reason could be the fact, that load piston and load saddle of the experimental device is rigid and creates circumferential local deformations of the membrane surface, while the numerical model is loaded with a free circular load, and thus the deformation result is concentrated to the centre of the membrane surface.

Obtaining experimentally determined values of the material characteristics of technical textiles could be the subject of further research to refine the numerical model of the experimental device and their subsequent comparison.

8. References

[1] Kmeť S 2012 Modern cable and membrane constructions *Building materials* **8** 48–52
[2] Del Grosso A E and Basso P 2010 Adaptive building skin structures *Smart Materials and Structures* **19** 12
[3] Kapolka L, Štulerová L and Kmeť S 2019 Adaptive membrane structure – experimental device *Assembly of steel constructions workers* **19** 79–86
[4] INOVA Prague s.r.o. 2017 *Documentation of experimental device – 17 120 part 1 – Membrane machine* Prague
[5] Kapolka L 2019 *IOP Conf. Ser.: Mater. Sci. Eng.* **566** 012014
[6] Wüchner R and Bletzinger K U 2005 Stress-adapted numerical form finding of pre-stressed surfaces by the updated reference strategy *Int. J. Numer. Meth. Eng.* **64** 143–166

Acknowledgments
This work was supported by The Slovak Scientific Grant Agency under the contract No.VEGA 1/0129/20.

IOP Conf. Series: Materials Science and Engineering **867** (2020) 012016 doi:10.1088/1757-899X/867/1/012016

Evaluation of frost resistance of various blended binders

E Kardosova[1]

[1] Technical university of Kosice, Faculty of Civil Engineering, Department of Environmental Engineering, Slovakia

Abstract. The use of blended binders using alternative raw materials is one of the key areas of sustainable development in terms of reducing CO_2 in the atmosphere. This approach reduces the proportion of energy-intensive clinker in cement when satisfying technical parameters. The paper presents a proposal of variants of blended binders with the application of fly ash, blast furnace slag and perlite for their use in building mixtures. The durability of the mixtures is assessed on the basis of testing the frost resistance of mortars. Test specimens were produced according to STN EN 196-1. The samples were subjected to 25 freeze-thaw cycles, and both the flexural and compressive strength were then determined, to express a durability factor. Except for slag as a standard component of blended binders, perlite appears to be a perspective material; it meets a limit for durability factor when used as a 30% substitution for cement, as well as when used in ternary composition with slag. Fly ash showed weak frost resistance, even when used with a suitable perlite.

1. Introduction

One of the most important properties of concrete that determines its durability is frost resistance. Freezing-thawing effect is one of the main problems that can lead to damage to materials like concrete or mortar. Durability problems of cement-based materials associated with cold weather conditions can take two main forms:

a) Internal cracking due to freeze-thaw cycles accompanied by loss of strength and stiffness.
b) Surface scaling due to freezing in the presence of de-icing salts.

Typically, damage occurs when the material is saturated and its pore structure is unable to withstand the stresses caused by ice formation. According to [1], concrete can be resistant to cyclic freezing and thawing provided it has sufficient strength and an adequate air-void system, and the aggregates are frost-resistant. The main factors affecting the above mentioned features of concrete, thus its frost resistance, are varieties of cement, water-cement ratio (directly linked to the porosity/permeability of matrix) and the density of aggregates [2]. In the case of mortars, the degree of saturation, total pore volume, pore size and distribution, frost velocity and tensile strength contribute to freeze-thaw resistance [3]. High porosity means higher water content and a higher frost force which reduces the resistance to freezing and thawing cycles.

According to [3], also aggregate size is a factor affecting frost resistance. Small grains are more favourable as they provide a larger specific surface area. As a consequence, there is more cement mortar between the grains. The bonding force is stronger and the resistance to freezing and thawing cycles is better.

Above mentioned varieties of cement are on its composition and on the possible application of mineral additives, including their type and quantity. A categorization of the effects from the mentioned additives is given by [4]; they can be by their type, their scale and the mechanism by which they improve the freeze-thaw performance of concrete.

IOP Conf. Series: Materials Science and Engineering **867** (2020) 012016 doi:10.1088/1757-899X/867/1/012016

Studies on application of ground granulated blast furnace slag (GGBFS) [5, 6] show 25%, 50%, 61% and 75% range of it usage in concrete presenting that GGBS has performed well under frost resistance tests [5, 6]. Tavasoli et al. point that the frost resistance of mixtures is highly influenced by percentage; by increasing slag replacement percentage the durability is decreasing. According to author, using 50% GGBS do not have impressive effect on the frost resistance concrete [7].

Perlite powder contains approximately from 70 to 75 % SiO_2 and from 12 to 18 % Al_2O_3. Due to its glassy structure and high SiO_2 and Al_2O_3 contents, perlite can be specified to be a pozzolan [8]. Some investigations report on the properties of cement composites with perlite powder [9,10,11,12], however there are limited results on the frost resistance.

On the other hand, there are quite lots of investigations on effect of fly ash on the frost resistance of cement-based mixtures [1, 13, 14, 15], however as mentioned by [14], there are often not compared on an equal footing. Generaly can be summarized that concrete containing fly ash may be less resistant when subjected to freezing and thawing [1]. As presented by [15], frost resistance decreases with dosage of fly ash, while the rate of decline from reference (cement) sample is different with different kinds of fly ash.

The frost resistance of cement composites is affected also by external factors, like the number of freeze-thaw cycles [15]. Liu et al. state, that as the number of freeze-thaw cycles increased, the quality of the cement mortar gradually decreased. Compressive strength, flexural strength and modulus of elasticity of the mortar were also reduced [16]. Glukhovsky et al. concluded that slag concrete activated by sodium silicate usually has the lowest porous structure, the highest strength, and the best frost resistance. These concretes were found to withstand between 300 and 1300 freeze-thaw cycles [17].
The measures to improve frost resistance of cement mixtures are to enhance density, reduce water-cement ratio and apply an air-entraining agent. Bubbles are able to accommodate water being pushed forward by the ice in the process of being formed [2]. Pogorelov and Semenyak report that one of the effective ways to improve the strength and resistance of a material to cracking, including frost aggression, is to introduce fibre reinforcement [18].

The principle of the frost resistance test is the freezing and thawing of test beam which are soaked with water. The number of freezing cycles, as well as applied temperatures, depends on the relevant standard or project [19].

The aim of this work was to verify the influence of selected mineral additives on the freeze-thaw resistance of cement composites. Perlite, fly ash and ground granulated blast furnace slag, together with cement in different proportions, were used in this work to prepare different blended binders. A reference sample was made up of cement, standard sand and water. The samples were subjected to 25 freeze-thaw cycles. The effect of individual blended binders on the frost resistance of mortars was evaluated by durability factor based on both the flexural strength and the compressive strength.

2. Materials
For testing the frost resistance of mortars, following materials were used:
- Standard sand CEN according to STN EN 196-1 [20]
- Cement CEM I 42,5 R (C): CRH Turňa nad Bodvou
- Mineral additives (Ad):
  - Fly ash (F): US Steel Košice
  - Ground granulated blast furnace slag GGBFS (S): US Steel Košice
  - Perlite powder (P): Lehôtka pod Brehmi
- Water (W)

3. Methods
Preparation of samples
To verify the effect of selected additives (F, S, P) on the freeze-thaw resistance of mortars, six different variants of blended binders were prepared in following models:
- 70 % C + 30 % Ad: C70S30, C70F30 and C70P30
- 70 % C + 15 % Ad + 15 % Ad: C70S15F15, C70P15F15 and C70P15S15
- 100% C - reference sample

Then, mortars were prepared in accordance with STN EN 196-1 [20]. The mixtures contained of 450 g ± 2 g of binder, 1350 g ± 5 g of standard sand and 225 g ± 1 g of water. After mixing by standard regime, testing beams of 40x40x160 mm were casted and cured in standard way (water curing, 28 days). Then, one series of samples was subjected to freezing-thawing testing, while the second one (comparative samples) was cured in the water under constant temperature of + 20 °C.

Parameter evaluated

To express the frost resistance of cement composites, durability factor based on the flexural strength is usually used, as given by formula (1). Then, the samples are considered to be frost resistant if the freeze-thaw durability factor is not less than 0.85 [19].

Freeze-thaw durability factor is calculated as follows:

$$x_f = \frac{f_{f,f}}{f_{f,0}} \tag{1}$$

where,

x_f freeze-thaw durability factor;

$f_{f,0}$ flexural strength of the comparative samples, in MPa (N/mm^2);

$f_{f,f}$ flexural strength of samples after freezing-thawing cycles, in MPa (N/mm^2).

In the paper, this model was also used for the evaluation of frost resistance in terms of compressive strength and relevant durability factor is marked as x_c. Both of durability factors are compared in figure 1.

To find the strength characteristics for calculation of durability factors, one series of samples was subjected to 25 freezing-thawing cycles, according to Slovak standard STN 73 1322: Determination of frost resistance of concrete [19]. The cycles consisted of freezing under -20 °C (2 hours of temperature rise, 2 hours of temperature holding) and defrosting under + 20 °C (0.5 an hour of temperature rise and 1.5 hours of temperature holding). One freezing-thawing cycle lasts 6 hours. After 25 freezing-thawing cycles, the flexural strength of both series of samples (frozen and comparative) was measured to calculate durability factor x_f, following by testing the compressive strength using the halves of the beams broken, to calculate durability factor x_c.

4. Results and discussion

The results of flexural strength of comparative samples ($f_{f,0}$) as well as samples after freezing-thawing cycles ($f_{f,f}$) are given in table 1. Table 2 shows the results of compressive strength of comparative samples ($f_{c,0}$) as well as samples after freezing-thawing cycles ($f_{c,f}$).

Table 1. Flexural strength of samples before and after freezing-thawing cycles.

Samples	Before freezing $f_{f,0}$ [MPa]	After freezing $f_{f,f}$ [MPa]
Reference sample	7.0	7.9
C70 S30	7.2	7.2
C70 F30	5.9	1.4
C70 P30	5.3	5.0
C70 S15 F15	6.8	2.2
C70 P15 F15	6.2	1.9
C70 P15 S15	5.9	5.6

Table 1 shows that the flexural strength of samples after freezing-thawing cycles decreases, excepting for reference sample (the strength increased from 7 MPa to 7.9 MPa) and C70S30 sample (no any change after the freezing-thawing process).

Table 2. Compressive strength of samples before and after freezing-thawing cycles.

Samples	Before freezing $f_{c,0}$ [MPa]	After freezing $f_{c,f}$ [MPa]
Reference sample	54.5	50.5
C70 S30	41.5	34.0
C70 F30	39.0	25.0
C70 P30	31.0	33.5
C70 S15 F15	44.5	31.5
C70 P15 F15	38.5	25.5
C70 P15 S15	38.0	36.5

Table 2 shows that the compressive strength of samples after freezing-thawing cycles decreases, excepting for C70P30 sample.

The results are evaluated deeper by durability factors, presented in figure 1.

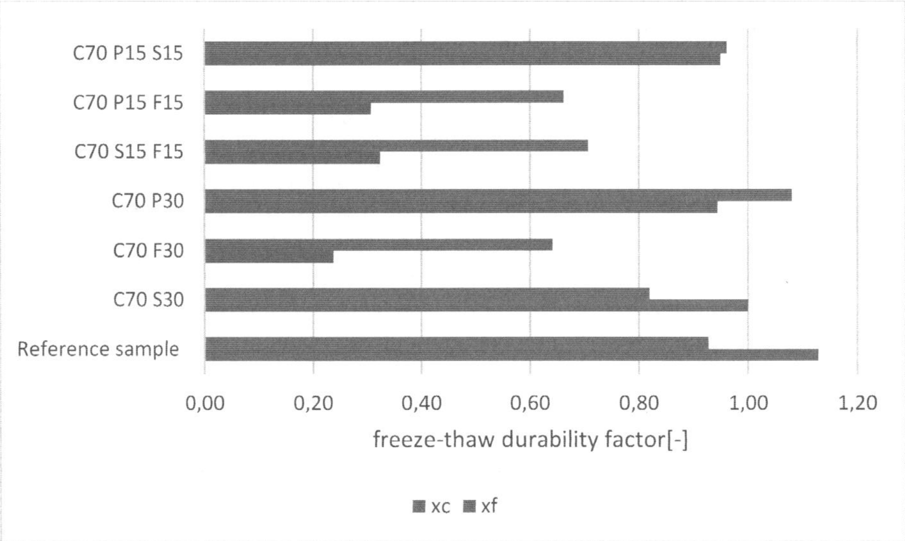

Figure 1. Freeze-thaw durability factors of samples.

The samples are considered to be frost resistant when durability factor is bigger than 0.85. In addition to the reference sample CEM 100, this limit is achieved by mixtures C70S30, C70P30 and C70P15S15 in terms of x_f, indicating good performance of GGBFS and perlite. This is supported also by durability factor x_c, which runs in acceptable (even better) values.

On the other hand, the samples containing fly ash (C70F30, C70S15F15 and C70P15F15) do not meet limit for frost resistance. But it was also observed that the durability factor x_c of these samples is more than twice as good as durability factor x_f.

The results are in line with [21], who states that samples containing GGBFS achieved better results than that of containing fly ash. The beneficial effect of GGBFS is also presented by [4], who associated this with the influence of slag dosage on the lower permeability.

Good performance of perlite is in line with results presented by [22]. They show that the frost resistance is improved when mixtures are prepared with higher concentration of perlite. Authors suggest this is partly attributed to the porous nature of perlite that helps reducing the disruptive expansive stresses caused by freezing-thawing process.

5. Conclusion

The results confirm both the perlite and GGBFS as satisfactory components of blended binders, either in combination with cement only (C70P30 and C70S30), or in ternary composition C70P15S15, while 30% of the dose seems to be optimal. Compared to GGBFS as a standard component of blended binders, perlite appears to be a perspective material – while keeping the limit for x_f durability factor and when used in separate combination with cement (C70P30), it provides the best performance in compressive strength out of all samples. This finding may be useful in the production of blended cements in regions suffering from the unavailability of GGBFS.

6. References

[1] Thomas M D A 2007 *Optimizing the use of fly ash in concrete* (United States: Portland Cement Association, Skokie, IL) pp 1-24
[2] Zhang H 2011 *Building materials in civil engineering* (Cambridge: Elsevier) pp 1-424
[3] Liu H et al 2018 *Appl. Sci* **8**(8)
[4] Ebrahimi K et al 2018 *Construction and Building Materials* **186**
[5] Li K L et al 2011 *Trans Tech Publications Ltd* 338-343
[6] Malhotra V M 1989 *Canadian Journal of Civil Engineering* **16**(2) 140-156
[7] Tavasoli S Nili M Serpoush B 2018 *Construction and Building Materials* **165** 717-722
[8] Erdem T K et al 2007 *Taylor and Francis* **13** 978-0
[9] Erdem T K et al 2007 *Cement and Concrete Composites* **29**(1) 13-21
[10] Ramezanianpour A A et al 2014 *Construction and Building Materials* **66** 222-228
[11] Bektas F Turanli L Monteiro P J M 200*Cement and Concrete Research* **35**(10) 2014-2017
[12] Yu L H Ou H Lee L L 2003 *Cement and Concrete Research* **33**(1) 73-76
[13] Gebler S H Klieger P 1986 *Special Publication* **91** 483-520
[14] Hasholt M T Christensen K U Pade C 2019 *Cement and Concrete Research* **119** 102-112
[15] Quan H Kasami H 2013 *The Open Civil Engineering Journal* **7**(1)
[16] Liu T Wang Y Zhou K Gao F Xie S 2019 *Advances in Civil Engineering* **2019**
[17] Ross C T F et al 2016 *Woodhead Publishing Series in Civil and Structural Engineering* (Cambridge: Elsevier) pp 1-740
[18] Pogorelov S N Semenyak G S 2016 *Procedia Engineering* **150** 1491-1495
[19] STN 73 1322 Determination of frost resistance of concrete
[20] STN EN 196-1 2016 Methods of testing cement Part 1 Determination of strength
[21] Toutanji H Delatte N Aggoun S Duval R Danson A 2004 *Cement and Concrete Research* **34**(2) 311-319
[22] El Mir A Nehme S G Assaad J J 2020 *Heliyon* **6**(1) e03165

Acknowledgement
This research has been carried out within the project of Slovak Scientific Grant Agency VEGA (Grant No. 1/0524/18).

IOP Conf. Series: Materials Science and Engineering **867** (2020) 012017 doi:10.1088/1757-899X/867/1/012017

Analysis of mechanical properties of I-beam with web from transparent wood

J Kanócz[1], V Bajzecerová[2] and V Karľa[3]

[1] Technical University in Košice, Faculty of Arts, Department of Architecture
[2] Technical University in Košice, Faculty of Civil Engineering, Department of Structural Engineering
[3] Technical University in Košice, Faculty of Civil Engineering, Department of Architectural Engineering

Abstract. This article contains a brief information about history of transparent wood, its fabrication process and its physical and mechanical properties. The main goal of this article, however, is to find whether transparent wood is a suitable material for load-bearing architectural constructions. This is achieved by using numerical analysis of I-beam with web from transparent wood (and various types of it) and comparing its results with the same sized beam with web from OSB (oriented strand board) and web from glass, which are both successfully used for such applications nowadays. Suggestions on how to use transparent wood as a load-bearing material are also presented in the article.

1. Introduction

Transparent wood was first mentioned in the article *Transparent Wood - A New Approach in the Functional Study of Wood Structure* by German scientist Sigfried Fink in 1992 [1]. In his research, he produced transparent wood samples in order to better observe specific three-dimensional wood structures. In his experiment, he was inspired by Spalteholz, who in 1914 introduced a technique to make coarse body organs transparent for analytical purposes using organic liquids with appropriate refractive indices.

Between 2015 and 2016, two teams at the Swedish KTH and the University of Maryland rediscovered Fink's article, began researching the properties of transparent wood and designed its possible applications for lightweight low-cost structures in light-transmitting buildings, solar cells, furniture or as construction material in cars and optoelectronics [2][3]. While in Maryland their research on transparent wood is part of the ongoing research on super-wood structures, in KTH they focus exclusively on transparent wood, thus have published more articles on this topic and have come up with interesting innovations with regard to its fabrication concepts and improvements of its properties.

In this paper, a numerical analysis of the mechanical properties of transparent wood with various parameters compared to other commonly used materials is published in a model example of a composite I-beam with flanges of natural wood in order to verify the suitability and efficiency of its use as a load-bearing building material.

2. Characteristics of transparent wood

2.1. Transparent wood fabrication

The fabrication of transparent wood consists of two steps. The first is removal or modification of lignin and the second is impregnation of the delignified sample with a transparent polymer. In KTH, delignification was achieved using a 1% sodium chlorite solution ($NaClO_2$) with an acetate buffer solution (pH 4.6) at 80 °C. The delignified samples were then purified in deionized water. They were then dehydrated. Initially, with pure ethanol and then with a 1:1 mixture of ethanol and acetone. After the delignification was complete, a prepolymerized methyl methacrylate (MMA) solution was infiltrated into the wood under vacuum [2]. In Maryland, delignification was a two-step process. The first step was to put the wood samples into a boiling solution of sodium hydroxide (NaOH) and sodium sulphate (Na_2SO_3). This was followed by placing of partially delignified samples into a solution of hydrogen peroxide (H_2O_2) to remove the remaining lignin. Subsequently, the epoxy resin was infiltrated into the delignified wood by repeated vacuum and de-vacuum [3]. In addition, in KTH they were further modifying the delignification process, which was criticized for the use of aggressive chemicals, leading to a not very ecological process for producing transparent wood. Instead of completely removing the lignin, the modification procedure was chosen using deionized water, sodium silicate $Na_2S_2O_3$ (3.0 wt%), sodium hydroxide solution NaOH (3.0 wt%), magnesium sulphate $MgSO_4$ (0.1 wt%). DTPA of diethylenetriaminepentaacetic acid (0.1 wt%) and finally hydrogen peroxide H_2O_2 (4.0 wt%). The wooden samples were then immersed in a solution heated to 70 °C until the veneers became pale. The samples were then thoroughly washed with deionized water and stored in water until further use. Subsequently, the PMMA infiltration process was identical to their previous procedure [4].

2.2. Mechanical and physical properties of transparent wood

So far, only thin samples of transparent wood have been produced. Scientists mainly examined their optical and mechanical properties. From these studies we know about the following findings.

Transparent wood of different thicknesses exhibits different optical properties. The thinner the sample, the higher its optical transmittance. This is because transparency depends on the volume fraction of cellulose. At KTH, the highest optical transmittance was achieved with a 0.7 mm thick sample and it was 90%. The lowest value was approximately 40% for a 3.7 mm thick sample [2]. In Maryland, researchers focused on how the properties of transparent wood depend on the cut of the original wood, meaning that they studied the properties of radially and longitudinally sawn wood. As radially sawn wood has a lower volume fraction of cellulose, it achieved 90% optical transmittance with a 2 mm thick sample and approximately 80% with the same thickness of longitudinally cut wood. Another optical feature studied by both universities is the optical haze of the samples. [3] At KTH they measured 60% to 80% haze values, with higher values in thicker samples. In Maryland, they measured a haze of 90% for longitudinally sawn wood and surprisingly close to 100% for radially sawn wood. Radially cut wood has a lower volume fraction of cellulose, but an important factor here is the arrangement of the wood cells that have produced a higher haze [3]. In a greener version of transparent lignin-modified transparent wood at KTH, they found that a 1.5 mm sample had an optical transmittance of 83% (equally thick delignified sample 86%) and a haze of 73% (equally thick delignified sample 68%) [4].

Transparent wood also has excellent mechanical properties. At KTH they found that they are better for transparent wood than for natural wood and PMMA. The modulus of elasticity for transparent wood with a 5% cellulose content was 2.05 ± 0.13 GPa (i.e. 1.80 ± 0.18 GPa for PMMA and 0.22 ± 0.08 GPa for delignified wood). Transverse tensile tests of transparent wood with a total volume fraction of 19% measured its tensile strength of 90.1 ± 10 MPa with a modulus of elasticity of 3.59 ± 0.27 GPa (for PMMA it was 44.1 ± 9.5 MPa) [2]. In Maryland, they also examined the mechanical properties of the samples and found that both radially sawn and longitudinally sawn woods are stronger than their natural counterparts and that transparent longitudinally sawn wood has approximately twice the strength of transparent radially sawn wood (table 1) [3]. In the case of transparent lignin-modified wood, the strength was found to be even higher than that of lignin-depleted samples (100.7 ± 8.7MPa) [4].

Table 1. Comparison of mechanical properties of natural radially sawn wood, natural longitudinally sawn wood, transparent radially sawn wood and transparent longitudinally sawn wood [3].

	Strength (*MPa*)	**Modulus of elasticity** (*GPa*)	**Toughness** (MJm^{-3})
Natural radially cut wood	4.46	0.19	0.10
Transparent radially cut wood	23.38	1.22	0.59
Natural longitudinally cut wood	42.72	5.78	0.33
Transparent longitudinally cut wood	45.38	2.37	1.20

3. Suggested usage of transparent wood in the applications of load-bearing constructions

If there is a chance to produce large pieces of transparent wood, we can simply use those and make some type of bearing structural elements, as beams, trusses etc... The problem is that elements produced from massive transparent wood will not transmit enough light and are difficult to produce. This is the reason to assume that different beams will have to be made using thin, transparent wood elements that can be fabricated and the appearance of the entire structure will be much more subtle and attractive at the end.

One possible option is I-beam, where the web of this beam is made of thin transparent wood and the flanges are made of either natural wood, laminated veneer lumber or thicker transparent wood elements. This is the lightest looking version of transparent wooden beams possible. Architecturally, a structure made with such beams almost seems to be floating or levitating in space, which may be highly desirable

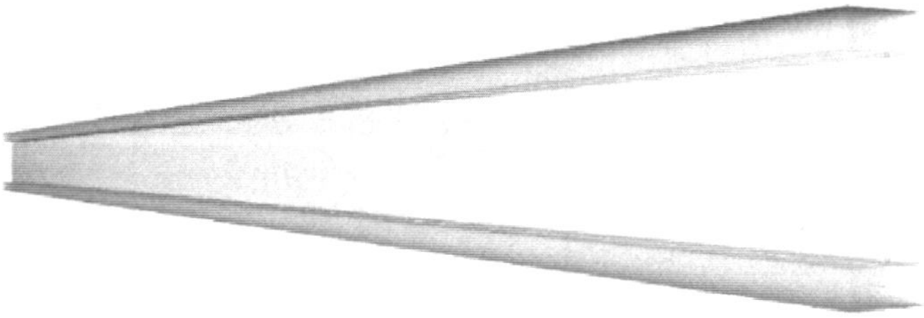

Figure 1. Rendering of I-beam from transparent wood [5].

Another possibility is to create box beams. They would be formed in a similar way as "I" beams. The flanges could again be made of natural wood, laminated veneer lumber or thicker transparent wood elements and the webs would be made of thin transparent wood panels. These beams are not so light in appearance (although they still look quite light), but can support more loads.

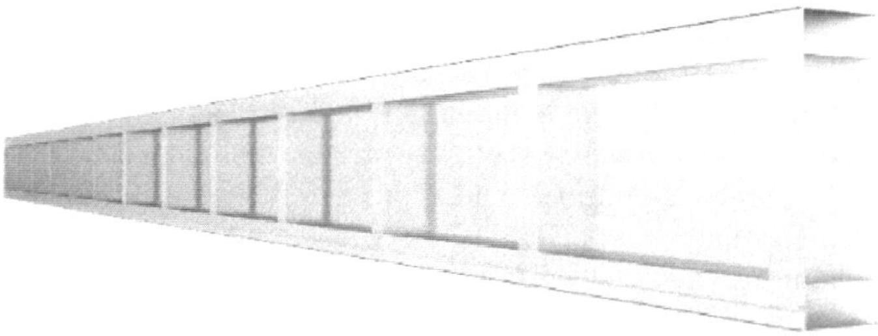

Figure 2. Rendering of Vierendeel box beam from transparent wood [5].

A variation of this type is the creation of Vierendeel beams (again with the aid of a natural, laminatef or thick transparent wood) with thin transparent wood panels added on both sides. Of all the above mentioned, it is the least light-appearing variant, but the strongest.

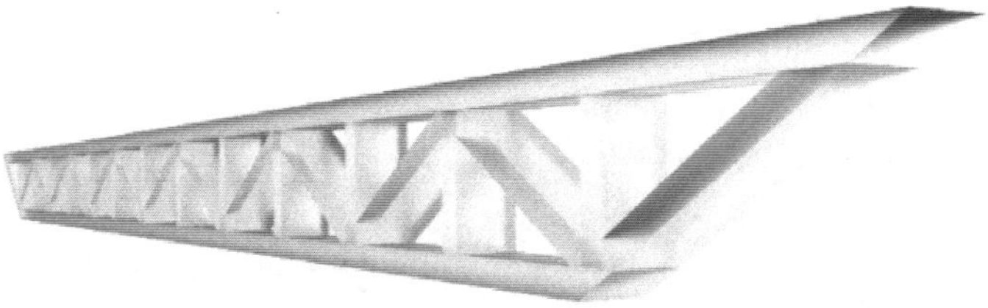

Figure 3. Rendering of transparent wood truss [5].

Transparent timber trusses are another way to use transparent wood in construction. Of course, thicker elements will have to be used to make them, but not as thick as one massive beam would require. Such trusses may be entirely of transparent wood, or only a flange would be made of transparent wood and then be coupled with a steel cable to achieve a much lighter construction overall.

There is also a chance to use massive transparent wood beams inspired by parallam beams. Research on such transparent wood elements has shown that their strength is comparable to conventional transparent woody samples (46.8 MPa), while it is easier to achieve greater transparency (10 mm thick – 68%, 1 mm thick – 92%) with lower haziness (10 mm thick – 90%, 1 mm thick – 56%) [6].

4. Numerical model of I-beam with web from transparent wood

ANSYS Workbench 17.2 was chosen for the theoretical analysis (figure 4). 7 types of beams were modelled, differing in the material of the beam web: (1) transparent wood (TW) single-layer (veneer) with fibre orientation in the direction of the beam axis, (2) single-layer TW with fibre orientation perpendicular to the beam axis, (3) 5-layer TW type CP with cover layer orientation in direction of beam axis, (4) 5-layer TW type CP with cover layer orientation 45° to beam axis, (5) 5-layer TW type Qi, (6) OSB, (7) glass. The flanges of the I-beam were of massive natural wood. In numerical analysis, natural wood, OSB and glass were modelled as linear elastic isotropic material, transparent wood as linear elastic orthotropic material. The material properties used are listed in table 2. Data for mechanical properties of transparent wood were obtained from the article *Transparent plywood as a load-bearing and luminescent biocomposite* [7].

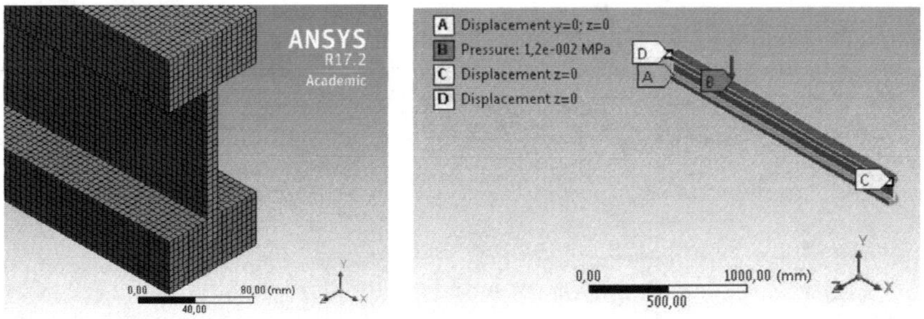

(a) (b)

Figure 4. Numerical model of I-beam; (a) Mesh; (b) Analysis parameters.

The contact between the web and the flanges was set as "bonded" for all nodes of the connection. Newton-Raphson's iterative method considering large deflection was used for the calculation. Half of the beam was modelled with the symmetry set at the centre of the span. The maximum size of the mesh element was 5.0 mm.

Table 2. Material parameters for numerical analysis.

Material	Direction	Modulus of elasticity E (*GPa*)	Shear modulus G (*GPa*)	Poisson's ratio μ (-)	Tensile strength f (*MPa*)
Timber	longitudinal	11.0	0.37	0.48	16.0
TW_veneer	longitudinal	4.3	1.40	0.37	62.5
	transverse	2.4	1.00	0.10	14.6
TW_CP	longitudinal	4.1	1.40	0.37	50.1
	transverse	3.9	1.00	0.10	44.9
TW_Qi	longitudinal	3.9	1.40	0.37	45.4
	transverse	3.5	1.00	0.10	42.0
OSB	-	3.8	1.08	0.30	9.9
Glass	-	70.0	28.70	0.22	20.0

5. Comparison between I-beams with web from transparent wood, OSB and glass

The beams were loaded with the uniform area load of 0.012 MPa which is corresponding to 1.0 kN/m of beam load. The calculated deflection values at the middle of the span of the timber-based beams are compared in figure 5. The differences between the deflections are relatively small, but the smallest deflection was calculated for sample no. 4 with a 5-layer transparent wood with a 45° inclination relative to the beam axis.

Further results of the numerical analysis are shown in table 3. The glass-web beams had a deflection up to 33% less than the wood-based beams, but the normal stress was relatively high due to the tensile strength of the glass.

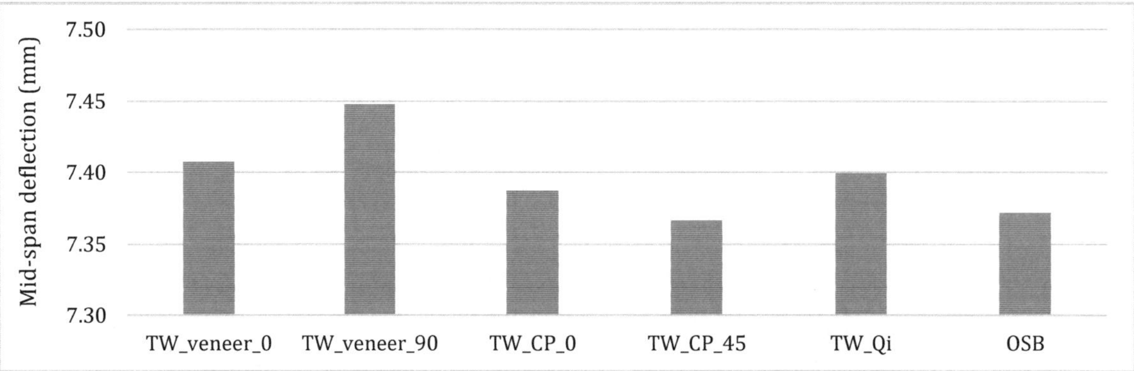

Figure 5. Calculated deflection of I-beams with various web type.

Table 3. Results of numerical analysis.

No.	Material of web	Deflection (*mm*)	Max normal stress σ_x (*MPa*)	tensile strength f (*MPa*)	σ_x /f (%)	Max shear stress τ_{xy} (*MPa*)
1	TW_veneer_0	7.408	1.2403	62.5	2.0%	1.805
2	TW_veneer_90	7.448	0.6745	14.6	4.6%	1.688
3	TW_CP_0	7.388	1.1628	50.1	2.3%	1.894
4	TW_CP_45	7.367	1.0931	44.9	2.4%	1.845
5	TW_Qi	7.400	1.1089	45.4	2.4%	1.897
6	OSB	7.372	1.0586	9.9	10.7%	1.741
7	Glass	4.929	15.231	20	76.2%	2.122

6. Conclusion

Numerical analysis has shown that transparent wood is suitable for use in load-bearing structures, as an I-beam with a transparent wood web of different types has a comparable deflection at the same web thickness as well as a normal stress with an OSB web I-beam. Although the glass-web beam has smaller deflection compared to transparent wood and OSB, its normal stress is already approaching a critical value, while transparent wood and OSB could still bear much heavier loads.

7. References

[1] Fink S 2009 Transparent Wood – A New Approach in the Functional Study of Wood Structure *Holzforschung International Journal of the Biology, Chemistry, Physics and Technology of Wood* **46** 403-408

[2] Li Y, Fu Q, Shun Yu, Yan M and Berglund L 2016 Optically Transparent Wood from a Nanoporous Cellulosic Template: Combining Functional and Structural Performance *Biomacromolecules* **17** 1358-1364

[3] Zhu M, Song J, Li T, Gong A, Wang Y, Dai J, Yao Y, Luo W, Henderson D and Hu L 2016 Highly anisotropic, highly transparent wood composites *Advanced Materials* **28** 5181–5187

[4] Li, Y., Fu, Q., Rojas, R., Yan, M., Lawoko, M., & Berglund, L. (2017). A new perspective on transparent wood: Lignin-retaining transparent wood. *ChemSusChem*. **10**. 3445-3451

[5] Katunský D, Kanócz J and Karľa V 2018 Structural Elements with Transparent Wood in Architecture *International Review of Applied Sciences and Engineering = IRASE* **2** 101-106

[6] Wang X, Zhan T, Liu Y, Shi J, Pan B, Zhang Y, Cai L and Shi S 2018 Large-Size Transparent Wood for Energy-Saving Building Applications *ChemSusChem* **5** 3855-3862

[7] Fu Q, Yan M, Jungstedt E, Yang X, Li Y and Berglund L A 2018 Transparent plywood as a load-bearing and luminescent biocomposite *Composites Science and Technology* **164** 296-303

Acknowledgments

This contribution is supported by the Scientific Grant Agency and project KEGA 046TUKE-4/2019 supported by the Cultural Grant Agency of the Ministry of Education, Science, Research and Sport of the Slovak Republic and the Slovak Academy of Sciences. This paper was elaborated with financial support project with ITMS code: 26220220182 (TECHNICOM).

IOP Conf. Series: Materials Science and Engineering **867** (2020) 012018 doi:10.1088/1757-899X/867/1/012018

Improvement of the backfilling characteristics by activation of halite waste for non-waste geotechnology

A M Khayrutdinov[1], Ch Kongar-Syuryun[2], T Kowalik[3] and V Faradzhov[2]

[1] Peoples' FriendshipUniversity of Russia (RUDNUniversity), Miklukho-Maklaya Str. 6, 117198, Moscow, Russia
[2] National University of Science and Technology MISIS, Leninsky Av. 4, 119991, Moscow, Russia
[3] Wrocław University of Science and Technology, Wyb. Wyspiańskiego 27, 50-370 Wrocław, Poland

Abstract. To reduce the frequency of man-made disasters it is recommended to use development of deposits systems with backfilling of working space at potash mines. The use of halite enrichment waste in the filing mixture is proposed in order to improve the technical and economic indicators of the mine and the environmental situation in the mining area. The factors influencing the choice of development technology with artificial maintaining of winning area are considered. The advantages and disadvantages of using the technology of the backfilling are indicated. Aspects of preparation of hardening backfill and questions of improvement of its quality by disintegration activation of components are revealed. The possibility of activation of halite enrichment waste used as an inert aggregate in disintegrators is considered.

1. Introduction

The frequency of geotechnogenic disasters during the extracting minerals from the depths of the earth increases. In 1986 - surface failure BKPRU-3 Uralkali, Perm Territory; 1999 - a rock blow at the Umbozero mine in the Murmansk region; 2007 - surface failure at BKRP-1 Uralkali, Perm Territory; 2014 - water breakthrough at the Solikamsk-2 mine, Perm Territory; 2017 - water breakthrough from a spent quarry into an underground mine at the Mir enterprise in Yakutia; 2019 - the collapse of the Korbalikhinsky mines in the Altai Territory, Sarylakh in Yakutia, Novo-Kalinskaya in the Sverdlovsk Region, etc. Many of these disasters led to the cessation of work at the mines, and accordingly entailed the loss ofreserves for mineral deposits [1].

The catastrophe, including the geotechnological one, arises as a result of the appearance of several causes and with the combination of a number of adverse factors. The most common causes leading to a geotechnological disaster are: errors made during design; errors made during construction, and they are often compounded by incorrect exploitation and an attempt to make design changes during exploitation.

All this indicates the need for a more detailed approach to the selection of development technology at the time of mine design and careful monitoring at the time of construction and exploitation.

2. Geotechnology with artificial maintenance of natural space

One of the ways to minimize geotechnogenic disasters is the use of technology with backfilling of working space [1], [2].

Analyzing the world experience, it can be concluded that about 35% of mines use systems with backfilling of working space when extracting minerals [3].

Factors that influence the choice of technology with backfilling:
- increasing the depth of mining;
- improving the completeness of the extraction of minerals;
- increase the service life of enterprises;
- mining under protected objects (the need to preserve the earth's surface);

Development systems with a hardening backfilling based on the cement binder component are used in deposits with a high value of the extracted raw materials [4], [5]. In coal deposits, due to the low value of the mineral, dry or hydraulic backfilling are used by coarse-ground rocks from sinking, finely crushed coal waste, ash and slag waste of thermal power plants or specially mined sand. Hardening backfilling in coal deposits is used only in exceptional cases:
- when extracting powerful, steeply dipping layers
- when it is necessary to reduce endogenous fire hazard
- when mining layers under protected objects

The use of backfilling of working space in underground geotechnology allows:
- manage rock pressure;
- improve the safety of mining operations;
- conduct simultaneous development of the deposits by underground and open pit methods;
- increase the rate of extraction;
- minimize losses;
- improve the quality of extracted ores;
- reduce the negative impact of mining on the environment;
- preserve the earth's surface;
- extract reserves previously considered off-balance or left in pillars;
- increase the life of the mine, which in addition to the economic aspect, solves the social issue in the regions where the mining enterprise is a city-forming.

Despite the large number of advantages of underground geotechnology with the backfilling of working space, it is not always implemented in a sufficient way and requires improvement of the filling mixture itself, its constituent parts and related processes, since it increases the cost of mining

A filling mixture is used, which is a composite material that can harden in underground conditions, when applying technology with the backfilling of working space. This mixture has in its composition: astringent; inert aggregate; mixing water and chemical additives. As an inert aggregate for the preparation of filling mixtures, both traditional building materials (sand, gravel, etc.) and production waste (slag of metallurgical processing, tailings of concentration plants, ash and ash and slag of thermal power plants, etc.) are used. The use of waste materials in the filling mixture obliges to take into account a number of environmental, economic and other restrictions when choosing the composition [6], [7].

The choice of the composition of the hardening mixture for various development systems is carried out taking into account [8], [9]:
- specified strength;
- method of transportation and good mobility for a given time (the filling mortar must remain in a fluid state for the entire time necessary for its transportation to the chamber to be laid);
- the cost of building an artificial array (to minimize the cost of backfilling works);
- the time to achieve the specified strength characteristics of the material, which determine the possible intensity of the actual mining (i.e., a rapid increase in the value of the elastic modulus E and the ultimate shear value T_0 are required);
- minimal water loss and segregation at rest;
- good adhesion to rocks.

Determination of the optimal rheological properties of the hardening mixture is provided by [10]:

- sustainable transportation along the route of the backfilling pipeline;
- lack of stratification at all stages of backfilling works;
- uniform spreading over the chamber;
- the required depth of penetration into the rock when using the combined method of laying cameras.

The use of various activation properties of the backfilling mixture and its individual components can improve the strength properties and uniformity of the artificial array, the use of new materials in the composition of the filling mixture, including production waste materials [11], [12].

3. Activation as a way to improve the quality of the filling mixture

By activation, it is necessary to understand one of the methods for processing the components of the filling mixture separately or the finished filling mixture, as a result of which the properties of the components are used to the fullest extent [12].

Purposesofactivation:

- increase the mechanical strength of cement stone;
- intensificationofitshardening;
- increase the mobility of mixtures;
- reducing binder costs without reducing strength;
- improving the quality of the inert aggregate, etc.

Ways to activate filling mixtures are divided:

a) for the intended purpose:
  - increased mobility of mixtures (transportability, spreadability);
  - increase in strength properties;
  - regulation of the hardening time.

b) on the technological stages of backfilling operations:
  - preparations;
  - transportation;
  - filling.

c) by objects of influence:
  - processing of individual components (binder, mixing water, inert aggregate, chemical additives);
  - processing of their combinations.

d) by methods of influence:
  - exposure to vibration;
  - electromagnetic processing;
  - heat treatment (firing) of components;
  - the introduction of chemical additives

IOP Conf. Series: Materials Science and Engineering **867** (2020) 012018 doi:10.1088/1757-899X/867/1/012018

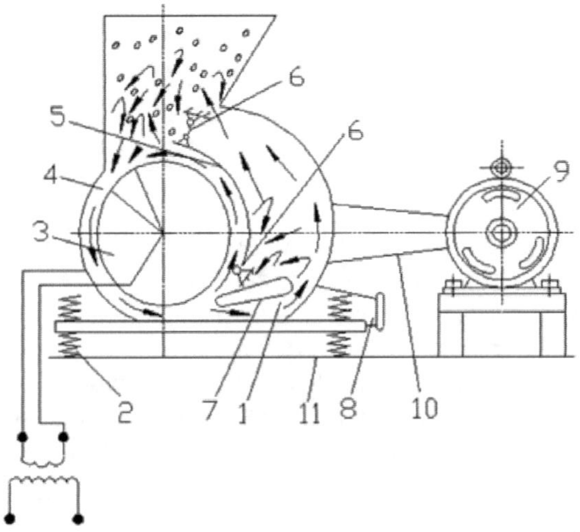

Figure 1. Disintegrator
1 – body; 2 – elastic supports; 3 – rotor; 4 – unbalance; 5 – hinged flap; 6 – hinges; 7 – divider; 8 –unloading device; 9 – electronic motor; 10 –V-belt transmission; 11 – frame [8].

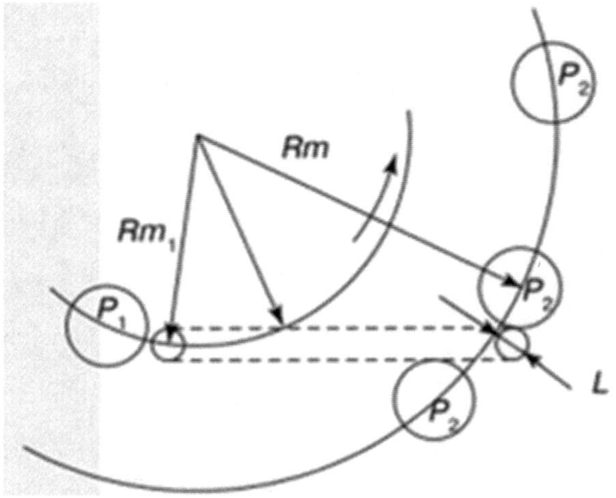

Figure 2. The movement of material in the working body
Rm, Rm_1 – radii of a circle of fingers; L – particle diameter of material; P_1P_2 – centers of adjacent fingers of a circle [Tyulyaeva Y.S].

Treatment of halite waste is performed before mixing it with other components. In the working body of the disintegrator, large internal stresses occur, which are caused by an uneven redistribution of speeds. In this case, the hydrate bonds on the surface of the inert aggregate are destroyed. The filling mixture obtained on the basis of activated halite waste is less prone to delamination, since it is more difficult to "give" water, is more mobile and uniform in composition (figure 3). The use of a disintegrating activation method allows to increase the strength by 25-30 % while reducing the cement by 40%.

Humanity faces a huge problem: the deterioration of the environment as a result of human activity, especially in the extraction of minerals [13], [14]. Recycling of mining waste is almost impossible due to its large volume and specific features. The main task of modern mining production is to create an effective environmentally balanced extraction and processing of minerals [18]. The use of industrial waste [15], [16], [17] along with mining waste [1], [4], [5], [6], [8], [9], [13], [18] is the most important environmental aspect in the world.

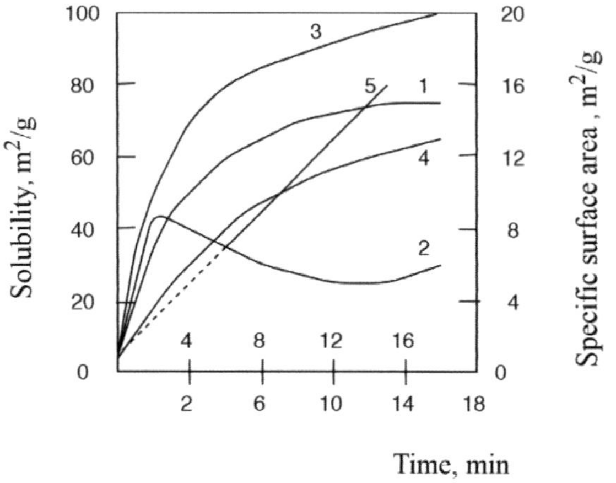

Figure 3. The effect of the duration of activation of halite enrichment waste
1 - solubility; 2–specific surface area of non-aggregated powders; 3 – specific surface area of aggregated powders; 4 – amorphization; 5 – comparison of specific surface growth and solubility [Tyulyaeva Y.S].

The use of disintegration activation of halite enrichment waste in disintegrators allows the use of enrichment waste in the filling mixture, improves the transportable properties of the filling mixture, increases the strength of the artificial array. This makes it possible to implement the principle of organizing mining, eliminating the generation of waste and calculating the use of intermediate products in cyclic production.

4. References

[1] Khairutdinov M M 2009 *The use of mining waste as a filling material to reduce the harmful effects on the environment* Gornyi Zhurnal no. 2 pp 64-66

[2] Khairutdinov M M *Waist improve development systems with laying the developed space*, Gornyi Zhurnal no. 11 pp 40-43

[3] Khairutdinov A M and Tyulyaeva Y S 2019 *Extraction of minerals on celestial bodies. Background, technological aspects and legal framework* Materials of the 14th International Scientific School of Young Scientists and Specialists (M: IPKON RAS) pp 280-283

[4] Khairutdinov M M and Karasev G A 2008 *The formation of diverse stowage arrays in the development of mineral deposits* Mining Information and Analytical Bulletin (In Russian) no. 3 pp 276-2283

[5] Khairutdinov M M 2008 *Technology for laying high-density mixtures (based on enrichment tailings) in underground ore mining* Technology of laying high-density mixtures (based on tailings) in underground mining of ores (Mining Information and Analytical Bulletin) (In Russian) no. 11 pp 276-278

[6] Tyulyaeva Y S and Khairutdinov A M 2019 *Gentle geotechnology*, *Problems of subsurface development in the 21st century through the eyes of young people(Moscow)* Materials of the 14th International Scientific School of Young Scientists and Specialists (M: IPKON RAS) pp 283-286

[7] Tyulyaeva Y S and Khairutdinov A M 2019 *Extraction of minerals on celestial bodies. Background, technological aspects and legal framework*, *Problems of subsurface development in the 21st century through the eyes of young people* Materials of the 14th International Scientific School of Young Scientists and Specialists (M: IPKON RAS) pp 280-283

[8] Khairutdinov M M and Votyakov M V 2007 *Development of compositions for hardening filling mixtures from wastes from potash ore processing* Mining Information and Analytical Bulletin (In Russian) no. 10 pp 220-222

[9] Votyakov M V 2009 *Increasing the completeness of the extraction of potash ore reserves based on the laying of the developed space of halite waste* Thesis (Moscow State Mining University)

[10] Khayrutdinov M M, Ivannikov A L, Arad V and Long V 2017 *Huang Problems of transporting filling mixture to the place of laying. Socio-economic and environmental problems of mining,*

construction and energy 13th International Conference on Mining, Construction and Energy In 2 vol. / T.1: proceedings of the conference (Tula: Publishing house of TulSU) pp 282-287

[11] Chistyakov Khairutdinov A N M and Artyukow E V 2009 *The effect of various activation methods on the physicomechanical properties of the bookmark* Mining Information and Analytical Bulletin (In Russian) no. 3 pp 232-24

[12] Congar-Syurun Ch B 2019 *The construction of a mathematical model for predicting the qualitative and quantitative indicators of processing plants* Problems of subsurface development in the 21st century through the eyes of young people. Materials of the 14th International Scientific School of Young Scientists and Specialists (M: IPKON RAS, October 28-November 01, 2019) 336-338

[13] Khayrutdinov M and Ivannikov A 2017 *The use of mining waste for backfill as one of sustainable mining activities* Proceedings of International Conferences on Geo-spatial Technologies and Earth Resources (GTER 2017: Hanoi, Vietnam, 5-6 October 2017) 715-717

[14] Kawalec J, Kwiecień S, Pilipenko A and Rybak J 2017 *Application of crushed concrete in geotechnical engineering - selected issues* IOP Conf. Ser: Earth Environ. Sci. 95 022057

[15] Herbut A, Marat M Khairutdinov, Cheynesh Kongar-Syuryun M and Rybak J 2019 *The surface wave attenuation as the effect of vibratory compaction of building embankments* IOP Conference Series: Earth and Environmental Science, World Multidisciplinary Earth Sciences Symposium (Prague, Czech Republic: (WMESS 2019) 9–13 September) Volume 362

[16] Rybak J and Schabowicz K 2010 *Survey of vibrations generated in course of geotechnical works* NDE for Safety: 40th int. conf. and NDT exhibition: proceedings (Brno University of Technology) pp 237-246

[17] Herbut A Rybak J 2017 *Guidelines and recommendations for vibration control in the case of rapid impulse compaction* Advances and trends in engineering sciences and technologies II (CRC Press, Taylor & Francis Group) pp 761-766

[18] Alexander L Ivannikov CheyneshKongar-Syuryun Rybak J and Yulia Tyulyaeva 2019 *The reuse of mining and construction waste for backfill as one of the sustainable activities* IOP Conference Series: Earth and Environmental Science (Prague, Czech Republic: World Multidisciplinary Earth Sciences Symposium (WMESS 2019) Volume 362

[19] Kaplunov D R, Melnik V V and Rylnikova M V 2016 *Integrated subsoil development/study guide* Ministry of Education and Science of the Russian Federation (Federal State Autonomous Educational Institution of Higher Education "National Research Technological University" MISiS " - Tula: Publishing House TulSU,. – 33)

IOP Conf. Series: Materials Science and Engineering **867** (2020) 012019 doi:10.1088/1757-899X/867/1/012019

Innovation in construction: trends and obstructions of adoption and implementation

D Klosova[1] and M Kozlovská[1]

[1] Department of Technology, Economy and Management in Civil Engineering, Faculty of Civil Engineering, Technical university of Kosice, Vysokoskolska 4, Kosice 042 00, Slovakia

Abstract. Innovation initiatives are crucial for any industry because their future growth depends on it. Innovative technologies that can have a great impact on the construction industry, such as building information modelling, automation and robotics, augmented reality, virtual reality, simulation, internet of things were reviewed in this article. In spite of that, innovation implementation in the construction sector has generally lagged behind other industry sectors affecting the productivity as well. The governments are aiming to improve this through increasing the level of investments into manufacturing and innovative research in this area. This article reviews the process of adopting innovative technologies and the reasons why there are obstacles to their implementation. It is expected that integration between machines and humans will enhance human capabilities, allowing them to concentrate on more creative and value added, socially important activities in the future. There is also need of changing thinking in order to transfer the results of these technological studies into production in the construction industry.

1. Introduction

Innovation in general can be understood as an actual use of any idea, technology or material that is perceived to be new to the respective adopting unit [1]. Innovation activities include all financial, development and commercial activities carried out by the company, which should lead to result in an innovation for the firm [2]. Successful management of innovations requires attentive studying of innovative activity. At start, there is a need to make a difference between innovations and insignificant modifications in technological processes or products like esthetic changes in colors or forms. Such changes do not have apparent impact on properties, parameters and product cost. Depending on the external parameters, innovations were classified as product innovations and process innovations [3]. The definition of product innovation is a new or improved service or good that is significantly different from the previous units and that has been introduced on the market. A process innovation is a new or improved process that is significantly different from the firm's previous processes and that has been brought into use by the firm [2].

Innovation processes play an important role in meeting the growing needs of industry and keeping a competitive advantage for enterprises. Huge amount of studies in construction sector shows that new technologies have a positive effect on overall productivity, efficiency and safety. For example, innovations in biotechnology and materials are allowing the development of new intelligent materials and processes and can reduce the impact of natural resources scarcity [4]. Despite the obvious advantages, the process of adoption new and innovative technologies is very slow and insecure because of the complexity, conservative nature of construction companies and high implementation risk. Suppliers make great efforts to overcome this and spread their technologies [1].

IOP Conf. Series: Materials Science and Engineering **867** (2020) 012019 doi:10.1088/1757-899X/867/1/012019

The construction industry is one of the main factors contributing to the growth of the economy and employment in each country, but it still has the lowest research and development intensity with less than 1% of net sales (according to the EU R&D table for 2015) [5]. In this regard, the following research questions were formulated:

1. Which technologies in the construction industry are associated with innovative technologies within the framework of the Industry 4.0 concept?
2. Which innovative technologies are widely used in the construction industry in recent years?
3. Which innovative technologies are still not very used?
4. What are the obstructions to the adoption and implementation of new innovative technologies?

The main purpose of this article is to make a comprehensive review of the existing literature, provide a structured and accomplishment vision of the topic of the article, by answering the above research questions.

2. Research methodology

First of all, an initial scan and review of existing literature and practices was made. Major innovative construction technologies commonly discussed in the literature were identified. To write this article with high-quality and up-to-date information, the following two sources were mainly used: Journal search from Web of Science and SCOPUS. High attention was payed to the impact factor of each journals and to the publication date. So by setting journal selection criteria were chosen the most recent articles from 2016 to 2019 years. Besides, were decided to include practical publications, even when most of them are not peer-reviewed and, therefore, can be considered as opinion-based and biased [5]. Withal, some of the none-scientific web pages', like open Google search, libraries and databases were reviewed too, for complete and more accomplish point of view in this problematic. At the same time, we included publications from all areas in order to get a comprehensive overview of the state of the art, but mainly from Civil engineering, Industrial and Manufacturing engineering. For searching engine, the following words combinations and keywords were used: "Construction industry" plus "Innovation in constructions", "Building information modeling", "Internet of Things", "Automation", "Product-Lifecycle-Management (PLM)", "Smart City", "Smart Building", "Robotics", "Simulation tools". The total volume of documents for analysis reaches 6,540 units. At the second stage of relevance assessment, we evaluated the content of articles by reading the title and analyzing the abstract. After evaluation and selection, 526 articles remained. In addition, precautions have been taken to eliminate of irrelevance and duplication. It is expected that with these data, current problems of constant importance and emerging issues will be identified with major transformative potential in the construction industry.

3. The concept of Industry 4.0 in the construction industry

Industry 4.0 is a popular term to describe the trend for the growing use of information data and automation technologies in the manufacturing environment [5]. But mainly, Industry 4.0 is a strategy that characterizes the fourth industrial revolution after the appearing of mechanization, electrification, and computerization [7]. Industry 4.0 is mentioned under other synonyms in the literature like Smart Production, Smart Manufacturing or Industrial Internet [5].

Twenty years ago, the architectural designs were represented in two dimensions on paper, or, in 3 dimensions, by means of models. Customer experience was limited to these dimensions until recently, when new ways of interaction and communication of space and architectural objects appeared using commercial modeling programs in 2D and 3D. The greater development of Building information modeling confirms a tendency towards the visual transformation of the architectural object and the transformation construction process into information [6]. BIM can be described as a platform that contains all the information about the process and the product throughout the entire life-cycle of the constructed object [8]. It does not represent a design methodology but a high-performance tool, particularly attractive to construction companies. The great interest of this methodology lies in some facilities to modify the architectural object together with its costs and its estimated energy demand.

IOP Conf. Series: Materials Science and Engineering **867** (2020) 012019 doi:10.1088/1757-899X/867/1/012019

The difficulty of visualizing the design object to the consumer facilitated the experience in 3 dimensions and demanded the immersion of the subject in the space prior to its actual materialization. Virtual reality (VR) brought the object closer to the render stage and facilitated 3D animations, however, the subject could not experience space. This spatial limit is overcome by what is known as "immersive reality" or "virtual immersion". This technology has continued to advance today to provide the latest in construction, Augmented Reality. This is the way to see a two-dimensional element in 3D and to share the information (in 4D, 5D and 6D) that it contains, just by having a smartphone [6].

Already in 1910, the French artist Villemard made a postcard that depicts his vision of a construction site in 2000. International Association of Automation and Robotics in Construction, continuing to organize annual symposia on this issue (since 1984) unknowingly shared Villemard's early vision that robotics will become commonplace at construction sites by the year 2000 in countries with developed economics [9]. But Villemard's vision has not materialized yet. It is expected that integration between people and machines will expand human capabilities, but will not replace them at all. This will allow people to focus on more creative, valuable and socially essential activities [10].

Figure 1. A Villemard's vision of using robotics at a construction site [9].

The ongoing 4th Industrial Revolution relates to the development and application of new innovative technologies such as advanced robotics, artificial intelligence, additive manufacturing, simulation, augmented reality, industrial Internet, cybersecurity, cloud computing, big data and analytics, smart and multifunctional materials, enabling the integration of physical and virtual worlds [11]. There are some benefits the construction industry can obtain through Industry 4.0:

- reduced labor costs through the use of automatic workflows and robotics;
- use of embedded sensors like RFID (Radio frequency identification) can reduce material costs through the automatic tracking of materials and equipment;
- time savings by using prefabrication and additive manufacturing;
- BIM-based platforms, cloud platforms or social media applications can significantly improve collaboration among companies and help to keep projects within budget and shorten project delivery time;
- improving work safety on-site by using wearable technologies like, Smart Helmets, Smart Glasses or Smart Work Clothes [7].

We decided to group the identified technologies, synonyms, concepts and terms into 3 main clusters and 17 concepts to build a concept list by reviewing existing literature. If an article is associated with two or more concepts, it has been grouped into all of them. The Table 1 presents the comparison of the statistical results of the scientific and technical resources about the concept of Industry 4.0 in years 2010-2015 identified by Oesterreich and Teuteberg [5] with the results in years 2016-2019 obtained by authors.

IOP Conf. Series: Materials Science and Engineering **867** (2020) 012019 doi:10.1088/1757-899X/867/1/012019

Table 1. The summary of statistical results on the concept of Industry 4.0 – comparison of the research results from 2010-2015 [5] and authors' research.

Cluster	Key technologies and concepts in the context of Industry 4.0	Num. of relevant scientific publications	Num. of relevant technical publications
		2010-2015 / 2016-2019	2010-2015 / 2016-2019
Smart Factory	Cyber-Physical systems (CPS)	2/8	0/4
	Radio-Frequency identification (RFID)	11/47	6/25
	Internet of Things (IoT)/ Services (IoS)	0/6	27/35
	Automation	21/71	3/28
	Modularisation/Prefabrication	7/18	4/15
	Additive Manufacturing	0/14	8/10
	Product-Lifecycle-Management (PLM)	2/7	3/7
	Robotics	6/29	13/20
	Human-Computer Interaction (HCI)	2/26	1/5
Simulation and modelling	Simulation tools/Simulation models	55/68	1/13
	Building information Modelling (BIM)	30/146	11/42
	Augmented/Virtual/Mixed reality	18/40	13/9
Digitisation and virtualisation	Cloud Computing	6/8	5/11
	Big Data	0/13	9/6
	Mobile Computing	2/8	1/7
	Social media	3/4	6/7
	Digitisation	4/13	8/11
	Total	169/526	119/255

Many publications have been found about Industry 4.0 in the construction industry, and their number is growing every year. In the other hand, authors admitted that existing studies focused on the possibility of adopting Industry 4.0 in the construction industry rather than on the developing theoretical plan for realizing the adoption [12].

Despite the given advantages of many technologies, their widespread adoption by construction companies has not yet taken place. The technologies mentioned are currently on different levels of maturity. For example, Cloud Computing, Mobile Computing, BIM or Modularization have reached market maturity therefore they are currently available. On the other hand, Additive Manufacturing, Augmented, Virtual and Mixed Reality technologies are still at the formative stage and don't ready for mainstream use [15]. Dallasegaa et al. say that according to the technical literature, modern technologies as GIS, RFID, E-Business, 3D printing and BIM are currently finding their way into practice in customer supply chain and facility management [7].

The development of new and unique technologies or materials is a complex and durable process. It may take decades for one new technology to reach industry level and years to completely replace old technologies and methods in the industry [10]. Furthermore, the new technologies and the growing level of automation and digitalization of design and manufacturing processes will require more technically qualified employees. In particular, it is expected that employees will need skills in computing, digital techniques, analytical thinking, manufacturing methodologies and machine ergonomics. The authors argue that technology connects and leads from a technological-instance to transcendental effects. This means that "invention stimulates invention." It has its own impulse, and even when undesirable side effects become obvious (for example, car pollution becomes a problem), it is almost impossible to stop the wave of innovation [13]. Since construction has its tendency to transit from manual control systems and equipment to automate and machine-dominated operations, it is recognized as an important area for future research. Additionally, many governments support corporations in implementing new technologies. Acceleration the pace of technologies diffusion is possible by facilitating their implementation, that's why it is crucial to know how, why and where construction companies could adopt new technologies [1].

4. Obstructions to the adoption and implementation of innovative technologies

Technological advancement of the new technology has the potential to improve quality, design, sustainability and productivity dramatically. However, reviewed literature shows that there are many barriers, which can impede the effective adoption of these technologies. Ramilo and Embi classified these challenges into six categories, such as technological, organizational, financial, psychological, governmental, and process barriers [15].

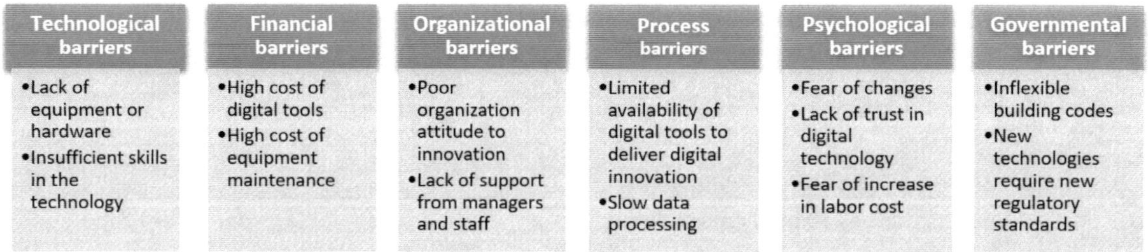

Figure 2. Obstructions to the implementation of innovative technologies [15].

There is some correlation between size of the organization and new technology implementation process. Small organizations are highly affected by financial, psychological and organizational barriers that result in technological problems with large consequences on the company. Medium-sized architectural organizations have the same barriers as well, but are less affected by psychological barriers. Large organizations have significant projects with good investments that can support digital innovation, so they are less affected by barriers compared to small and medium-sized firms. Thus, financial barriers are the most important among the all six barriers presented [15]. Currently, the various stages of the construction process can be described as opaque, slow, uncertain. According to Ermolli and De Toro the main problems are:

- the lack of a structured activities to update professional data for managing the entire process;
- the inadequacy of the current organizational models of the process, since they no longer correspond to new tools and new organization and management methodologies aimed at increasing synergies between main contractor and client;
- the low use of IT tools for the management and collection of the huge amount of data generated during construction process;
- the complexity of integration and exchange of information between numerous process participants;
- the widespread use of remuneration criteria based on the lowest price form, which takes economic aspect as a priority, ignoring the quality aspect [14].

There are two main gaps in the European manufacturing domain, that need to be addressed with funding mechanisms and specific instruments. The first lies between fundamental research and applied research while the second lies between applied research and market development [10].

5. Conclusion

The main purpose of this article was to investigate the current state of Industry 4.0 related to innovative technologies in the construction industry, and to identify obstacles to the adoption and implementation of these technologies in construction by providing an overview of existing literature. It can be concluded from the preceding analysis that the specific definition of the Industry 4.0 in the concept of construction industry include plenty of interdisciplinary technologies which enable the automation, digitisation and integration of the construction processes at all stages of the construction value chain. Withal the process of adoption new advanced and innovative technologies is very slow and insecure because of the complexity, conservative nature of construction companies and high implementation risk. In spite of it suppliers are making determined efforts to overcome this and spread their technologies. Moreover, it was found that the size of the organization has a significant relationship with the obstructions to digital innovation. This conclusion means that smaller organizations face more barriers to digital innovation whereas larger organizations face fewer barriers.

6. References

[1] Sepasgozar S M E and Davis S 2018 Construction Technology Adoption Cube: An Investigation on Process, Factors, Barriers, Drivers and Decision Makers Using NVivo and AHP Analysis *Buildings* 8 74

[2] OECD/ European Union *2018 Oslo Manual 2018: Guidelines for Collecting, Reporting and Using Data on Innovation the Measurement of Scientific, Technological and Innovation Activities* vol 4 (Paris: OECD Publishing/Eurostat: Luxembourg) p 258

[3] Gureva M A Kirillov A V Vinichenko M V Melnichuk A V and Melnychuk Y A 2016 Management of Innovations and Innovative Process: Concept, Essence, Classification and Diffusion *International Review of Management and Marketing* 6(S6) pp 147-153

[4] Ezell S 2018 Why Manufacturing Digitalization Matters and How Countries Are Supporting It *Information Technology & Innovation Foundation* p 66

[5] Oesterreich T D and Teuteberg F 2016 Understanding the Implications of Digitisation and Automation in the Context of Industry 4.0: A Triangulation Approach and Elements of a Research Agenda for the Construction Industry *Computers in Industry* vol 83 (Amsterdam: Elsevier) pp 121–139

[6] Ramos-Sanz A I 2019 Transformación de la Construcción y la Arquitectura en los Ultimos Veinte Años: Prospectivas y Perspectivas. Análisis Bibliométrico de los Tópicos más Desarrollados en Revistas Internacionales de alto Impacto *Arquitecturas del Sur (Argentina) vol 37 pp107-25*

[7] Dallasegaa P Raucha E and Linderb Ch 2018 Industry 4.0 as an Enabler of Proximity for Construction Supply Chains: A Systematic Literature Review *Computers in Industry* vol 99 (Amsterdam: Elsevier) pp 205-225

[8] Samuelson O and Björk, B.-C 2013 Adoption Processes for EDM, EDI and BIM Technologies in the Construction Industry *Journal of Civil Engineering and Management* vol 19(1) p16

[9] Skibniewski M J and Zavadskas E K 2013 Technology Development in Construction: A Continuum from Distant Past in the Future *Journal of Civil Engineering and Management* vol 19(1) pp 136–147

[10] Manufuture High-Level Group 2018 Manufuture Vision 2030: Competitive, Sustainable and Resilient European Manufacturing *Report from ManuFUTURE High-Level Group p74*

[11] Schwab K 2017 *The Fourth Industrial Revolution* (United Kingdom: Portfolio) p 192

[12] Maskuriy R Selamat A Ali K N Maresova P and Krejcar O 2019 Industry 4.0 for the Construction Industry—How Ready Is the Industry? *Applied science* p 26

[13] Woodheada R Stephensona P and Morrey D 2018 Digital construction: From point solutions to IoT ecosystem *Automation in Construction* vol 93 (Amsterdam: Elsevier) pp 35-46

[14] Ermolli S R and Toro P 2017 Innovazioni di processo per la digitalizzazione degli appalti pubblici: sinergie tra BIM e analisi multicriterio *Ricerca e Sperimentazione* p 9

[15] Ramilo R and Embi M 2014 Critical analysis of key determinants and barriers to digital innovation adoption among architectural organizations *Frontiers of Architectural Research* vol 3 pp 431-451

Acknowledgements

The activity presented in the paper is part of the research grant VEGA 1/0557/18 "Research and development of process and product innovations of modern methods of construction in the context of the Industry 4.0 principles".

Fungicidal properties of cement composites based on waste sludge water from concrete plant

L Klus[1], T Dvorsky[1], J Svoboda[1], J Charvat[1], K Macalova[1], L Chlebkova[1] and V Václavík[1]

[1] VSB - Technical University of Ostrava, Faculty of Mining and Geology, Department of Environmental Engineering, Czech Republic

Abstract. The present article deals with the fungicidal properties of cement composites based on Waste Sludge Water (WSW) from a concrete plant (CP). The article compares the ability of standard cement composites and cement composites based on WSW to withstand microscopic fungi *Aspergillus niger, Trichoderma nereus* and *Penicillium glabrum*. The results of the tests show that the use of WSW as a substitute for pure mixing water in the preparation of cement composites does not have a significant positive or negative effect on the fungicidal properties of the prepared composites.

1. Introduction

Cement composites are frequently used in construction practice due to their very good price-performance ratio. For example, concrete pipes are used for waste water drainage or as process water supply systems to industrial plants due to their favourable properties. However, the presence of microorganisms together with nutrients leads to algae growth on the concrete surface. They are capable of affecting the flow rate [1]. The growth of filamentous fungi poses another problem in the use of cement composites. There are research projects and studies on concrete biocorrosion [2-4] with various surface treatments in controlled conditions as protection against fungal growth [5] or with the addition of various additives as protection against biocorrosion [1,3,6,7]. Recently, there has been increasing pressure to use various waste materials from metallurgical, energy and other industries as secondary raw materials in the production of cement composites. Several studies and research papers address this issue as well [8-12]. The use of secondary raw materials in cement composites has also been studied in terms of added value in the form of improvement of the fungicidal properties of the resulting composites [13,14]. One of the possibilities of the utilization of waste generated in the construction industry is, for example, the utilization of waste sludge water from concrete plants as a partial or complete replacement of mixing water in the production of cement composites. Waste sludge water in a concrete plant is produces during the cleaning of agitation trucks from the remaining transported concrete or other cement-based materials.

This research is currently very popular in terms of environmental protection, thanks to the reduction of waste sludge water and its economic benefits. The aim of these research is compares the ability of standard cement composites and cement composites based on WSW to withstand to most frequently occurrence microscopic fungi in civil engineering industry which are *Aspergillus niger*, *Trichoderma nereus* and *Penicillium glabrum*.

2. Materials and Methods

The cement composites were prepared and tested according to CSN EN 196-1 [15]. A precisely determined amount of pure mixing water and waste sludge water from the concrete plant (see table 1)

was used in the preparation of the cement composites. The parameters of the water are given by the CSN EN 1008 standard [16]. The test specimens with the dimensions of 40×40×160 mm were produced according to formulas R1 and R3 for the purpose of the tests. They were subsequently cut into partial sections with the dimensions of 40×40×10 mm and used for the testing of fungicidal properties.

2.1. Composition of cement composite formulas

Two experimental formulas with the designations of R1 and R3 were designed to verify the possibility of replacing the mixing water in the production of cement composites with waste sludge water from the concrete plant. The composition of the individual formulas is described in table 1 and is based on the standard requirements for the production of cement composites. The mixing water was replaced with waste sludge water from the concrete plant in the amounts of 25%, 50%, 75% and 100% in both formulas. Portland cement CEM I 52.5 R EN 197-1 [17] manufactured by Cement Hranice a.s. as the binder for mixture R1 was used and Portland mixed cement CEM II / B-LL 32.5 R EN 197-2 [18] manufactured by Cement Hranice a.s. for R2 mixture was used. Standard sand according to EN 196-1 as the filler was used [15].

Table 1. Composition of designed formulas.

Formula			**R1**					**R3**		
Mixing water repl. [%]	**K**	**25**	**50**	**75**	**100**	**K**	**25**	**50**	**75**	**100**
CEM I 52.5 R [g]	450.0	450.0	450.0	450.0	450.0	-	-	-	-	-
CEM II/B-LL 32.5 R [g]	-	-	-	-	-	450.0	450.0	450.0	450.0	450.0
Pure mixing water [g]	225.0	168.8	112.5	56.3	-	225.0	168.8	112.5	56.3	-
Waste sludge water [g]	-	56.25	112.5	168.8	225.0	-	56.25	112.5	168.8	225.0
Sand [g]	1,350	1,350	1,350	1,350	1,350	1,350	1,350	1,350	1,350	1,350

2.2. X-ray diffraction of used cements

A Bruker Advance D8 (Bruker, USA) powder diffractometer equipped with a LynxEye linear semiconductor detector and a SOL-XE energy dispersion detector was used for the X-ray diffraction analysis. This system was suitable for quantitative analyses of mineralogical-phase composition of the examined samples. During the sample preparation, we ensured the samples had not come into contact with other components to prevent their contamination with other materials. The samples were ground to an approximate size of 10-50 μm in order to determine the parameters. Subsequently, the cement samples were placed in sample tubes in which they were stable and free from any interactions.

2.3. Mixing water

Drinking water from the water supply system together with waste sludge water from the concrete plant were used for the preparation of the cement composites. The properties of pure mixing water and waste sludge water from the concrete plant are presented in table 2. The properties were determined according to CSN EN 1008 standard [16].

2.4. Testing the resistance against fungi

For the testing of the fungicidal properties of cement composites were selected microscopic fungi *Aspergillus niger*, *Trichoderma nereus* and *Penicillium glabrum* because they are most frequently occurring microscopic fungi in civil engineering industry. These microscopic fungi were obtained as new isolates from metal-contaminated environments of sludge beds after ore extraction [19–22]. The strains were inoculated in a standard manner on Malt Extract Agar (M137, HiMedia Laboratories, Mumbai, India) in the following composition: malt extract 30.0 g, mycological peptone 5.0 g, agar 15.0 g, final pH (at 25 °C) 5.4 ± 0.2. The samples of composites in the form of beams with the dimension of 40x40x10 mm were sterilized in closed beakers in Stericell sterilizer (BMT Medical technology, Czech Republic) at 180 °C for 60 min. The sterilized composite specimens were then placed in sterile Petri dishes and bathed in sterile growth medium for the cultivation of microscopic filamentous fungi, see figure 1. Sabouraud Dextrose Agar (M063, HiMedia Laboratories, Mumbai, India) was selected as the culture medium containing: dextrose (glucose) 40.0 g , mycological peptone 10.0 g, agar 15.0 g, final pH (at 25 °C) 5.6 ± 0.2. To verify the activity of the composites, the beams were overlaid with a thin layer of the medium. Spores of cultured microscopic fungi (15-day cultivation) were evenly applied on the prepared dishes using moistened transport swabs (Transsystem, Dispolab, Czech Republic). Each fungus was placed on a separate Petri dish with growth medium. The duration of the experiments was 30 days.

Table 2. Properties of mixing water for the preparation of cement composites.

Tested Properties	Pure Mixing Water From A Concrete Plant	Waste Water From A Concrete Plant	Limit Concentrations According To ČSN EN 1008
pH	6.84	12.31	> 4
Temperature [°C]	17.0	17.0	-
Conductivity [μS/cm]	335.20	15.30	-
Humus substances	acceptable	acceptable	paler than yellow-brown
Chlorides μS/cm	12.30	156.5	< 600 < 2,000 < 4,500
Sulphates [mg/l]	40.268	875.3	< 2,000
Nitrates [mg/l]	9.4	-	< 500
COD_{Cr} [mg/l]	5.236	-	-
Na [mg/l]	31.326	135	< 1,000
Pb [mg/l]	0	0.1821	< 100
Zn [mg/l]	0.0655	0.0232	< 100
Ca [mg/l]	33.47	1,108.5	-
Mg [mg/l]	5.27	935	-
K [mg/l]	3.96	691	-
SS [mg/l]	2.0	134,900	

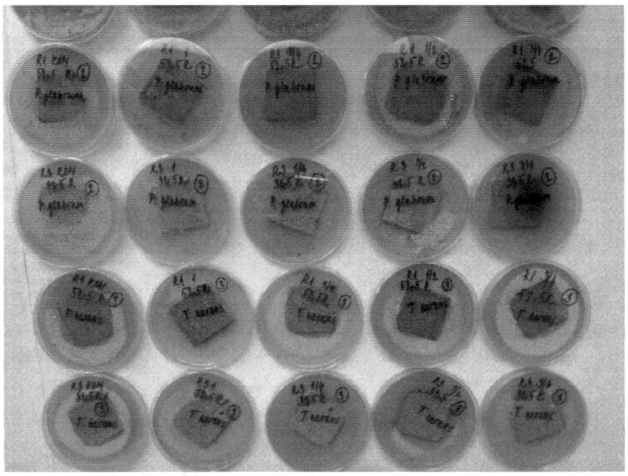

Figure 1. Cement composite samples at the beginning of fungi resistance test.

3. Results and Discussion

3.1. X-ray diffraction of the used cements

Table 3 shows the X-ray diffraction results of the individual cements used in the production of the cement composites. The results show that the composition of cements is not mineralogically different, only the percentage of the individual components is different. Cement CEM II / B-LL 32.5 R contains a higher proportion of limestone, a lower proportion of dicalcium silicate (C2S - Belite), tricalcium aluminate (C3A - Celite), tricalcium silicate (C3S - Alite) and calcium sulphate.

Table 3. Results f X-ray diffraction of the used cements.

Cement type	X-ray analysis	[%]
Portland cement CEM 52.5 R EN 197-1	C_3S monoclinic (NISHI) Hatrurite	65.752
	Gypsum	1.815
	C_3A cubic	6.198
	C_4AF Brownmillerite(Si,Mg)	9.850
	C_2S beta Larnite	10.268
	Calcite	4.096
	Portlandite	2.019
Mixed Portland cement CEM II/B-LL 32.5 R EN 197-2	C_3S monoclinic (NISHI) Hatrurite	63.390
	Gypsum	0.718
	C_3A cubic	5.554
	C_4AF Brownmillerite(Si,Mg)	6.843
	C_2S beta Larnite	3.596
	Calcite	17.883
	Portlandite	2.016

3.2. Results of fungi resistance test

After the fungi resistance test, the individual beam samples were evaluated according to CSN 724310 [23] as shown in table 4. The results of the evaluation are summarised in table 5 below.

Table 4. Scale of evaluation of the mould proofness properties of building materials according to CSN 72 4310: 1977 [23].

Degree of fungi growth	Description
0	No growth of fungi
1	The growth of fungi is negligible (the colonies of fungi are dispersed)
2	The growth of fungi is gradual (numerous small fungi colonies which cover 25% of the sample surface)
3	The growth of fungi is intensive (fungi colonies cover up to 50% of the sample surface)
4	The growth of fungi is very intensive (fungi colonies cover up to 75% of the sample surface)
5	The sample surface is fully covered by fungi (fungi colonies cover 100% of the sample surface)

Table 5. Antifungal efficiency of cement composites with waste sludge water.

Sample		Degree of fungi growth		
		Penicillium glabrum	*Aspergillus niger*	*Trichoderma nereus*
R1	comp.	0	0	1
	25%	0	0	1
	50%	0	0	1
	75%	0	0	1
	100%	0	0	1
R3	comp.	1	1	1
	25%	1	1	1
	50%	1	0	1
	75%	1	0	1
	100%	1	0	1

The results clearly show that the samples of R1 mixture containing waste sludge water corresponded to the results of the comparative mixture containing only pure mixing water and showed high resistance to *Penicillium glabrum* and *Aspergillus niger* fungi. The samples of mixture R1 showed a slight growth of scattered colonies at the edges of the beam in case of *Trichoderma nereus* fungi, see figure 2.

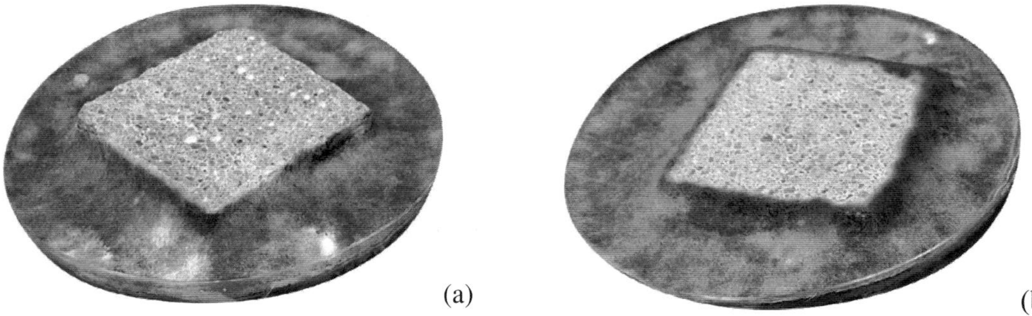

(a) (b)

Figure 2. Illustration of slight growth of *Trichoderma nereus* fungi on the edges of beams in mixtures R1(a) and R3 (b).

4. Conclusion

The aim of this paper was to evaluate the effect of waste sludge water, used as the substitute for pure mixing water in the preparation of cement composites, on the resistance to fungal growth. The fungicidal properties have been studied in three strains of filamentous fungi, namely *Penicillium glabrum, Aspergillus niger* and *Trichoderma nereus*. The test specimens were exposed to fungi for 30 days and the results suggest that:

- the use of waste sludge water as a substitute for pure mixing water does not in any way impair the properties of the cement composites compared to the comparative sample prepared using pure mixing water;
- 50% or higher replacement of pure mixing water with waste sludge water showed a slight improvement in fungicidal properties against the growth of *Aspergillus niger.*

It can therefore be stated that the use of waste sludge water as a substitute for pure mixing water does not impair the fungicidal properties of cement composites.

5. References

[1] Vaquero J M et al 2016 *Cem. Concr. Compos.* **74** 109–19
[2] Gu J-D, Ford T E, Berke N S and Mitchell R 1998 *Int. Biodeterior. Biodegrad.* **41** 101–9
[3] Kong L, Zhang B and Fang J 2017 *Constr. Build. Mater.* **149** 1–8
[4] O'Connell M, McNally C and Richardson M G 2010 *Cem. Concr. Compos.* **32** 479–85
[5] Giannantonio D J et al 2009 *Int. Biodeterior. Biodegrad.* **63** 252–9

[6] Erofeev V, Rodin A, Rodina N, Kalashnikov V and Irina E 2016 *Procedia Eng.* **165** 1448–54
[7] Hoang C P et al 2010 *Int. Biodeterior. Biodegrad.* **64** 104–13
[8] Junak J and Sicakova A 2017 *Procedia Eng.* **180** 1284–91
[9] Števulová N and Junák J 2014 *Chem. Listy* **108** 620–3
[10] Stevulova N, Schwarzova I, Hospodarova V and Junak J 2016 *Chem. Eng. Trans.* **50** 367–72
[11] Junakova N and Junak J 2017 *Sustain. Switz.* **9**
[12] Jursová S 2010 *METAL 2010 - 19th Int. Conf. Metall. Mater, Conf. Proc.*115–20
[13] Strigáč J, Števulová N, Mikušinec J and Sobolev K 2018 *Constr. Build. Mater.* **159** 351–60
[14] Smolakova M et al 2018 *IOP Conf. Ser.: Mater. Sci. Eng.* **385**
[15] CSN EN 196-1. Methods of testing cement – Part 1: Determination of strength. Prague: Czech office for standarts, metrology and testing 2016
[16] CSN EN 1008. Mixing water for concrete–Specification for sampling, testing and assessing the suitability of water, including wate recovered from processes in the concrete industry, as mixing water for concrete. Prague: Czech office for standards, metrology and testing 2003
[17] CSN EN 197-1 ed.2. Cement - Part 1: Composition, specifications and conformity criteria for common cements. Prague: Czech office for standards, metrology and testing 2012
[18] CSN EN 197-2 Cement - Part 2: Conformity evaluation. Prague: Czech office for standards, metrology and testing 2014
[19] Šimonovičová A et al 2017 *Chemosphere* **171** 89–96
[20] Šimonovičová A et al 2016 *Water. Air. Soil Pollut.* **227** 336
[21] Smolakova M and H Vojtkova 2019 IOP *Conf. Ser.: Mater. Sci. Eng.* **566**
[22] Urík M et al 2019 *Sci. Total Environ.* **664** 683–9
[23] CSN 72 4310 Testing of mould proofness of building products and materials. Prague: Czech office for standards, metrology and testing 1975

Acknowledgments
This research was funded by:
- VŠB–TUO, Faculty of Mining and Geology—grants numbers SP2020/30 and SP2019/29;
- Moravian-Silesian Region—grant number RRC/10/2018.

IOP Conf. Series: Materials Science and Engineering **867** (2020) 012021 doi:10.1088/1757-899X/867/1/012021

Alternatives of solar systems and their analysis

J Kosican[1]

[1] Technical university of Kosice, Faculty of Civil Engineering, Institute of Architectural Engineering, Slovakia

Abstract. This study negotiates with percentage fraction for heating and hot water of various solar systems. Research building of the Technical University of Košice was selected. For this analysis, we used simulation appropriate simulation software, which replicates solar systems and calculates the production of the proper solar system. If we consider DHW solar ratio, the fifth alternative, which includes a combination tank system (heating buffer tank) with an outside heat exchanger and a swimming pool, is the most productive. Nonetheless, in accordance of overall solar fraction, the top choice is alternative 4, which is a mix of DHW tank system with a heat exchanger inside the tank. At the end, the outcome demonstrates that, mix of DHW tank and buffer tank for space heating with a swimming pool, that is alternative 6, seems to be the worst case. Efficient usage of different systems based on renewable energy sources and their applications secures Investigated centre mentioned before.

1. Introduction

Solar energy, among other forms of renewable energy sources, is promising and freely available; it can address the energy crisis in the long time [1]. Population growth rate and industrial evolution are the pre-dominant factors towards increasing world's energy demand. Primary energy needs are met using conventional energy such as coal, petrol, and diesel; and by using non-conventional energy sources such as solar energy, bio energy, wind energy, geothermal energy [2]. The selection of the best solution for a given project requires a full understanding of the system performance and the interaction of its components rather than the technology efficiency only [3]. Design and installing solar photovoltaic panels in the hard environment is based on the safety of the solar system from dust and wind. From the measurement of different climatic factors, which are humidity, temperature, solar irradiation, and the available data, we are able to produce hydrogen and electricity in the same time from photovoltaic solar energy systems [4]. The majority of the installation of solar power capacity is seen in Europe and China, whereas countries with the highest potential for solar energy still lag [5]. Unexpected climatic changes can affect power prediction significantly. Due to the variable nature of solar power, predicting the power output of solar panel installation is important for its optimal use. Using analytical method, certain months show more errors than other months because the correction factor does not fit all the months equally [6]. The global solar industry is constantly evolving; there is considerable demand for energy, but the main source of energy — fossil fuels — is limited, and other sources are expensive. The risks associated with solar panels' life cycle must be addressed by setting clear regulations. One related environmental impact is land use, given that utilisation of solar energy requires a considerable area of land to be allocated mainly for generating electricity [7]. As the world population increases in the coming decades, the International Energy Agency has estimated that global energy demand will increase by an additional 30% between 2016 and 2040, with 40% alone coming from an expected increase in electricity usage [8]. Investigated experimental apartment is constructed as a passive requirement model and is situated in a chamber that can reproduce outdoor conditions, ignoring the sunshine. Apartment consists of living room, area with a kitchen hygienic space. Solar panels are not situated on the roof of house but on roof of the research building. House in present time produces hot water in a 120-liter tank, situated in the kitchen on the ground. For cold and hot water networks were applied three-layer pipes, arranged in the floor. For this analysis work was used an auxiliary heat source of wooden fire boiler (15kW). Source of energy is a mass storage tank situated under the investigation centre. The sewerage is composed of polyvinyl chloride (PVC).

2. Materials and methods

Investigated experimental apartment is constructed in a passive requirement and is situated in a chamber that can reproduce outdoor conditions, except sunshine. For cold and hot water networks were applied three-layer pipes, arranged in the floor.

Figure 1. Experimental house.

Main house attributes, important DHW and heating requirements are displayed in table 1.

Table 1. Attributes and requirements for DHW and heating.

House attributes	Unit	Amount
Occupied house area	m^2	53 m^2
Outside temperature	°C	-14.3
Calculated house thermal losses	kW	3,1
DHW consumption	L/day	120
Annual energy demand for DHW	kWh	2031
Solar preheating tank	litres	200
Auxiliary heat source - wooden fire boiler	kW	15

For this case of study, we chose solar panels from company Thermosolar Žiar TS 300 with a surface of 2 m^2. In all alternatives are designed five solar collectors. One panel has surface of 2 m^2, is located on the roof of the investigation centre. Inclination of solar panels is set up to 45° and their occupied area of roof is 10.15 m^2. Active surface of all panels is 8.9 m^2. The direction of the competent panels is to the south. Acknowledges to the calculations of simulation program we found out external conditions - outside temperature for Košice is 14.3°C. Wood boiler with 15 kW performance is designed as the auxiliary heat source. We figured out that, heat loss a day of the reported apartment is 3.1 kW. Total living area is 53 m^2. All required equipment for all alternatives are described in table 2.

Table 2. Equipment for all alternatives.

Equipment	Alternative					
	1	2	3	4	5	6
Solar panel	•	•	•	•	•	•
Wood boiler	•	•	•	•	•	•
DHW120 litres/day	•	•	•	•	•	•
Space heating	•	•	•	•	•	•
Dual coil water tank 100 litres				•	•	
Heat excharger			•		•	•
Dual coil water tank 120 litres	•					•
Dual coil water tank 300 litres			•			
Space heating buffer tank 550 litres	•					•
Buffer tank 550 litres			•			
Swimming pool					•	•
Solar preheating tank 200 litres		•				
DHW Standby tank 120 litres		•				
Buffer tank 500 litres		•				

In this case we want to get a proper result for heating and DHW. For simulation of solar schemes for heating and DHW in experimental house located in the research centre was used proper software - TSOL 2018. This program is an expert device for designing a solar thermal structure. The simulation results based on construction known input data can be interpreted in a chart or board form. Results are also considering the performance of the system corresponding to various natural conditions, such as the construction's temperature area, the value of heat loss of designed building, heating types or hot water system. The outcome of given simulation can be figured out by considering the numerical design with variable time values up to almost 6 minutes. Current weights of outcome can be diverse from these amounts due to particular temperature areas, consumption and much more factors and elements. The simplified diagram of the system mentioned above cannot change and cannot replicate fulfilment of technological design these solar systems.

3. Results of simulations

Table 3 and table 4 presents solar fraction performance of the year. Analysis occurs for various combination for solar systems. Experiments for production and useful analyses for methodology verification have been done. The simulation confirms on function contribute relevant knowledge, and in few instances can be the reason for not reaching the demands of the Thermosolar Ziar brand. As a outcome of presented simulation of the six mentioned alternatives. Research shows, that in DHW solar fraction the most effective is alternative 5, which contains a combination tank system (heating buffer tank) with additional heat exchanger and swimming pool. According to examination of the overall solar fraction, the top choice is alternative 4, which includes combination tank for DHW system with heat exchanger embedded to the tank. The alternative with the lowest efficiency in total and DHW solar fraction our research implies alternative 6 - combination tank DHW and buffer tank for heating with swimming pool.

Table 3. Sorting alternatives by DHW solar fraction.

Alternative number	DHW solar fraction [%]
1 - combination tank DWH system and space heating without preheating tank	64
2 - combination DHW system and space heating with solar preheating tank	75
3 - DHW system with heating buffer tank with external heat exchanger with a combination of solar panel and a wood boiler	77
4 - combination tank DHW system with heat exchanger inside the tank	18
5- combination tank system (heating buffer tank) with external heat exchanger and swimming pool	100
6 - combination tank DHW and buffer tank for space heating with swimming pool	15

Table 4. Sorting alternatives by total solar fraction.

Alternative number	Total solar fraction [%]
1 - combination tank DWH system and space heating without preheating tank	18
2 - combination DHW system and space heating with solar preheating tank	20
3 - DHW system with heating buffer tank with external heat exchanger with a combination of solar panel and a wood boiler	20
4 - combination tank DHW system with heat exchanger inside the tank	25
5- combination tank system (heating buffer tank) with external heat exchanger and swimming pool	23
6 - combination tank DHW and buffer tank for space heating with swimming pool	15

4. Conclusion

This report presents the results for solar systems in small range and their effect for DHW and also heating. All necessary simulations were done in TSOL and in plan to complete the efficiency of energy and its production and following optimization of the technology. Construction of solar collector systems, and further accurate simulation tool established with solid analysis data of the mentioned systems is needed. For optimization the solar thermal system, each unit (solar collectors, pipelines, tanks, boilers, etc.) can be measured independently. And with the help of the diversity, it is too desirable to produce many different simulations together in one time, e.g. to automatically design the collector space and storage tank size.

5. References

[1] Rozentale L 2018 Accelerating power generation with solar panels. Case in Latvia *Energy Procedia* 147 pp 600-606
[2] Kalidasan B 2020 Phase change materials integrated solar thermal energy systems: Global trends and current practices in experimental approaches *Journal of Energy Storage 27*
[3] Ayadi O 2019 Comparison of solar thermal and solar electric space heating and cooling systems for building in different climatic regions *Solar Energy Volume 188* pp 545-560
[4] Menad C A, Gomri R, Bouchahdane M 2018 Data on safe hydrogen production from the solar photovltaic solar panel through alkaline electrolyser under Algerian climate *Data in Brief 21* pp 1051-1060
[5] Arabian Gulf University 2019 Public perception toward residential solar panels in Bahrain *Energy Reports 5* pp 253-261
[6] Nageem R R J 2019 Predicting the Power Output of a Grid- Connected Solar Panel Using Multi-Input Support Vector Regression *Procedia Computer Science 115* pp 723- 730
[7] Lauka D First solar power plant in Latvia. Analysis of operational data 2018 *Energy Procedia 147* pp 162-165

[8] Kafka J L and Miller M A 2019 A climatology of solar irradiance and its controls across the United States: Implications for solar panel orientation *Renewable Energy* 135 pp 897-907
[9] Dr. Valentin G, T*SOL basic version 5.0 2012 *Design and Simulation of thermal solar systems User Manual*
[10] http://web.tuke.sk/kpt/aktualita_vukonze.html
[11] Nfaoui M and El-Hami K 2018 Extracting the maximum energy from solar panels *Energy Reports 4* pp 536-545
[12] Soloha R, Pakere I and Blumberga D 2017 Solar energy use in district heating systems. A case study in Latvia *Energy* 137 pp 586-594
[13] Dinniyah F S, Wahab W and Alif M 2017 Simulation of Buck-Boost Converter for Solar Panels using PID Controller *Energy Procedia* 115 pp 102-113
[14] Xu Y, Li J, Tan Q, Peters A L and Yang C 2018 Global status of recycling waste solar panels *Waste Management* 75 pp 450-458

Acknowledgments
This work was financially supported by the Grant Agency of Slovak Republic to support project No. VEGA 1/0512/20: “Analysis of new approaches and certification of sustainable office buildings from the perspective of wellbeing and performance of employees.”

Compressive strength of concrete specimens with inclusions

M Kovac[1], P Sabol[1], S Kusnir[1], D Kusnirova[1] and M Rovnak[1]

[1] Technical university of Kosice, Faculty of Civil Engineering, Institute of Structural Engineering, Vysokoskolska 4, 042 00 Kosice, Slovakia

Abstract. Large diameters and grain aggregates and/or reinforcing bars can have a negative impact on the strength of concrete elements. Tests of concrete specimens with inclusions under uniaxial compression should contribute to a better understanding of the process of failure of concrete elements. The identification of the unstable phase of the cracking and quantification of the effect of diameters, positions, distances, number and material of inclusions, as well as the relative size of inclusions and matrix on compression strength could provide information for an explanation of the causes of the size effect in compressed structural elements. The introduction of the article provides a brief overview of the problem of concrete structures subjected to uniaxial compression. Attention is paid to the failure mechanisms in the process of loading the test specimens from initial crack formation to specimen collapse. The second part is devoted to the investigation of the size effect of varying dimensions of the test specimens with one or three inclusions.

1. Introduction

In 1925, Gonnerman was the first to investigate axial-compressed concrete elements. The influences of the sizes of concrete blocks and cylinders on their nominal compressive strength were analysed based on experimental results [1]. Compression failure is a complex process because the number of non-critical cracks is much greater than tensile failure [2]. It was found that there were cracks in the concrete specimen before the mechanical load was applied. One of the main causes of this phenomenon is cracks due to concrete shrinkage [3]. Concrete is a complex heterogeneous system. Wittmann [4] was the first to introduce three different levels of concrete models: macro-level, meso-level and micro-level. Concrete in the meso-level model can be considered as a material consisting of grains of aggregate, cement putty and transition zone. It has been found that the aggregate grain and the cement putty exhibit almost linear elastic brittle behaviour under axial compression. However, the curve of the σ - ε of ordinary concrete has a typical curvature, which is caused by the interface between the aggregate grain and the cement putty, where the first cracks arise under load. As the load increases, cracks penetrate into the matrix. In this way, an interconnected network of cracks forms in the matrix and at the matrix/aggregate grain boundary until the specimen fails [5].

When designing concrete structures, the compressive strength of the concrete, which is determined from the standard dimensions of test specimens tested in the laboratory, is very important. However, these dimensions do not correspond to the proposed dimensions of concrete elements in real constructions. According to experimental research and analytical studies performed so far, it is possible to generalize the principle of dimensional effect in the design of concrete structures. The strength of the specimens decreases with increasing specimen size and thus larger concrete specimens show lower strength and vice versa, smaller specimens show higher strength [2].

2. Concrete failure mechanisms

Depending on the approach of the concrete model at the macro- and meso-level, we distinguish the different failure mechanisms in the concrete with the acting axial compression. Basically, three types of microcracks can be categorized [2]:

- Cracks at the interface of the cement matrix and aggregate.
- Cracks in the material matrix.
- Cracks passing through the aggregate.

2.1. Meso-level concrete model

Cracks in the concrete can develop and spread due to the concentration of tensile and compressive stresses in the specimen, which occur through the interaction between large aggregate grains and the matrix (figure 1a). A weak connection between the aggregate grain and the cement matrix leads to the initiation of a crack at their interface. The soft matrix flows around the stiffer grains, and thus the lateral deformation in the relatively soft matrix must be greater than in the solid grains of the aggregate, which can lead to transverse tensile and cracks at the interface. The shear stress will develop at the top and below the aggregate grain. Spatial compression is created in these cone-shaped regions, causing the matrix to fail in these regions. The failure then proceeds to grow the crack along the side of the cone (figure 1b). E_a indicates Young´s modulus of aggregate and E_m Young´s modulus of the matrix. The type of failure and the course of crack propagation depend largely on the ratio of stiffness between matrix and inclusion. The inclusions may be represented by aggregate grain, reinforcing bars or cavities and macropores in the matrix. The cleavage of the lightweight aggregate grain results in a lower stiffness value compared to the stiffness of the surrounding matrix (figure 1c). Fragmentation of the matrix occurs in the case of stress concentrations around large air inclusions (figure 1d). It is generally assumed that in the field of compressive stress, the formation and development of cracks are parallel to the direction of the compressive load. However, the real propagation of micro-cracks caused by axial compression does not correspond to the above assumption. Cracks at the interface between the aggregate grain and the matrix often divert to the sides of the matrix (figure 1e) [6]. The local crushing mechanism of concrete occurs in the interface transition zone (ITZ) when pore concentrations occur above and below the aggregate (figure 1f) [7].

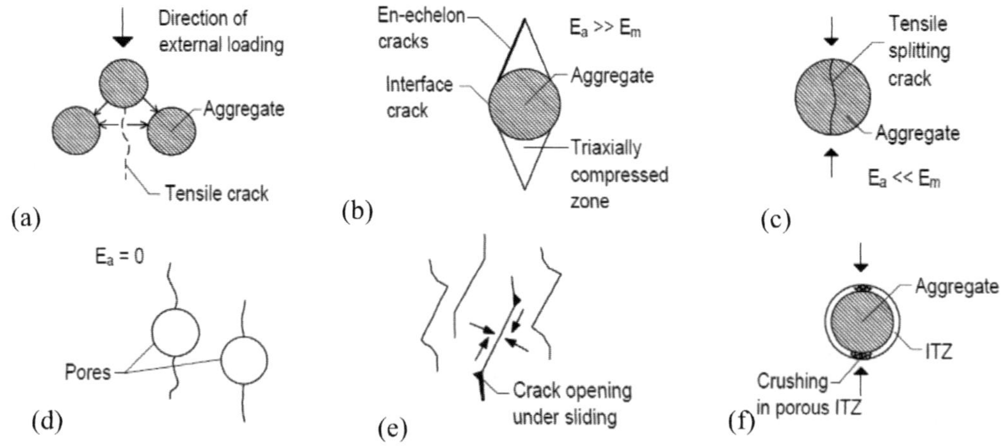

Figure 1. Mechanisms of crack formation and propagation in concrete.

2.2. Macro-level concrete model

The crack propagation process and the failure pattern of the test specimen depend on the design of the experimental equipment, the mode of loading, the boundary conditions and the slenderness (ratio *h/d*) of the concrete elements. The propagation of macrocracks in the specimen depends on the orientation and magnitude of the shear stress on the contact between the specimen and the load plate. The rigid load plates cause shear stress on the contact which prevents horizontal displacement of the specimen in the transverse direction and thus creates a region of compressive stress under the plate. If a flexible load plate is used, the lateral displacement is greater than in concrete and the shear stresses are applied outwards from the centre of the specimen. The result is the tensile stress in the matrix. If the stress exceeds the concrete tensile strength, there is a split - tensile cracks in the transverse direction. The mentioned types of boundary conditions in combination with the specimen size ratio (slenderness = height/diameter) *h/d* determine the mode of failure of the test specimen. In the case of greater slenderness of concrete elements with a ratio *h/d* > 2 and using a rigid steel plate, the specimen may be cutted off due to a slanted crack. On the other hand, the use of an elastic plate causes a transverse tension, i.e. splitting. At the ratio *h/d* ≤ 1 and the resilient plate, cleavage also occurs, but with a higher density of cracks in the specimen. At a smaller *h/d* ratio and a rigid plate, the crushing of the concrete body prevails (figure 2a and figure 2b) [6].

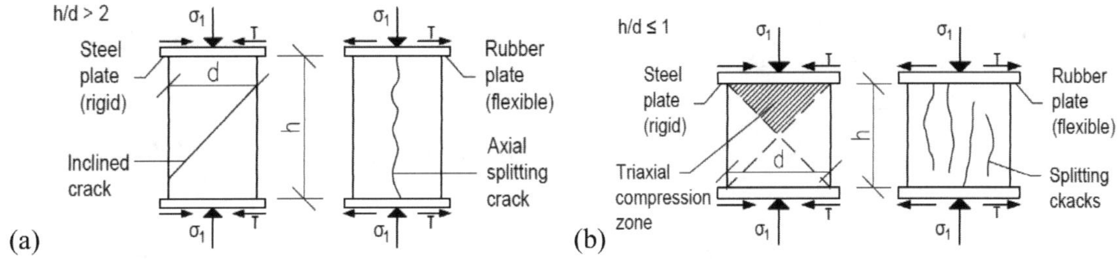

Figure 2. Different mechanisms of concrete failure under uniaxial compression.

3. Experimental research

3.1. Material and specimens

The compression tests were performed on specimens of three different sizes (100×100×50; 150×150×50 and 200×200×50 mm), with 1 or 3 inclusions and two different types of inclusion/matrix interface: steel inclusion/concrete matrix (*S*) and steel inclusion/epoxy/concrete matrix (*SI*). To prepare the specimens, dry concrete mix class C16/20 was used. The test specimens were cast into the steel framework where the inclusions were previously glued in the desired position. At the same time, the prisms were concreted to determine the material characteristics of the concrete. The specimens were tested 7 days after the concreting. During the storage, relative humidity and temperature of the environment were measured every 15 minutes. The diameter of the inclusions was 20, 30 and 40 mm. Their height was identical to the depth of the specimens 50 mm. Figure 3 shows uncoated steel rollers, which were used as the steel inclusion in the specimen, intended to test the inclusion-concrete matrix interface (*S*). Figure 4 shows steel rollers with an epoxy layer with an approximate thickness of 1 mm, which were used as the steel inclusion in the specimen intended to test the inclusion-epoxy-concrete matrix interface (*SI*).

Figure 3. Steel inclusions (ϕ = 20, 30, 40 mm).

Figure 4. Steel inclusions with epoxide layer (ϕ = 20, 30, 40 mm).

The inclusions strength class was S235. For the coating of the steel roller, epoxy Sikadur 30 was used. The producer declares the tensile strength of 26 MPa, the compressive strength of 75 MPa, the tensile modulus of 11.2 GPa, and the compressive strength of 9.6 GPa. The strength characteristics of the concrete were measured on the test day. Mean values of concrete compressive strength, modulus of elasticity, flexural strength and splitting tensile strength are given in table 1.

Table 1. Material characteristics of concrete

Inclusion	Specimen	f_{cm} (MPa)	E_{cm} (GPa)	$f_{cf,m}$ (MPa)	$f_{ct,sp,m}$ (MPa)
Steel	SQ-7-100-20-S-2	18.79	18.57	4.01	3.13
	SQ-7-100-20-S-9	19.49	20.32	4.04	3.3
	SQ-7-150-30-S-2	18.57	16.5	4.3	3.03
	SQ-7-150-30-S-9	18.47	19.17	4.2	3.05
	SQ-7-200-40-S-2	19.3	19.7	4.6	2.5
	SQ-7-200-40-S-9	18.5	18.1	4.5	2.4
Steel with epoxide layer	SQ-7-100-20-SI-2	19.51	17.52	3.2	2.27
	SQ-7-100-20-SI-9	14.44	14.59	3.19	2.07
	SQ-7-150-30-SI-2	20.22	19.73	4.5	3.16
	SQ-7-150-30-SI-9	19.9	18.86	4.05	3.18
	SQ-7-200-40-SI-2	15.8	15.2	3.6	2.5
	SQ-7-200-40-SI-9	20.5	17.6	4.4	3.3

3.2. Test arrangement and procedure

Tests were performed by using a loading device consisting of rigid steel plates, threaded rods and a hydraulic jack (figure 5). Test specimens were subjected to uniaxial compression. The tests were controlled by force, the load step was about 9.3 kN. During the test, the vertical deformation of the samples was measured using two indicator gauges. Before the test, the surface of specimens was coated with a white colour and then, the chaotically arranged black dots were applied for non-contact measurement of deformation by photogrammetry GOM Aramis system. Figure 6 (a) shows the specimen before the test and figure 6 (b) shows the typical failure of the specimen after the test.

Figure 5. Arrangement of the uniaxial compression test.

(a)

(b)

Figure 6. Specimen (a) before (b) after the test.

3.3. Results and discussion

Figure 7 shows the $\sigma/f_{cm} - \varepsilon$ relationship of the test specimens with one inclusion in the middle. Test curves with 3 inclusions are shown in figure 8. In the early stages of loading, an elastic phase was identified that changed into the plastic phase and later to overall specimen failure. Rigid steel plates and a load-controlled test caused, that the curves did not show a dominant peak and the post-peak branch. It was found that smaller test specimens showed higher strength and larger specimens achieved less concrete compressive strength. This tendency can be observed for both single-inclusion and 3-inclusion specimens. In the second case, the impact of significant size effect occurred. The compressive stress at the failure of specimens decreased as the number of inclusions increased from one to three. This phenomenon was evident in specimens of 200×200×50 mm and 150×150×50 mm. The test specimens with dimensions of 100×100×50 mm achieved approximately the same strength values. It could be caused by a triaxial compression zone, which represents a large part of the total sample volume. The comparison of the different inclusion and matrix interface depending on the specimen size revealed changes in the achieved strength values of concrete specimens. Test specimens with 1 and 3 inclusions of 200×200×50 mm and 150×150×50 mm with the *SI*-interface exhibited higher strengths than *S*-interface.

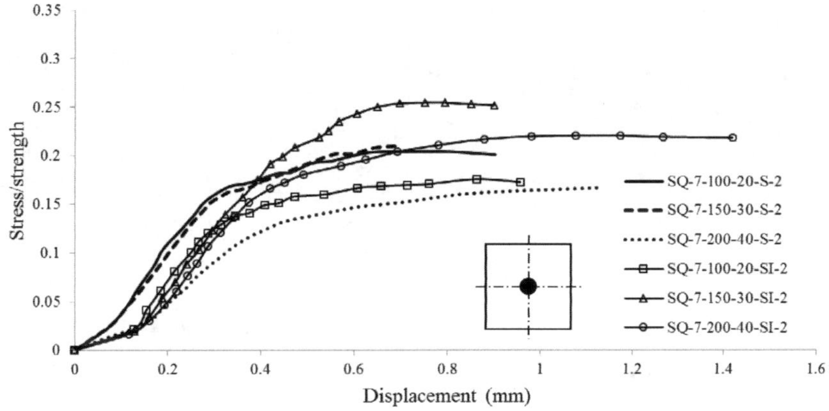

Figure 7. Stress-concrete strength ratio vs displacement of specimens with steel inclusions and steel inclusions with epoxide layer.

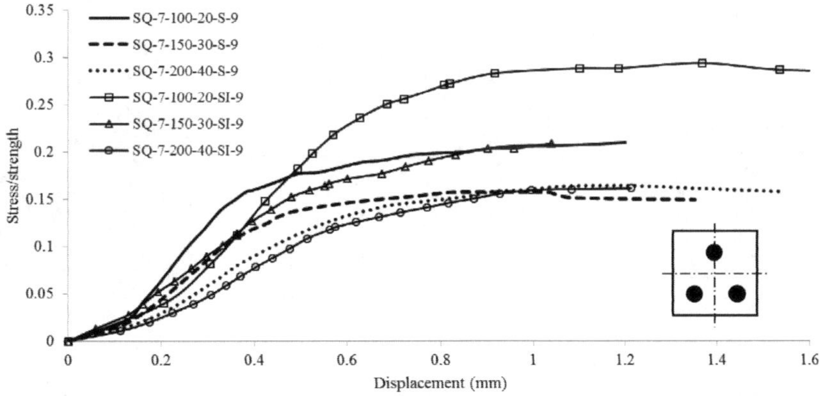

Figure 8. Stress-concrete strength ratio vs displacement of specimens with three steel inclusions and steel inclusions with epoxide layer.

Figure 9 shows the failure of specimens of three different dimensions with 1 inclusion and *S*-interface, and figure 10 shows the failure of test specimens with 1 inclusion and *SI*-interface.

IOP Conf. Series: Materials Science and Engineering **867** (2020) 012022 doi:10.1088/1757-899X/867/1/012022

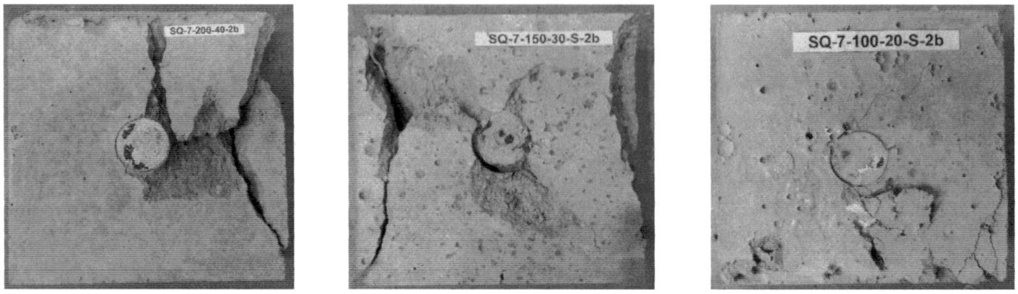

Figure 9. Failure of specimens with *S*-interface (Images size ratio 1:0.75:0.5).

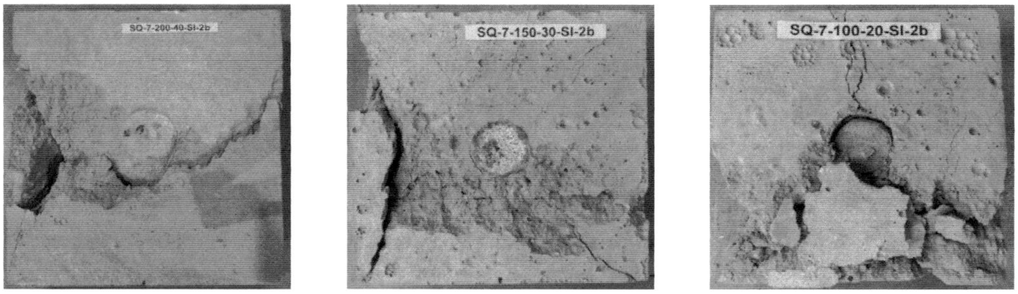

Figure 10. Failure of specimens with *SI*-interface (Images size ratio 1:0.75:0.5).

The failure was related to the development and opening of cracks in the vertical direction. Larger test specimens showed a more fragile failure while the smallest test specimens showed ductile behaviour. The crack development detected by the Aramis system before the collapse of the 3-inclusion test specimens and the *S*-interface is shown in figure 11 and for the *SI*-interface in figure 12. Based on the performed tests is possible to identify the initiation of cracks at the interface of inclusion and concrete matrix. As the load increased, the cracks subsequently spread into the matrix.

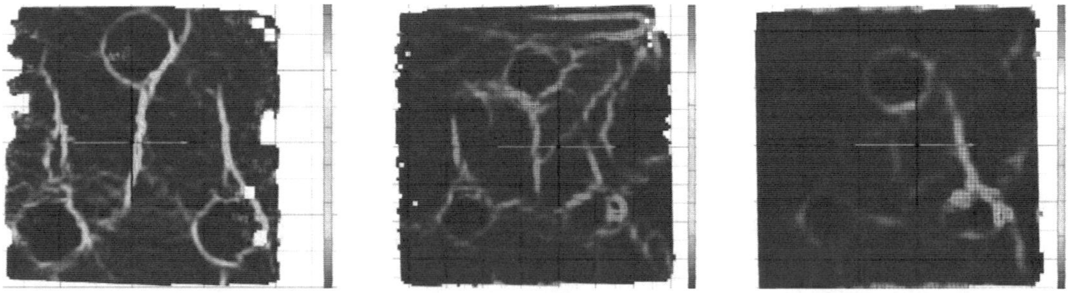

Figure 11. Crack development with *S*-interface (Images size ratio 1:0.75:0.5).

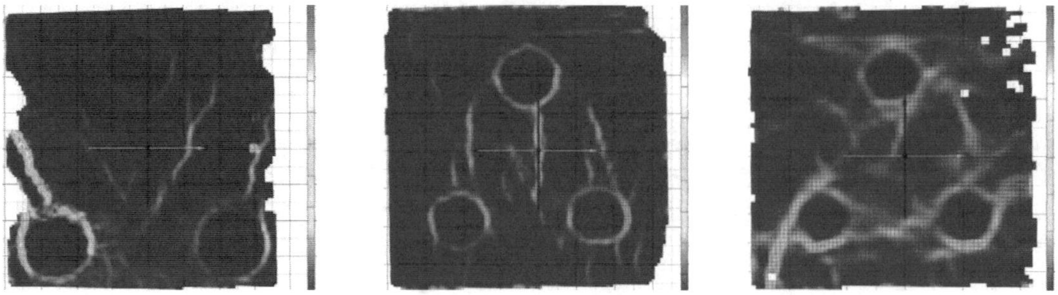

Figure 12. Crack development with *SI*-interface (Images size ratio 1:0.75:0.5).

4. Conclusion

The tests performed and comparison of the test results led to the conclusion that the inclusions contained in the specimens significantly affected their strength to uniaxial compression. The strength of specimens with inclusions was not only affected by the number of inclusions but they largely depended on the properties of the inclusions, their number and arrangement in the specimen. This experiment confirmed the phenomenon of the size effect - the compressive strength of specimens decreased with the increasing size of the specimen. The decrease in the strength of specimens was found with an increasing number of inclusions placed in the specimens.

Based on an overview of the factors and results of tests performed, it can be concluded that further investigation of the mechanisms of failure of the concrete elements stressed by axial compression is necessary for the formation of cracks at the interface and their spread into the matrix depending on stiffness, size, number and arrangement of inclusions. It is necessary to define the relationship between the strength of concrete test specimens and the number and arrangement of inclusion with different material interfaces. It is also useful to look at ways of formation and propagation of cracks during loading, which in turn affects the failure mode of the specimens and have a major impact on the value of strength.

5. References

[1] Gonnerman H F 1925 Effect of Size and Shape of Test Specimen on Compressive Strength of Concrete *Proc. Am. Soc. Testing Malts* vol 25 part II pp 237-250

[2] Van Mier J G M 1984 *Strain-softening of concrete under multiaxial loading conditions* PhD Thesis (Netherlands: Eindhoven University of Technology)

[3] Hsu T T C and Slate F O 1963 Tensile bond strength between aggregate and cement paste or mortar *Journal of the American Concrete Institute Proc.* **60** 465-86

[4] Wittmann F H 1983 Structure of Concrete with respect to Crack Formation *Fracture Mechanics of Concrete,* ed F H Wittmann (Amsterdam: Elsevier Science Publishers) pp 43-74

[5] Neville A M 1997 Aggregate bond and modulus of elasticity of concrete *ACI Materials Journal.* **94** 71-74

[6] Van Mier J G M 1998 Failure of concrete under uniaxial compression: An overview *3rd Int. Conf. on Fracture Mechanics of Concrete and Concrete Structures: Proc. FraMCoS* (Gifu, Japan) ed H Mihashi and K Rokugo (Freiburg: AEDIFICATIO) pp 1169-1182

[7] Bongers H 1998 Multilevel analysis of concrete in multiaxial compression *Computational Modelling of Concrete Structures,* ed R de Borst et al pp 347-354

Acknowledgments

This work was supported by the Slovak Research and Development Agency under the contract No. APVV-15-0777.

A batch study of Ni(II) sorption on natural Slovak zeolite

Z Kovacova[1] **and C Pla**[2]

[1] Technical University of Kosice, Faculty of Civil Engineering, Institute of Environmental Engineering, Vysokoskolska 4, 042 00 Kosice, Slovakia
[2] University of Alicante, Department of Civil Engineering, Carr. de San Vicente del Raspeig, s/n 03080, San Vicente del Raspeig - Alicante, Spain

Abstract. The adsorption mechanism of nickel on natural Slovak zeolite, a regional low-cost material, was studied in a batch system. The effect of various initial concentrations of metal ions in the adsorption process was tested. The highest adsorption efficiency (almost 70%) was reached at a concentration of 50 mg/L. The lowest efficiency was determined at a concentration of 350 mg/L (approx. 19%). Sorption data have been interpreted in terms of Langmuir and Freundlich isotherms. The Freundlich model with regression coefficient 0.83 better fitted for the batch experiments. The change of pH in equilibrium showed to ion-exchange reaction. The conductivity increases in all cases. Due to relatively low efficiency in higher concentration of aquatic solutions, further research and modification of zeolite is needed.

1. Introduction

The aquatic ecosystem is frequently contaminated by heavy metals from diverse sources like industrial wastewater, urban runoff, agricultural processes and others. Nickel belongs to toxic heavy metals and is non-biodegradable in environment. The industries such as connecter, electroplating, metal finishing, nickel-cadmium batteries, lead frame, plastics manufacturing, fertilizers, pigments, mining and metallurgical processes are the source of nickel in wastewaters. Higher concentration of nickel causes severe health problems like pulmonary fibrosis, renal edema, damage to lungs, kidneys, gastrointestinal distress, e.g., abdominal discomfort, nausea, vomiting, diarrhoea and skin dermatitis. It is also a carcinogenic and in teratogenic to live beings in its metallic form [1]-[4][4].

There are many processes dedicated to water treatment, such as biological methods, electrocoagulation and electrodialysis, coagulation and flotation, membrane separation. Adsorption onto zeolites has advantages such as a simple process, easy maintenance, large adsorption capacities, appropriate selectivity and relatively low cost. Zeolites belong to aluminosilicates, they have a framework silicate with a three-dimensional cage structure. They represent a proper material for removing heavy metal ions from wastewater because their structure provides negative charges, which can balance by adsorption of exchangeable cations [4]-[7].

Zeolites were investigated to have high selectivity for heavy metal ions, like Pb(II), Cd(II), Zn(II), and Cu(II) [6],[8],[9].

The aim of this study was to examine the effectiveness of natural zeolite as a sorbent for nickel removal. In the experiments, the influence of pH, conductivity and various concentration of nickel in aqueous solutions on adsorption behaviour of zeolite was studied. To evaluate sorption process onto Slovak zeolite, the data from batch experiments were evaluated using Langmuir and Freundlich isotherms [2],[5].

2. Material and methods

2.1. Preparation of material
In the experiments, the natural zeolite (particle size 0.2-1.0 mm) from the Nizny Hrabovec (Zeocem, Inc.; Bystre, Slovakia), was used.

The chemical composition of studied sorbent was determined by X-ray fluorescence (XRF) spectrometry Philips-Magix Pro (Philips Electronics Co. Netherlands). XRF analyser determines the chemistry of a sample by measuring the fluorescent (or secondary) X-ray radiation emitted from a sample when it is excited by a primary X-ray source [10],[11].

Adequate amounts of $Ni(SO_4) \cdot 6H_2O$ were dissolved in de-ionized water to obtain the stock solutions of concentration 1000 mg/L of Ni. Concentrations of solutions (50, 100, 200, 350 and 500 mg/L) used in the batch studies were prepared by dilution of stock solution with de-ionized water.

2.2. Batch experiments
5 g of zeolite and 100 mL of single-component solution was added into 100 mL bottle vial. The vials were shaken at a regular rate in a magnetic shaker at 200 rpm. After 24 h reaction time, sorbent was removed by filtration through a laboratory filter paper. Residual concentration of nickel in filtrate was determined by ICP-MS Spectrometer PQ-ExCell (VG-Thermo Elemental, Winsford, Cheshire, UK).

In the filtrates, pH (Hach sensION+ PH1, USA) and conductivity (Sension+ EC5, Hach, USA) were measured. All experiments were performed at laboratory temperature (23±0,2°C).

The removal efficiency (E) was calculated by following formula as

$$E = \frac{(C_i - C_e)}{C_i} \cdot 100\ \% \qquad (1)$$

where C_i is the initial concentration (mg/L) and C_e is the equilibrium concentration (mg/L) [3][12],[13].

2.3. Sorption models
Adsorption isotherm is a function that associates the amount of adsorbate on the adsorbent. Several models can represent placement of metal ions between the liquid and the solid phase. Ones of most using are Langmuir and Freundlich.

The Langmuir model assumes monolayer adsorption onto a surface containing a limited quantity of adsorption sites of uniform strategies with no transmigration of adsorbate in the surface [12] . This model is represented by

$$Qe = Q_{max} \cdot \frac{K_L \cdot C_e}{1 + K_L C} \qquad (2)$$

where Q_{max} (mg/g) is the Langmuir constant corresponding to the uptake capacity, and K_L (L/mg) is the adsorption energy as well as the equilibrium constant of the adsorption-desorption process. The constants in the Langmuir model, Q_{max} and K_L, were evaluated using the slope ($1/Q_{max}$) and the intercept ($1/K_L Q_{max}$), respectively, of the plot of C_e/Q_e vs. C_e [14].

The Freundlich isotherm is recognised as an empirical model and represents a direct relationship between concentration and adsorption. The formula of this this model is:

$$Q_e = K_F \cdot C_e^{1/n} \qquad (3)$$

IOP Conf. Series: Materials Science and Engineering **867** (2020) 012023 doi:10.1088/1757-899X/867/1/012023

where Q_e (mg/g) represents the amount sorbed per amount of adsorbent at the equilibrium, Ce (mg/L) represents the equilibrium concentration, and K_F is Freundlich constant corresponding to adsorption capacity and n value the degree of nonlinearity between solution concentration and adsorption. If $1/n$ approaches 1, isotherm equation becomes linear [15].

3. Results and discussion

3.1. XRF analysis

A series of XRF analyses were carried out at the Department of Civil Engineering at University of Alicante – Polytechnic School, for the chemical characterization of zeolite. The obtained XRF analysis together with declared parameters from manufacturer is shown in table 1.

Table 1. XRF Analysis of raw zeolite.

Chemical component	Declared content (mass %)	XRF analysis (mass %)
SiO_2	62.25-68.28	60.62
Al_2O_3	12.22-12.62	10.81
CaO	2.67-2.98	3.50
K_2O	-	3.05
Fe_2O_3	1.31-1.66	1.39
MgO	-	0.58
Na_2O	-	0.41
BaO	-	0.20
WO_3	-	0.16
TiO_2	-	0.15

This analysis determined that the content of silica and alumina is the main components of natural zeolite. Calcium and potassium ions can be exchanged with metal or non-metal ions [11],[16]. Ratio of Si/Al in the natural zeolite is 4.95 which is little lower like in [17].

3.2. Sorption efficiency

The sorption efficiency, change of pH and conductivity are presented in table 2. The maximum amount Ni(II) removal (69,51%) was achieved at concentrations of 50 mg/L. The sorption efficiency decreased with higher concentration (at 500 mg/L was only 20.3%). The results indicated that nickel uptake rate was relatively high at the beginning of the process due to abundance possible adsorption sites. However, the number of active sites decreases towards to the formation of the adsorption equilibrium [5].

The pH of the solution is one of the most important parameters in the sorption process of heavy metals by zeolites, for its effect on the charge state of metals and adsorbent surface [5].

It was observed that pH of solutions 50 and 100 mg/L increased and 200, 350, and 500 mg/L decreased during experiments. The values of conductivity of all solutions increased during 24 h of reaction time.

Table 2. The pH, conductivity and efficiency of Ni(II) sorption.

Concentration (mg/L)	Initial parameters		Final parameters		
	pH	γ (μS/cm)	pH	γ (μS/cm)	E (%)
50	6.1	200	6.6	251	69.51
100	5.9	427	6.5	489	53.62
200	6.4	765	6.2	862	33.01
350	6.6	1312	6.3	1393	18.89
500	6,7	1815	6.4	1841	20.30

3.3. Isotherm models

To find the most appropriate model, data were fitted to Langmuir and Freundlich isotherms (figure 1). Both isotherm models indicated that during adsorption process monolayer ion-exchange was formed [7].

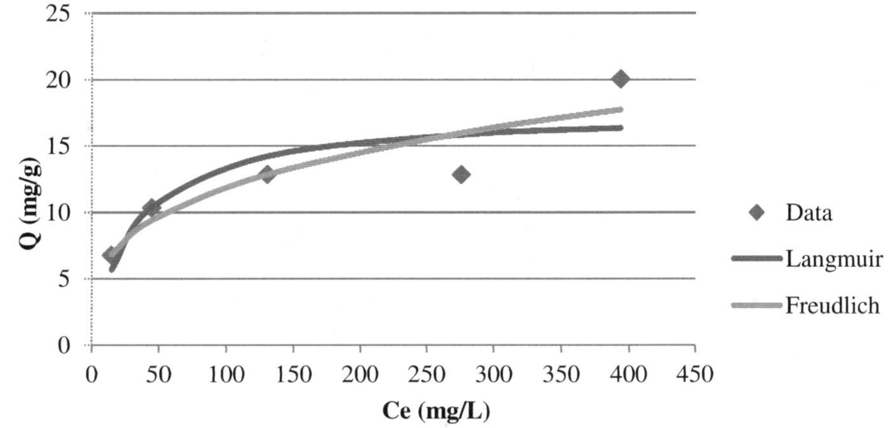

Figure 1. Lagmuir and Freundlich isotherm models for the Ni(II) adsorption.

The isotherm coefficients are presented in table 3. The coefficient of determination.(R^2) values derived from the Langmuir fitting was 0.728 for nickel and 0.830 was obtained from the Freundlich model. The statistical correlation using the Freundlich isotherm was better than the Langmuir isotherm. Freundlich isotherm is realized on heterogeneous surfaces. This model defines surface heterogeneity and exponential distribution active sites and energy [18].

Table 3. Isotherm parameters for adsorption of Ni(II).

	Langmuir constants			Freundlich constants		
	Q_{max} (mg/g)	K_L (L/mg)	R^2	K_F (mg/g(L/mg)$^{1/n}$)	1/n	R^2
Ni(II)	17.6236	0.03196	0.728	3.1170	0.2907	0.830

4. Conclusion

In this paper, the sorption of Ni(II) on a natural Slovak zeolite was investigated. The chemical composition of tested sorbent was characterized by XRF spectroscopy. The maximum removal efficiency (almost 70%) was achieved in solution with concentration of nickel 50 mg/L. Change of the pH can be caused ion-exchange process. Results from experiment fit to the Freundlich isotherm, which means that the sorption happens on heterogeneous surface. It can be concluded that zeolite is promising for the removal of metal ions from aqueous solution but had relatively lower efficiency in higher concentration. To improve the efficiency of sorption, next study will be aimed to modification of zeolite.

5. References

[1] Singovszka E, Balintova M, Demcak S and Pavlikova P 2017 Metal pollution indices of bottom sediment and surface water affected by acid mine drainage *Metals* **7** pp 1-11
[2] Can Ö, Balköse D and Ülkü S 2010 Batch and column studies on heavy metal removal using a local zeolitic tuff *Desalination* **259** pp17-21
[3] Spurthi L, T Brahmaiah, K S Sai Prasad, Chandrika K, Kausalya Chandra L, and Yashas S 2015 Removal of nickel from wastewater using low cost adsorbent *European J. Biomed. Pharm. Sci.* **2** pp 188-192
[4] Dhokpande S R, Kaware J P and Kulkarni S J 2013 Research for removal of nickel from wastewater–A review *International Journal of Science, Engineering and Technology Research* **2** pp 2162-2166
[5] El-Azim H A and Mourad F A 2018 Removal of Heavy Metals Cd (II), Fe (III) and Ni (II), from aqueous solutions by natural (clinoptilolite) zeolites and application to industrial wastewater *Asian J. Environ. Ecol.***7** pp 1-13
[6] Oter O and Akcay H 2007 Use of natural clinoptilolite to improve water quality: sorption and selectivity studies of lead (II), copper (II), zinc (II), and nickel (II) *Water Environ. Res.***79** pp 329-335
[7] Irannajad M and Haghighi H K 2017 Removal of Co2+, Ni2+, and Pb2+ by manganese oxide-coated zeolite: equilibrium, thermodynamics, and kinetics studies *Clays Clay. Miner.* **65** pp 52-62
[8] Holub M and Balintova M 2014 Using of zeolite for copper and zinc removal under acidic conditions *Pollack Period.* **9** pp 141-149
[9] Holub M, Balintova M, Demcak S and Hurakova M 2016 Characterization of natural zeolite and determination its adsorption properties *Journal of Civil Engineering, Environment and Architecture* **63** pp 113-122
[10] Samarghandi M R, Al-Musawi T J, Mohseni-Bandpi A and Zarrabi M 2015 Adsorption of cephalexin from aqueous solution using natural zeolite and zeolite coated with manganese oxide nanoparticles *J. Mol. Liq.* **211** pp 431-441
[11] Yulianis Y, Muhammad S, Pontas K, Mariana M and Mahidin M 2018 Characterization and Activation of Indonesian Natural Zeolite from Southwest Aceh District-Aceh Province *IOP Conf. Ser.: Mater. Sci. Eng* **358** pp 1-7
[12] Desta M B 2013 Batch sorption experiments: Langmuir and Freundlich isotherm studies for the adsorption of textile metal ions onto teff straw (Eragrostis tef) agricultural waste *J. Thermodyn.* **2013** pp 1-6
[13] Ehrampoush M Ghanizadeh and Ghaneian M 2011 Equilibrium and kinetics study of reactive red 123 dye removal from aqueous solution by adsorption on eggshell *Iran. J. Environ. Health Sci. Eng.* **8** pp 101-106
[14] Langmuir I 1916 The constitution and fundamental properties of solids and liquids. Part I. Solids *J. Am. Chem. Soc.* **38** 2221-2295
[15] Freundlich H M F 1906 Over the adsorption in solution *J. Phys. Chem.* **57** pp 1100-1107
[16] Bilgin A and Sanin S L 2014 Remediation of endosulfan using zeolite biobarrier *Fresen. Environ. Bull.* **23** pp 3272-3277
[17] Wirawan S K, Sudibyo H, Setiaji M F, Warmad I W and Wahyuni E T 2015 Development of natural zeolites adsorbent: chemical analysis and preliminary TPD adsorption study *J. Eng. Sci. Technol.* **1** pp 87-95
[18] Ayawei N, Ebelegi A N and Wankasi D 2017 Modelling and interpretation of adsorption isotherms *J.Chem.* **2017** pp 1-11

Acknowledgment
This work has been supported by the Slovak Grant Agency for Science (Grant No. 1/0419/19).

IOP Conf. Series: Materials Science and Engineering **867** (2020) 012024 doi:10.1088/1757-899X/867/1/012024

Environmental assessment of existing pavements asphalt layers for production of new asphalt mixtures

R Kovács[1] and J Mandula[1]

[1] Department of Geotechnics and Traffic Engineering, Technical University of Košice, Vysokoškolská 4, Košice, Slovakia

Abstract. Road construction requires a large amount of construction materials. During asphalt pavement construction it is natural aggregate and asphalt binder in particular. Their ratio varies depending on whether they are applied in an asphalt pavement as a wearing, binder or base course. Nowadays, pressure of public opinion on responsible environmental behaviour and effective use of financial resources create an optimal space for assessment of building materials from pavement layers. According to the EAPA (European Asphalt Pavement Association) data from 2017 the wearing courses accounted for 90 % of all pavement layers in European Union. In Slovakia their representation is also dominant. While in Western Europe reassessment of existing asphalt layers by recycling technologies has been used to the maximum, in Slovakia the reassessment was out of interest of construction companies. With the arrival of new machinery, legislation amendment and social pressure on responsible environmental behaviour, the strategy of some construction companies in Slovakia is also changing. The reassessment of existing asphalt layers for production of new asphalt mixtures goes to the forefront of the interest. This article summarizes necessary steps to achieve an increase in the re-use rate of R-material by hot-mix asphalt recycling technology in asphalt mixing plants.

1. Introduction

Since 2006, when EN 13108-8 came into force, the hot-mix asphalt recycling technology in asphalt mixing plants in Slovakia has begun to develop. After the initial enthusiasm of new technology introduction, its use has been minimalized over time. Crucial influence, apart from economic efficiency, was the difficulty of extracting materials from an asphalt pavement. Demolition, loading and transport to a reprocessing site using inefficient mechanisms was complicated and largely ineffective. Asphalt mixing plants were equipped only with cold batching technology, which limits them in adding a higher percentage of R-material in the production of new asphalt mixtures. In principle, they copied the requirement of TKP 6 for batching of R-material in quantities 10% for wearing courses and 20% for binder and base courses.

In Western Europe, the hot-mix asphalt recycling technology in asphalt mixing plants has progressed by big stride. For instance, in Germany the use of R-material achieves 85-90% of reprocessing in pavements. Other Western countries in Europe like Switzerland or France are in similar position. The Czech Republic has also made a great progress in reprocessing of R-material in new asphalt mixtures. By Decree 130/2019 the Ministry of the Environment set environmental criteria, after which achievement a milled asphalt mixture become a by-product. Technical limitation of the highest acceptable amount of R-material is set by ČSN 73 6121/2019.

IOP Conf. Series: Materials Science and Engineering **867** (2020) 012024 doi:10.1088/1757-899X/867/1/012024

2. Current use of R-material in Slovakia

The legislation in Slovakia, Act no. 79/2015 on Waste and on amendment to certain acts in § 2, paragraph 2 states „Waste is not“ according to letter c) waste which go through a preparation process for re-use and meets requirements for the launched product set by special separation. Technical specification limiting the acceptable amount of R-material do not dictate them.

Recently, the Government of the Slovak Republic has committed itself to support circular economy in order to reduce negative environmental aspects in relation to natural resources, but also to increase economic efficiency of spending financial resources. Appropriate management with R-material can help to achievement both of these goals. It requires a change in approach and an awareness of the potential in using a high amount of re-used R-material. Current approach to obtaining high-quality R-material is practically impossible. The way, when the individual courses of a pavement are milled to the maximum thickness, damage high-quality wearing course, e.g. SMA type or AC type containing modified asphalt binder. The sorting and storage of the milled asphalt layer has also large reserves – figure 1.

Figure 1. The current way of R-material storage.

Figure 2. Cold batching technology.

Figure 3. From hopper to elevator.

The key to the successful application of this challenge is an investment to technological equipment of asphalt mixing plants and change the way R-material is obtained and managed.

The main barrier to increasing a ratio of batching is the cold batching technology of R-material in production process of a new asphalt mixture – figure 2 and 3.

3. Conditions for using a higher ratio of R-material in production of a new asphalt mixtures

To meet the goal of maximum efficiency in the process of re-use existing asphalt layers, it is through the management with R-material. As a good manager, it is necessary to plan entire process of assessment of existing pavements asphalt layers for production of new asphalt mixtures, which include the following:

- selection an appropriate compact section
- diagnostic research of individual layers
- milling in layers
- sorting and storage by fractions
- design of compositions with R-material
- production, laying and compaction of asphalt mixtures with R-material.

3.1. Selection an appropriate compact section

The final quality of obtained R-material depends on this decision. In ideal cases, an administrator of pavement is able to provide relevant data about age, thickness and identification built-in materials in a pavement. If these inputs are missing or they are insufficient, in these cases he relies on a result from a diagnostic research.

3.2. Diagnostic research of individual layers

This operation has to be made in any case. The difference in delivery of input data from an administrator of pavement can only be at the individual steps (distances) of diagnostic well compared to the condition, when relevant input data from an administrator of pavement are missing. Very important element is a visual inspection of a selected section. This operation is mainly about a research of a pavement surface and determination of an overall section homogeneity. If repairs are identified in the section (e.g. filled potholes), they inevitably reduce a quality of a milled material.

3.3. Milling in layers

It is essence of effective use of a milled material. Based on the results of the diagnostic research, the composition and thickness of individual layers is known. The result of the milling in layers is effective obtained material with maximum added value.

a) Example of the right practice – milling in layers:

1. a layer in thickness 40 mm is diagnosed on AC 11 wearing course containing 5,3% of PMB binder,
2. a layer in thickness 60 mm is diagnosed on AC 22 binder course containing 4% of asphalt binder.

The fraction of R-material thus obtained can be re-used with minimal addition of asphalt binder in layers.

b) Example of the incorrect practice – milling in single layer:
a layer in thickness 100 mm with mixed granulometry and approximate content of mixed binder 4,5%.

The fraction of R-material thus obtained is already considerably degraded and cannot be re-used as wearing course with a higher batching.

3.4. Sorting and storage by fractions

It is the next step to ensure a high-quality R-material. Under the proviso that the previous step of the milling in layers was strictly observed, it is not necessary to sort a milled material. Storage of milled material under a roofed landfill is a necessary condition its effective use – figure 4. Subsequently, sampling is performed and based on the results of prescribed laboratory tests the classification of R-material according to the STN EN 13108-8 take place. Examples of R-material label:

32 RA 0/16 or 22 RA 0/11

Note: 32 RA 0/16 is R-material whose largest aggregate size is 16 mm and pieces of asphalt mixture have a maximum size 32 mm. The material thus classified and certified is prepared for re-use.

Figure 4. Storage of classified R-material.

3.5. Design of compositions with R-material

STN EN 13108-8 defines R-material as a component that can be used in asphalt mixtures manufactured in asphalt mixing plants in accordance with the specifications for these mixtures. The requirements for asphalt mixtures are the same for asphalt mixtures with R-material or without R-material and therefore the level of material homogeneity is determined by the maximum amount of R-material that can be used. When design asphalt mixtures with a ratio of R-material higher than 20%, its properties have to be taken into the account. In general, the mentioned material contains an aged asphalt binder with which an asphalt mixture has stiffness, but it is prone to formation of fatigue and low-temperature failures. Therefore, the material has to be adjusted. There are two options. One of them is that a rejuvenator is added to the R-material to achieve a recovery of aged asphalt binder or the second one is the use a soft asphalt binder. Recent researches focused in this field suggests more stable results with using soft asphalt binders with which is possible from an aged asphalt binder make a stable material. With rejuvenators similar results can be obtained. However, they are very prone for exact batching and their lifespan is questionable.

Worldwide, there are cases for use of 100% R-material only with the help of rejuvenators. Last year, in the Czech Republic the norm for specification the maximum ratio of using R-material in manufactured asphalt mixtures was published. In table 1, there is a possible proposal of using R-material in Slovakia.

IOP Conf. Series: Materials Science and Engineering **867** (2020) 012024 doi:10.1088/1757-899X/867/1/012024

Table 1. Proposed batching of classified R-material in percentage by weight of asphalt mixture.

Wearing courses		Binder courses		Base courses	
AC 8 II	25	AC 16 I	30	AC 16 I	50
AC 11 I	15	AC 22 II	40	AC 22 II	60
AC 16 II	20				

The proposed batching of R-material takes into the account a quality criterion for asphalt mixtures according to the TKP 6. Asphalt mixtures of quality class I are used to construction pavements asphalt layers with traffic load classes I.-VI and asphalt mixtures of quality class II are used to construction pavements asphalt layers with traffic load classes IV.-VI.

3.6. Production, laying and compaction of asphalt mixtures with R-material

The current production facilities in Slovakia are not able batching R-material with ratio higher than 20%. In order to achieve this, it is necessary to invest in parallel drying drum technology – figure 5.

Figure 5. Asphalt mixing plants equipped with parallel drying drum technology.

Laying and compaction of asphalt mixtures with R-material does not require modification of laying and compaction technology. The attention should be taken only when the amount of used R-material is higher than 50% depending on whether rejuvenators or only soft asphalt binders have been added to the mixtures.

IOP Conf. Series: Materials Science and Engineering **867** (2020) 012024 doi:10.1088/1757-899X/867/1/012024

4. Conclusion

In respect to effective pavement management in Slovakia it is time to increase the ratio of recyclation of construction layers from existing pavements by hot-mix asphalt technology in asphalt mixing plants. By responsible behaviour towards to natural resources, which also include aggregates, will also bring an economic return on the mentioned R-material batching in increased quantity. This model is not experimental, it has been successfully working in Western Europe for years and it has been introduced in the Czech Republic in recent years. This article should be an impulse for effective processing of R-material which is a valuable feedstock for the production of new asphalt mixtures and also motion for construction companies that operate in Slovakia.

5. References

[1] TP 043: 2011 Recyklácia asfaltových zmesí za horúca v obaľovacích súpravách, MDVRR SR
[2] TKP 6: 2019 Hutnené asfaltové zmesi, MDV SR
[3] STN EN 13108-8: 2020 Asfaltové zmesi. Požiadavky na materiály. Časť 8: R-materiál
[4] ČSN 73 6121: 2019 Stavba vozovek. Hutněné asfaltové vrstvy. Provádění a kontrola shody
[5] Rubio MC, Martínez G, Baena L a Moreno F 2012 Warm mix asphalt: an overview *J. Clean. Prod.* **24** 76-84
[6] Woszuk A 2018 Application of fly ash derived zeolites in warm-mix asphalt technology *Materials* **11** 1542
[7] Lagos-Varas M, Movilla-Quesada D, Arenas J P, Raposeiras A C, Castro-Fresno D, Calzada-pérez M A, Vega-Zamanillo A a Maturana J 2019 Study of the mechanical behavior of asphalt mixtures using fractional rheology to model their viscoelasticity *Constr. Build. Mater.* **200** 124-34
[8] Estaji M, Coleri E, Harvey J T a Butt A A 2019 Predicting excess vehicle fuel use due to pavement structural response using field test results and finite element modeling *Int. J. Pavement Eng.* **200** 8436
[9] Singh H, Chopra T, Jain S, Kaur A a Kamotra S 2019 Effect of aggregate type and polymer modification on the performance of bituminous concrete mixes *Int. J. Appl. Sci. Eng.* **16** 1-13

IOP Conf. Series: Materials Science and Engineering **867** (2020) 012025 doi:10.1088/1757-899X/867/1/012025

3D printers as part of Construction 4.0 with a focus on transport constructions

P Krupík[1]

[1] CTU Prague, Faculty of Civil Engineering, Department of Construction Management and Economics, Thakurova 7/2077, 166 29 Prague 6 Dejvice, Czech Republic

Abstract. Digital and technological innovations that have grown rapidly in recent years have been referred to as the 4th Industrial Revolution or Industry 4.0. The paper comprehensively deals with the description of the 3D printers as part of Construction 4.0 with a focus on transport construction. The aim of the article is description of this issue comprehensively and with the help of SWOT analysis identify weaknesses, strengths, opportunities and threats with focus on the transport constructions. The content is complemented by an assessment of the current situation, indicating the possibility of use in practice.

1. Introduction

Autonomous self-guiding and self-learning systems are embarking on a number of activities. This process seems to be taking on the rapid acceleration. Dynamics these changes and their importance in shaping of the whole company was named 4th industrial revolution or Industry 4.0.

In the 21st century, there are four major changes in construction:

- Technology is the driving force. Virtual and augmented reality, 3D printers, advanced building materials, prefabricated and modular designs, Big data and predictive analytics, 3D scanning and photogrammetry, autonomous vehicles and drones, cloud collaboration (real-time collaboration) or wireless monitoring and interconnection devices, thanks to mass expansion they become cheaper and affordable.
- The needs, requirements and attitudes of end users have changed drastically
- Construction focus is also shifting towards mass adaptation, particularly in the urban area, on the one hand; and mass production at a very high speed in the countryside (for housing and other infrastructure).
- Due to climate change, particular emphasis is placed on sustainability and green construction, which has not been a priority in the past [3].

These changes increase the demand for construction, cost pressure and cost-effectiveness not only of the construction itself, but of the whole life cycle of buildings with regard to sustainability. The construction industry is beginning to run into the limits of the technologies used so far and the methods of planning and managing buildings. Figure 1 shows the need for construction 4.0, where construction productivity in EU is stagnating compared to most other sectors.

The construction industry thus requires a fundamental change, which seems to be the transition to the fourth industrial revolution, which began after 2010, building on the use of so-called cyber-physical systems and the digitization of the entire product life cycle. A necessary condition for the transition of the construction industry to such a system seems to be digitization of the entire life cycle of buildings, i.e. design, implementation and operation [5].

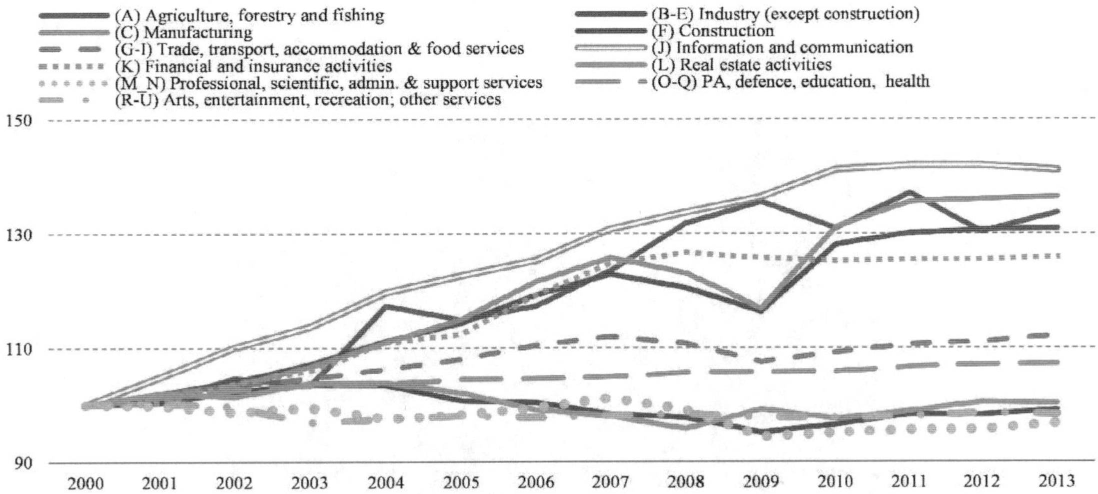

Figure 1. Evolution of labour productivity for the EU-28 (2000=100) [4].

2. 3D printers as a digitized production method

3D printers or 3D printing as manufacturing technology was raised in the late 1980s. Connection with construction industry was established in 1997 [1]. 3D printing, also known as additive manufacturing, is process based on a three-dimensional digital model, using automatic technology to create a physical objects layer-by-layer without human intervention [2].

The materials currently used in 3D printing are mostly either Portland cement-based [6] or geopolymer-based [7]. The possibilities are wide. Most materials are in the research phase. Thus, very diverse materials such as resin mortar or clay are used [8]. Another example of material is stainless steel [10].

2.1. Examples of using 3D printers with a focus on transport construction

Below are examples of how to use 3D printers. The possibilities are very wide, so it is only a cross-section of the most important ones.

2.1.1. Stainless steel bridge. The Dutch company MX3D developed the technology for 3D printing from metal and in 2018 printed stainless steel in Amsterdam with this patented technology a fully functional footbridge [10].

Figure 2. 3D printing a fully functional stainless steel bridge [10].

2.1.2. Concrete bridge. The Marines used an ACES (Automated Construction of Expeditionary Structures) printer to print a 3D footbridge that can carry the intended load. The printer uses concrete instead of paste or mortar. With this printer it is also possible to create other types of structures such as obstacles, protective structures, arches, culverts and spokes [11].

Figure 3. 3D printed concrete bridge [11].

3. SWOT analysis

Possibilities and possible problems of 3D printers or 3D printing in relation to swot analysis have been discussed with fourteen other researchers and experts. The following information was established on the basis of the questioning. In addition, an extensive research of previously published articles was also carried out.

3.1. Weaknesses/ Threats

- Possible errors in digital model creation.
- Inappropriate material.
- Inconsistency of possible use in technical standards and legal regulations / regulations.
- Lower production speed.
- Poor surface quality - need for additional treatment.
- Different mechanical properties due to material layering.
- For tall buildings up to tens of meters or even hundreds of meters, first of all there must be a printer that matches the height of the building, which means the higher the building, the higher the cost [8].
- High price of new technology (price of 3D printers and printing material) compared to established procedures including convective construction equipment and standard organization of human labour [9].

3.2. Strengths/ Opportunities

- New possibilities of design both practical and aesthetic. It is possible, for example, to design noise barriers which, due to the materials and profiles used, have a better ability to absorb or withstand sound. The advantages of 3D printing technology are also detailed construction accuracy. But that depends a lot on the 3D printing technology used.
- 3D printing technology facilitates the production of curved concrete components, allowing construction companies to create better structural sound structures. Unlike conventionally manufactured components, 3D components can be made from hollow centres, which is very practical, for example, when installing electrical wiring.
- This technology also produces less waste than conventional technologies. For example, lost, pruned and increased consumption due to the way the material is used.

- Printing building components requires less human work, saving not only time but, above all, the cost of the work itself. This is associated with another indisputable advantage, which is the fact that this technology reduces the risk of injury to construction site workers.
- 3D printing also has great potential for the use of environmentally friendly materials, such as bio-plastics. It is also possible to combine different types of materials.
- It is possible to print mechanically connected parts. It is not necessary to connect them directly on site.

4. Conclusion

Currently, 3D printing technology is in the exploratory research and development phase. It will take some time to overcome the current shortcomings and obstacles and become a commonly used means for use in construction. As such, however, it should play an important role in Construction 4.0.
The ideal picture for the future is shown in the following figure. However, this is still a long way off, even though research in this area is moving forward by leaps and bounds.

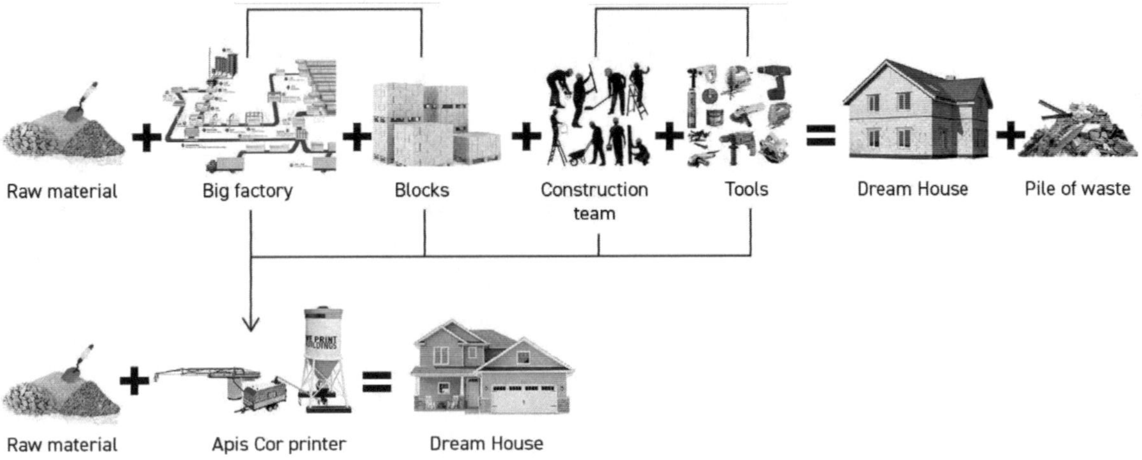

Figure 4. Future of 3D print [12].

5. References

[1] Pegna J 1997 Exploratory investigation of solid freeform construction *Automation in Construction* **5(5)** 427-437

[2] De Schutter G, Lesage K, Mechtcherine V, Nerella V N, Habert G and a Agusti-Juan I 2018 Vision of 3D printing with concrete — Technical, economic and environmental potentials *Cement and Concrete Research* **112** 25-36

[3] Subhash R Construction 4.0: The 4th Generation Revolution Online: https://www.researchgate.net/publication/331131645_CONSTRUCTION_40_THE_4_th_GENERATION_REVOLUTION

[4] European commission 2015 Report on Single Market Integration and Competitiveness in the EU and its Member States: Brussels Online: http://publications.europa.eu/resource/cellar/93a2b2cc-7d5a-11e5-b8b7-01aa75ed71a1.0001.03/DOC_2

[5] Bock T 2015 Construction Robotics Enabling Innovative Disruption and Social Supportability *Proceedings of the 32nd ISARC* (Oulu, Finland)

[6] Marchment T, Sanjayan J and Ming X 2019 Method of enhancing interlayer bond strength in construction scale 3D printing with mortar by effective bond area amplification *Materials & Design* **169** 107684

[7] Ming X and Sanjayan J 2016 Method of formulating geopolymer for 3D printing for construction applications *Materials & Design* **110** 382-390

[8] Gaojie Liu *et al* 2019 *IOP Conf. Ser.: Earth Environ. Sci.* **267** 032014

[9] P I Ospishchev 2019 *IOP Conf. Ser.: Mater. Sci. Eng.* **698** 066009
[10] MX3D Bridge 2020 Online: https://mx3d.com/projects/mx3d-bridge/
[11] Wagman D 2019 Marines 3D print a bridge in the field Online: https://insights.globalspec.com/article/11128/marines-3d-print-a-bridge-in-the-field
[12] Cheniuntai N 2018 What is construction 3D printing: perspectives and challenges Online: https://medium.com/@Nik_chen/what-is-construction-3d-printing-perspectives-and-challanges-5b57170c2a29

Acknowledgments
This work was supported by the Grant Agency of the Czech Technical University in Prague, grants No. SGS19/010/OHK1/1T/11 and No. SGS20/100/OHK1/2T/11.

IOP Conf. Series: Materials Science and Engineering **867** (2020) 012026 doi:10.1088/1757-899X/867/1/012026

Evaluation of microwave drying effects on historical brickwork and modern building materials

V Kvapilová[1]

[1] Technical university of Brno, Faculty of Civil Engineering, Czech Republic, e-mail: kvapilova.projekce@gmail.com

Abstract. One of the most common reasons for any construction work requirements is an increased humidity of the building. Increased humidity does not affect only buildings with basement, but also buildings without it, which were not properly hydro-isolated or the hydro isolation was damaged due to aging or to adverse environmental effects. Such situations are solved with designing suitable remediation measures, which necessarily includes drying of damp masonry as well. This almost always includes historical buildings, where protection against groundwater was never planned. Such buildings are usually very sensitive to any physical intervention into construction itself and destructive methods of revitalization are very often completely out of the question, therefore it is necessary to develop methods, which can be used particularly for historical buildings. This article focuses on evaluating effects of microwave drying historical brickwork as well as modern building materials. First, the method of microwave drying is described in general, then we cover the experiment, which was performed at the Faculty of Civil Engineering at Brno University of Technology. The experiment was applied on two pillars made of historical brickwork. Those samples were being thoroughly analysed during the drying process and final outcomes were compared to already existing experiments on modern materials. The goal of this work is to find out, if the method of microwave drying is more suitable for described conditions and also to set such process of drying, which would provide the highest effectivity.

1. Introduction

In the long-term high humidity of building materials, which, in addition, degrades building materials and structures. Excessive humidity can cause volume changes as well as deterioration of mechanical and physical properties of building materials and structures. Other negative effects include a deterioration in the thermal properties of the objects. There are currently several ways to artificially reduce excess moisture in building structures. One of these options is microwave drying, which so far seems to be a relatively quick and cheap solution. The knowledge gained so far in the field of microwave drying will now be applied and developed for research on historical objects. With excessive dampness constructions and building materials can change in volume and their mechanical and physical properties can worsen. Among other negative effects is degradation of thermo-technical properties of given objects. Currently we know more options how to artificially reduce content of excessive dampness in building structures. One of these options is microwave desiccation, which currently appears as relatively fast and inexpensive solution. So far collected knowledge in the area of microwave desiccation will be now applied on and evolved in the research focused on historical buildings.

2. Overview of the problematic

Followed by a theoretical reason for the issue of increased humidity of building structures and its reduction by microwave drying. Following is theoretical introduction of the problematic of excessive dampness of building structures and its reduction using microwave desiccation.

2.1. Water and moisture in building structures

In every building material there is a certain amount of free water, which we call equilibrium moisture content. Moisture enters the building materials already during their production, then it is practically incorporated into the building structures and is created by usage of already completed buildings as well. Each building has to be evaluated separately in terms of moisture, as many factors influence the final moisture of the structure. For historical buildings, it is especially important to evaluate all the input parameters and to suggest an optimal solution [1].

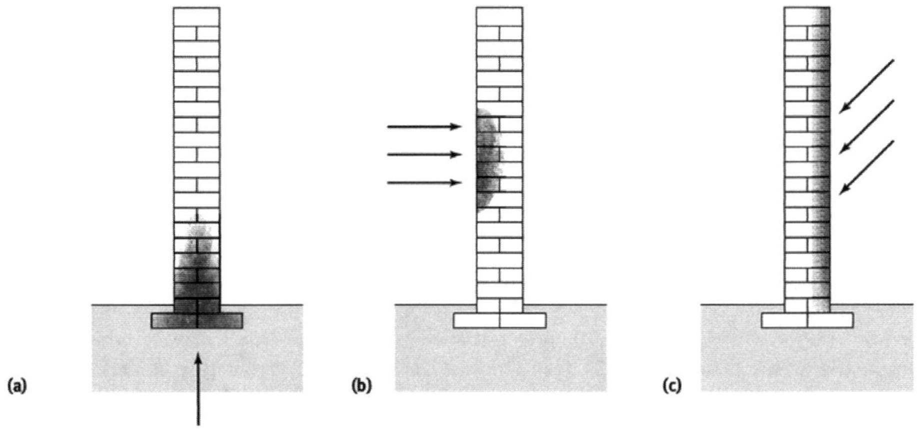

Figure 1. Water distribution in masonry depending on the moisture.

The amount of water in the porous material is expressed as volume or mass moisture. The definitions are based on the names given above, where the density is determined as the volume of water to the solid of the sample ratio, as shown in Equation (1), and mass moisture is defined as the weight of water to the solid of the sample ratio (2).

$$w_v = \frac{V_w}{V_d} \times 100, \ [\%] \tag{1}$$

where V_w is volume of free water in m^3 and V_d is volume of dry material in m^3.

$$w_h = \frac{m_w - m_d}{m_d} \times 100, \ [\%] \tag{2}$$

where m_w is weight of wet material in kg and m_d is weight of dry material in kg.

2.2. Electromagnetic microwave radiation

Electromagnetic microwave radiation is in fact electromagnetic waves with a frequency in the range from 300 MHz to 300 GHz. In free space microwave radiation spreads from the source in the form of waves at the speed of light. The most often used frequency in construction is 2450 MHz [2].

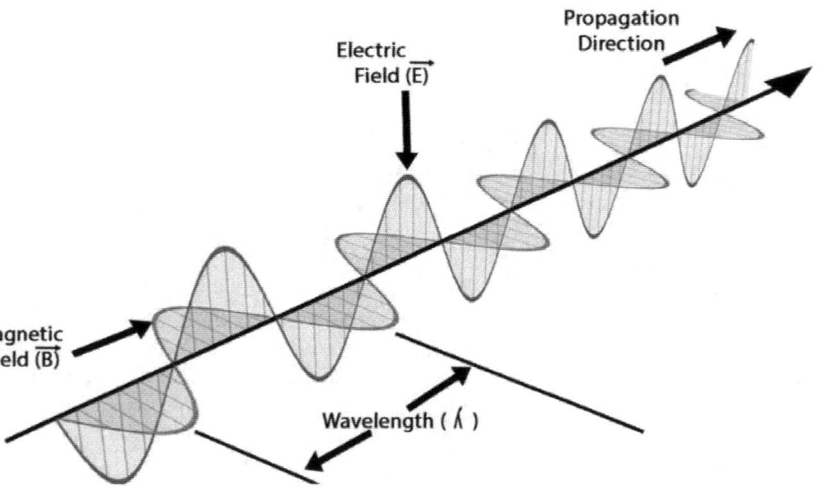

Figure 2. Electromagnetic waves.

2.3. Drying-out of building structures using microwave radiation

The use of microwave masonry desiccation usually requires synergy with other measures, since remediation is generally the summary of activities and measures that eliminate existing moisture, and also prevent the penetration of additional moisture. The principle of microwave drying-out is the ability to vibrate water molecules in the masonry mass and the subsequent generation of heat, which is created by their movement and friction, causes their conversion into water vapour. The water vapour thus converted is then only vented from the structure. The water temperature is higher than the temperature of the building material itself and it has been found that the heating increases from the centre of the material. The heated area then spreads so the moisture already removed allows the radiation to extend its effectiveness [3], [4].

3. Experimental measurement

This is one of the experiments carried out at the Faculty of Civil Engineering in Brno. The aim was to obtain input information on the reaction of historical masonry to microwave drying-out and to compare the results of measurements with already performed experiments on selected building materials.

3.1. Description of the experiment

The experiment was carried out on two sets of three pillars made of historical brickwork. These pillars were made of a brick tripod with age reaching 350 years. Walling up of the individual pillars was done on cement lime mortar. The samples were then left for 30 days in constant laboratory conditions. The first set of samples was prepared for drying-out using a microwave rod antenna by drilling holes in the pillars to fit the antenna. The diameter of the drill holes was set to 24 mm. The holes were designed so that the dried area was evenly distributed over the width of the test pillar. The second set of samples was prepared for drying-out using a microwave funnel antenna, such desiccated pillars do not need any special treatment. Initially, the samples thus prepared were weighed on digital scales and the approximate equilibrium initial humidity of each given sample was determined as well. Afterwards, these samples were left for one week in a soaking tub with water, where the water level reached height of approximately 150 mm. In the next phase the samples were removed from the water and weighed individually, so their moisture was determined by gravimetric method.

3.2. Microwave drying process

A 650 W microwave drying device was used for this experiment. Desiccation of samples was done in cycles. The length of each cycle was 20 minutes, with 17 minutes of drying, and the remaining 3 minutes was a break between drying to avoid overheating of the samples. Individual samples were weighed

IOP Conf. Series: Materials Science and Engineering **867** (2020) 012026 doi:10.1088/1757-899X/867/1/012026

during these pauses. In total, 12 drying cycles were performed on each sample. During the experiment three basic characteristics were observed - the weight of individual pillars, the weight loss of moisture at the time of drying and the weight moisture - using the gravimetric method.

4. Results and discussion

The experiment was carried out on two sets of samples, each set consisting of 3 pillars, and from the individual results from each set the average of measured values was calculated for better results' clarity. The weight loss of individual samples at the time of desiccation is shown below table 1. A set of samples dried using a microwave device with a plug-in antenna has a number 1 and a set of samples numbered 2 represents piles dried by a microwave device with a funnel antenna.

Table 1. Table of average weight loss of individual samples at the time of desiccation.

Drying cycle	Marking of fittings	KC1	KC2
	Starting weight [g]	15626.0	15791.0
	Moistened pattern [g]	17752.5	17826.0
1	20	17635.0	17710.5
2	40	17503.5	17590.0
3	**60**	17366.5	17465.0
4	80	17232.5	17336.0
5	100	17101.5	17203.0
6	**120**	16971.5	17067.5
7	140	16843.0	16931.0
8	160	16716.5	16798.5
9	**180**	16591.5	16664.5
10	200	16466.0	16528.0
11	220	16341.0	16398.5
12	**240**	16218.0	16267.5

Referring to above table, following figure 3 is shown, which is a diagram of the average weight loss on each set of samples at the time of drying.

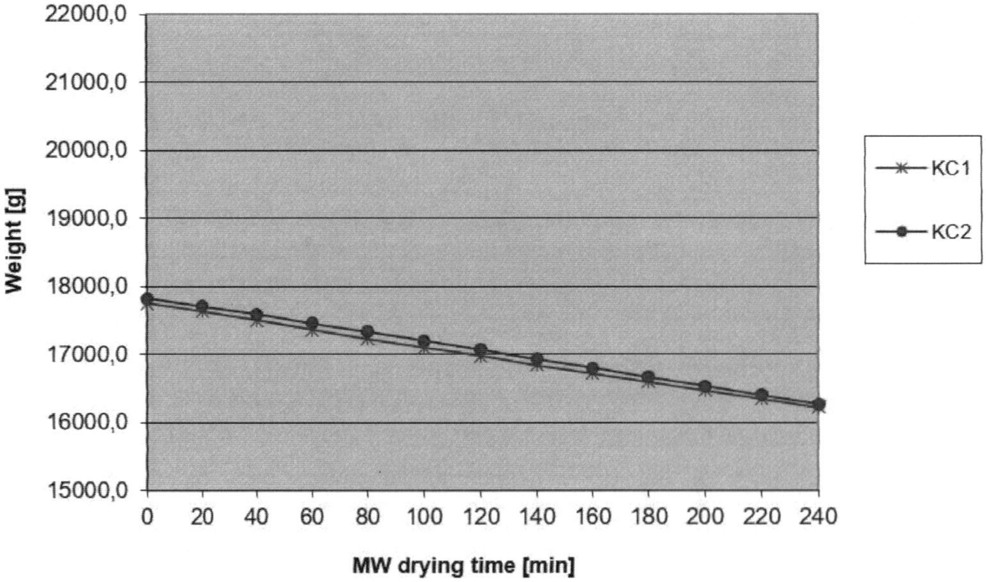

Figure 3. Graph of average weight loss of individual samples at the time of desiccation.

It can be seen from the figure 3 that the desiccation using the attachable funnel antenna, or the plug-in antenna was approximately the same, or to be exact - the moisture losses were comparable. We also monitored the weight loss of moisture at the time of drying on the individual samples and according to the type of drying device, that is using a plug-in antenna and a funnel antenna. This is shown in table 2.

Table 2. Table of average weight loss of moisture at the time of drying on the individual samples and according to the type of drying device.

Drying cycle	Marking of fittings Amount of moisture [g]	KC1 2126.5	KC2 2035.0
1	20	117.5	115.5
2	40	131.5	120.5
3	**60**	137.0	125.0
4	80	134.0	129.0
5	100	131.0	133.0
6	**120**	130.0	135.5
7	140	128.5	136.5
8	160	126.5	132.5
9	**180**	125.0	134.0
10	200	125.5	136.5
11	220	125.0	129.5
12	**240**	123.0	131.0
Total weight loss [g]		**1534.5**	**1558.5**
Residual moisture after MW drying [g]		592.0	476.5

Again, the above table 2 is supplemented with figure 4, which is a diagram of the weight loss of moisture in individual sets at the time of drying using two types of antennas.

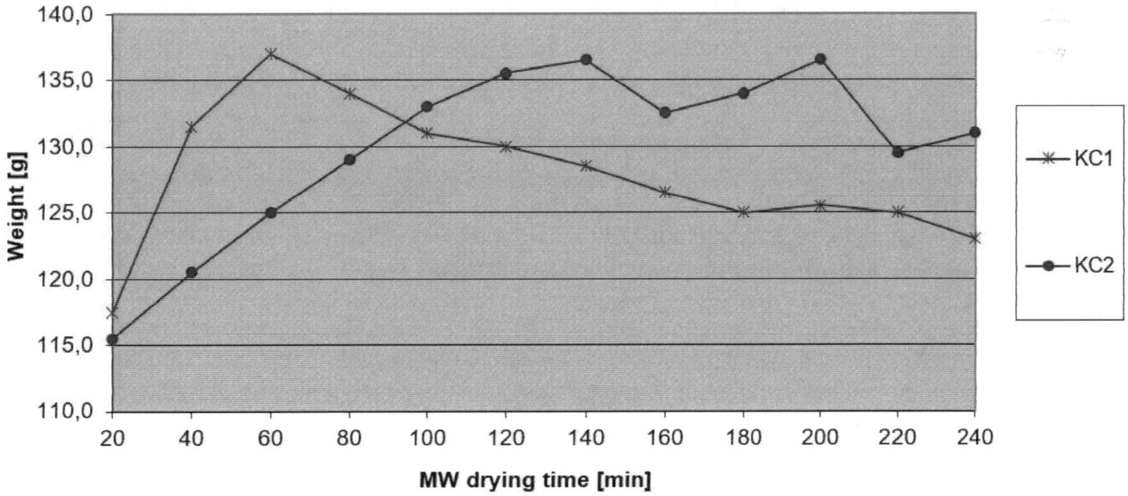

Figure 4. Graph of the weight loss of moisture in individual sets at the time of drying using two types of antennas.

Furthermore, the mass moisture of the samples was monitored, respectively its decrease during the drying time. The course of weight humidity decrease is shown in the following table 3. This table is also supplemented by the results of a similar experiment, which was performed on other types of masonry. These were samples of ceramic blocks of the Therm type with the KT designation and samples of air-entrained concrete blocks with the PT designation, while the numerical designation again indicates the type of drying device. This way it is possible to monitor the drying results using two different devices on different types of masonry with different structure and age.

Table 3. The course of weight humidity decrease.

Marking of fittings	KT1	KT2	PT1	PT2	KC1	KC2
Equilibrium humidity [%]	2.5	2.3	3.8	4.1	2.9	2.7
Moisture after dampening [%]	24.3	24.5	52.9	54.5	16.1	15.4
Moisture after drying [%]	11.7	12.5	25.1	24.9	5.8	5.2
Residual moisture [%]	12.6	11.8	27.8	29.6	10.3	10.2

5. Conclusion

It was found that the weight loss was almost the same when either desiccated with the plug-in rod antenna or the attachable funnel antenna. According to table 3, this is true for all types of masonry, but for historical brick pillars, the weight loss of moisture was almost identical for both types of drying antennas used.

Generally, it can be said that the weight loss when drying-out with the plug-in rod antenna starts to rise more slowly, up to a certain maximum value of the weight loss. From this limit on, the weight loss again decreases slowly until the value of the moisture weight loss value fluctuates around zero. The course of the moisture loss during desiccation using a funnel antenna shows a steeper increase in the weight loss values, that are even during the drying time approximately constant or fluctuate around same value. According to table 3, the efficiency of microwave drying was best reflected in samples of historical full burnt clay bricks, so it can be concluded that historical masonry reacts well to microwave desiccation and this experiment can be the foundation for further research.

6. References

[1] Solař J 2013 *Odstraňování vlhkosti: sanace vlhkého zdiva* (Praha: Grada Profi & hobby) p 104

[2] James Clerk Maxwell 2019 *Archiv historie matematiky MacTutor* [online] (Skotsko) [cit. 2019-10-13]. Available: https://www-history.mcs.st-andrews.ac.uk/Biographies/Maxwell.html

[3] Novotný M 2011 *Problematika využití mikrovlnného záření v pozemním stavitelství: Problems of utilization of microwave radiation in building constructions: thesis of a lecture on professorship appointment in the field of Building construction* (Brno: Vutium) p 29

[4] Alexa T 2010 *Odstraňování vlhkosti ve spojovací maltě mw ohřevem* (Brno: Faculty of civil engineering department of building structures) p 33

Acknowledgments

I would like to thank doc. Ing. Karel Šuhajda, Ph.D. for the expert assistance in the experiment and I am also grateful for being part of a specific research team. The research was created with the support of specific research FAST-J-19-5970.

Comparison of Mechanical and Adhesive Joints for Structural Glass – A review

M Lavko[1], V Kvočák[1]

[1] Technical university of Kosice, Faculty of Civil Engineering, Department of Steel and Timber Structures, Slovakia

Abstract. The presented paper compares types of connections mostly used for glass structures. Two groups are recognised and it is mechanical connections and adhesive joints. Bolted connections are mostly used as it is relatively simply to design and to provide, but to be able to create such a connections holes needs to be drilled in glass panes, which leads to discontinuities in geometry, weakening the glass panes and to creation of local stress peaks in the area where the bolt is inserted. On the other hand, adhesive joints don't lead to peak stresses as the stress is redistributed along larger area. Huge variety of accessible adhesives enable the joints to be used in almost each conditions. Main disadvantage of adhesives the change in behaviour over time, manly decrease in strength. Also, mechanical properties are influenced by UV radiation, humidity, temperature, etc. The most common adhesives used for structural glass and their performance is shown and discussed in the paper.

1. Introduction

As the dimensions of glass panes are considerably limited, it is necessary to combine the individual components to create large dimensional elements, often designed in modern architecture in order to achieve a safe transfer of loads to foundations and then to soil. Whether it is the attachment of glass facades to a supporting part, usually a steel structure, or the joining of individual segments of pillars or beams in large dimensional elements, it is necessary to carefully focus on detail design to avoid stress accumulation at these points leading to brittle fracture. For this reason, the requirements not only for the design but also for the construction of the joints are high. To prevent the creation of peak stresses plastic, resin, polyamide, and light metal materials [1] are placed in the contact area of glass and metals [5].

2. Mechanical connections

The most widespread types of glass joints are bolted joints. However, the architecture of nowadays is increasingly focusing on the minimal visual effect of connections, so research takes place in the development of glued joints that create a sense of uninterrupted construction. Thereby the attractiveness and impressiveness of such structures are increased. The most common joints of glass constructions are described in more detail in this chapter.

2.1. Supports of glass slabs

Linearly supported slab around its perimeter is one of the simplest supports for glass panes. In this type of support, glass is fitted into a frame made of aluminium, steel, plastics or timber. The disadvantage of this solution is the uneven distribution of stresses along with the support which must be taken into account in design [2]. Other types of panel support include local edge support, structural silicone seal and local point support [3].

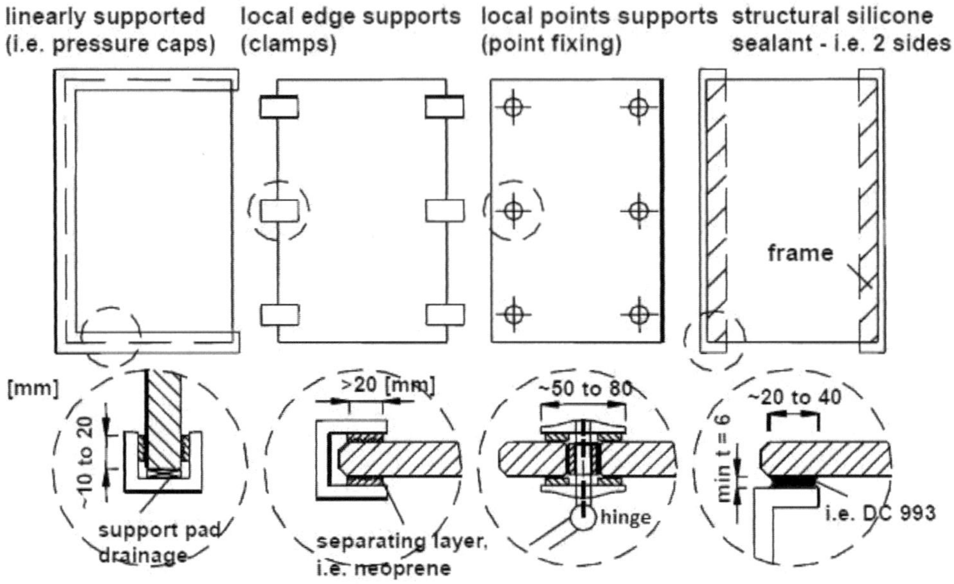

Figure 1. Types of supports of glass slabs [3].

2.2. Standard bolt connection – shear bolt

This is the simplest joint of the glass structures, where the bolt's head protrudes from the glazing plane. The load is transmitted from the glass to the screw through the inner flexible material. The contact surface is very small, therefore the load capacity is smaller, compared to other types of joints. Because the glass element is firmly attached to the support structure, this makes it almost impossible to move. Therefore, local peak stresses occur around the bolts [5].

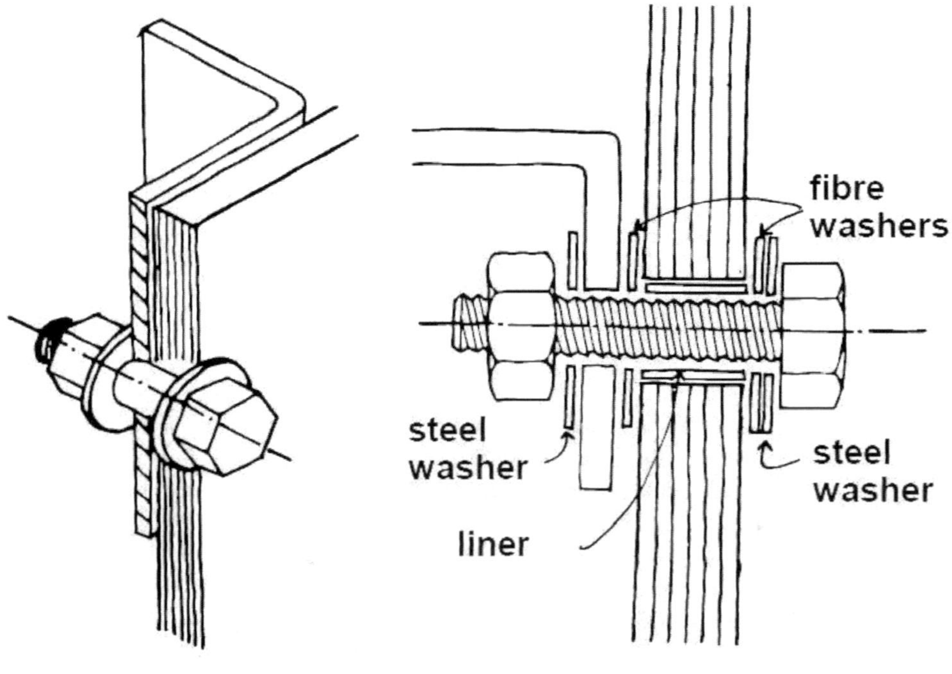

Figure 2. Standard bolt connection – shear bolt [3].

2.3. Simple countersunk-head bolt

This type of joint is similar to the previous one. The main difference is that the bolt's head is recessed in a plane with a glass plate. Also in this connection, local peak stresses are generated around the bolt. The described connection does not allow ratification between glass and steel.

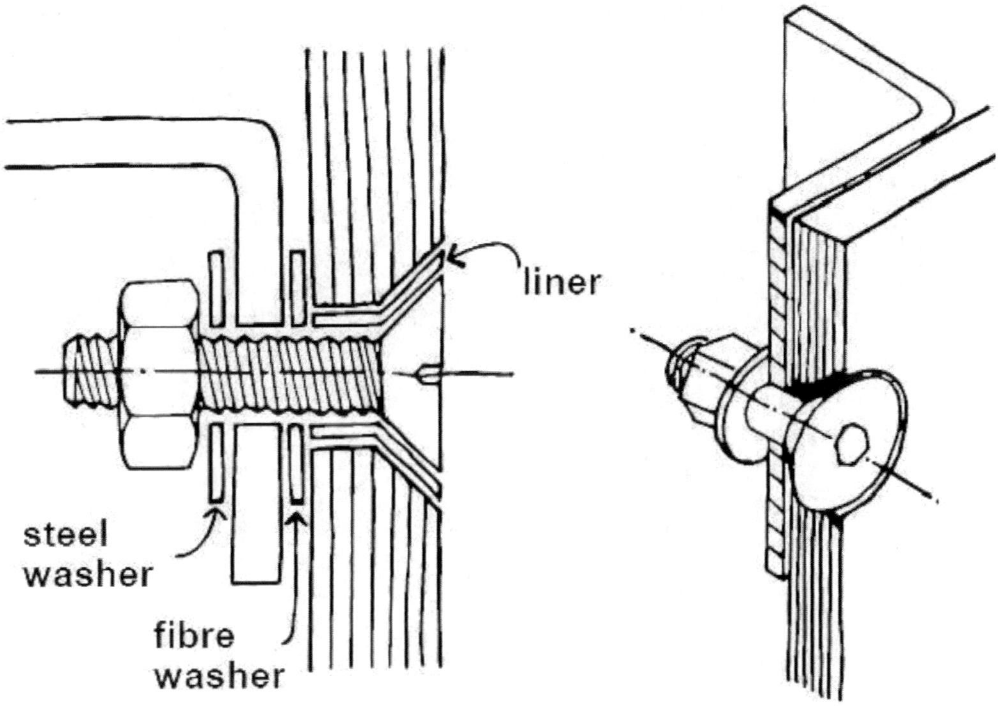

Figure 3. Simple countersunk-head bolt [3].

2.4. Bolted connection with steel splices

In this type of joint, steel splices are added on both sides of the glass pane, which are firmly connected by the bolts, so there is no mutual displacement between the glass and the steel plates. The opening for bolt may be larger than the diameter of the screw, which makes it possible to align the glass sheet when assembled. In practice, the joint is made using pre-tensioned screws tightened to the prescribed stress by a torque wrench. Load transfer occurs through friction between the steel and the glass, therefore the contact surfaces must be suitably adjusted and flexible material (plastics, neoprene, etc.) must be inserted between the glass and steel.

The latest research, however, points to the negative impact of the friction joints used in the laminated glass, where the interlayer is squeezed out at the joint site and consequently the mechanical properties of the laminated glass are changed [5].

Bearing bolts with steel splices are suitable only for tempered or heat-strengthened glass. In this type of connection, the load is transferred directly from glass pane through interlayer to bolt. Load bearing capacity depends on a number of bolts and their diameter. The scheme and behaviour of such connection are illustrated in the figure below [3].

IOP Conf. Series: Materials Science and Engineering **867** (2020) 012027 doi:10.1088/1757-899X/867/1/012027

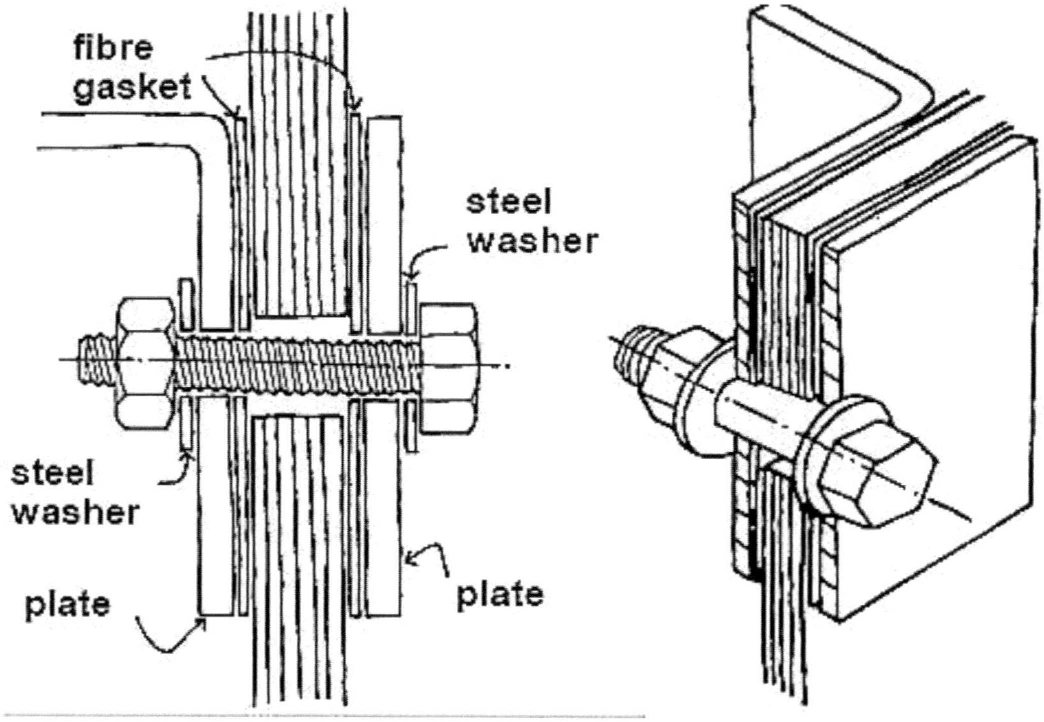

Figure 4. Bolted connection with steel splices [3].

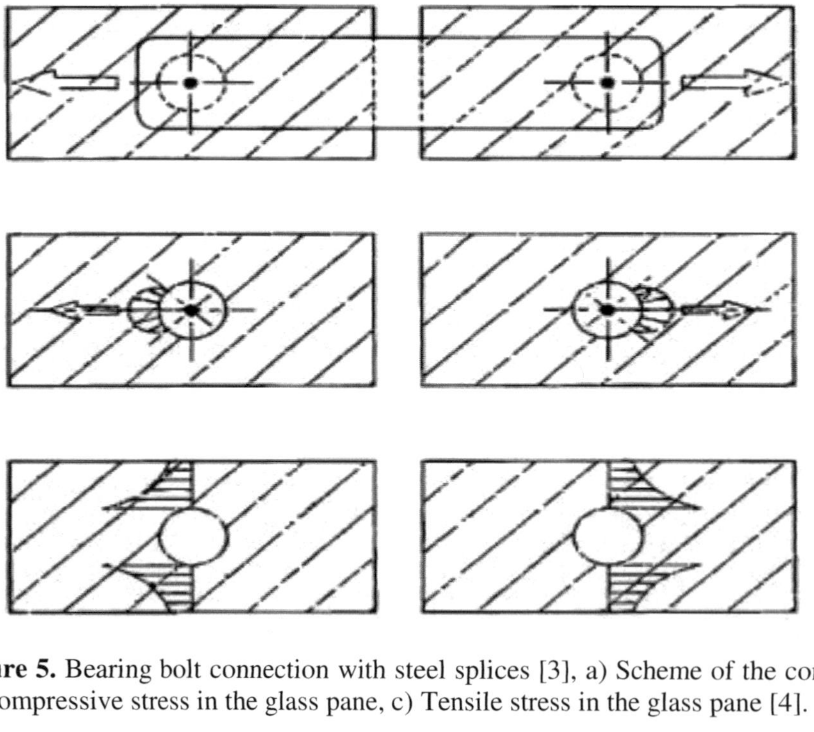

Figure 5. Bearing bolt connection with steel splices [3], a) Scheme of the connection, b) Compressive stress in the glass pane, c) Tensile stress in the glass pane [4].

2.5. Bolted connection with stud assembly

The main advantage of this type of connection is the transfer of vertical load by a stud that is part of the connecting steel plate. This stud is larger in size than the bolt, which leads to the higher pressed area and to lower stresses in the glass. The horizontal load is transferred by countersunk-head bolts. The diameter of holes for bolts is larger than the bolt's diameter to prevent the in-plane load from being transferred by them. The disadvantage of the joint is the need to drill more holes in close proximity [5].

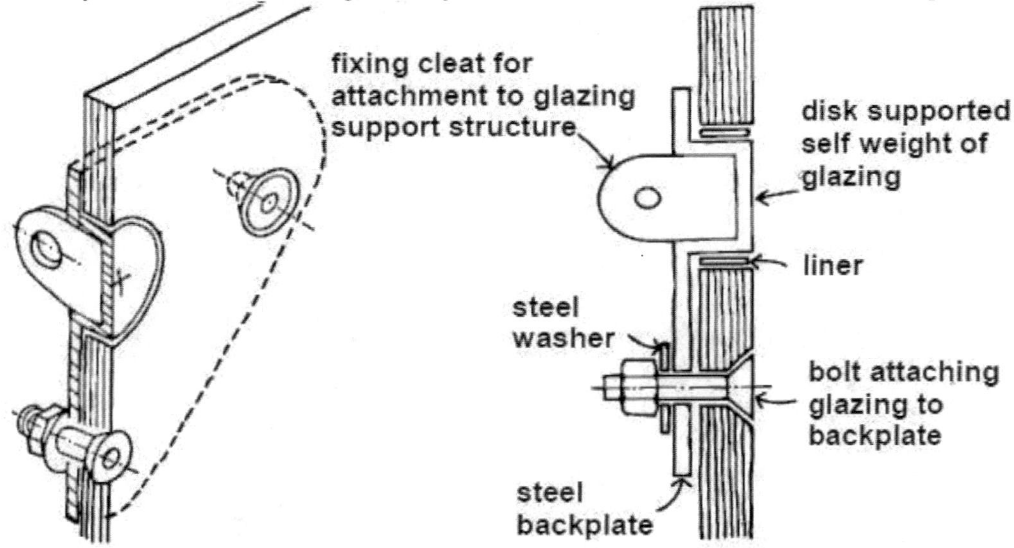

Figure 6. Bolted connection with stud assembly [3].

2.6. Spider connection

Spider joints are used for connecting the individual glass panes of facades to supporting structures. This joint may have one, two, three or four cantilevered arms that extend from the center. The steel part of the connection is attached to supporting structures. The advantage of this type of joint is the possibility of partial rotation of the glass pane using flexible profiles not only in the opening itself but also in brackets. This adjustment reduces the moment transmitted by the glass pane, resulting in a more economical design of the joint [5].

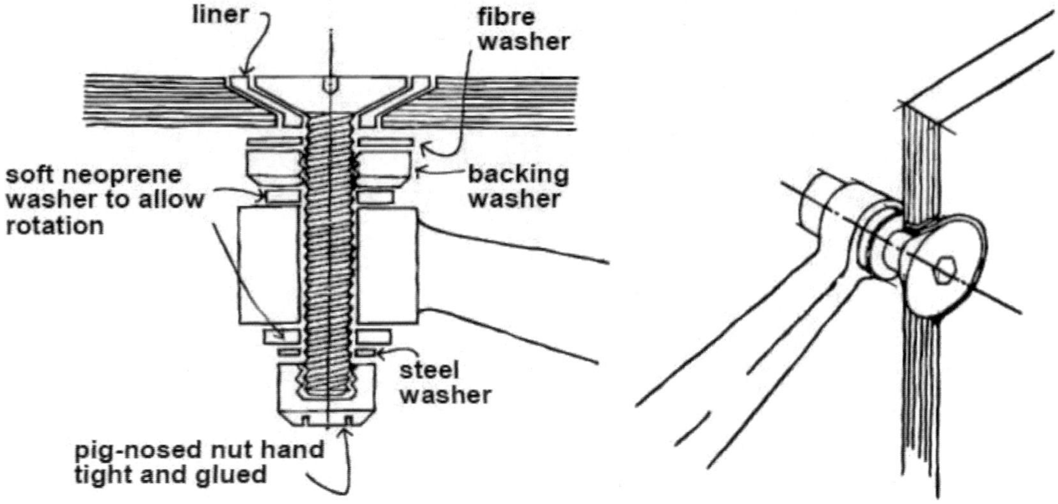

Figure 7. Spider connection [3].

3. Adhesive joints

The glued joints according to the type of adhesive used are divided into elastic (silicone) or hard (resin). The main advantage of glued connections is the distribution of stresses in the vicinity of the joint where, unlike the bolted connections, the local stress peaks do not occur [2]. Further advantages such as weight reduction, the ability to connect thinner materials, the possibility of bonding and, at the same time, sealing and aesthetic enhancement contribute to the development of these less researched areas of glass connections [3]. The design should take into account the minimum shear stiffness of the flexible adhesives, as well as the influence of the temperature and the length of the load to its strength. Also, the decrease in stiffness shall be noticed due to the aging of the adhesive.

It should be also said that the behaviour of the adhesive bond strongly depends on a type of adhesive and the thickness of the adhesive layer. The following are the most used adhesives [4]:

- Cyanoacrylates
- Modified epoxies
- Polyurethane resin
- Structural silicones

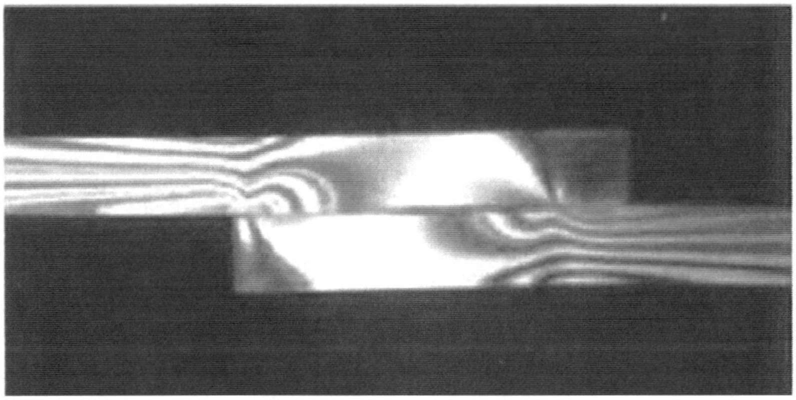

Figure 8. Stress distribution in the connection made with hard adhesive [3].

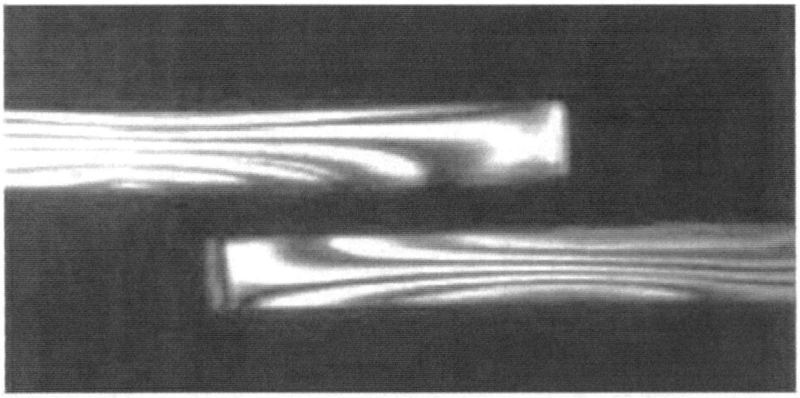

Figure 9. Stress distribution in the connection made with thick elastic adhesive [3].

The adhesive material has to be strong and stiff enough to transfer the design loads for long periods of time without degrading. However, it must also be flexible to redistribute stresses and to compensate dimensional tolerances and thermal expansions without breaking or causing the breakage of the glass elements. A progressive breakage is desirable; the failure of a connection must not result in a total and instantaneous collapse of the whole structure. There are some design considerations that may help in reaching a safe breakage. For example, designing structures with alternative load paths, in order to redistribute loads in case of local breakage, and implementing secondary support structures with the sole purpose of preventing the collapse of the main glass structure [2].

In one-part component adhesives, the curing process is generally done by exposure to UV. In two-part component adhesives, the curing happens as a result of a chemical reaction when the two components are mixed. The curing process may be affected by the thickness of the material. Some materials are specially formulated to cure for thin or thick sections, depending on its intended application. The thickness of the adhesive also affects its mechanical behaviour: thicker adhesive layers have a better capacity to redistribute stresses and to absorb shear strain caused by differential thermal expansion. However, thicker adhesives tend to have lower strength, especially for more rigid adhesives [7].

Table 1. General comparison of different adhesive systems, + good, ++ better, +++ the best [6].

	Tension and shear strength	Stiffness	Ductility	Viscosity	Temperature resistance	Ageing behaviour	UV resistance	Transparency and colour
Epoxy resin	+ + +	+ + +	+	+ + +	+ +	+ +	+ +	+
Polyurethane	+ +	+ + +	+ +	+ +	+ +	+ +	+ +	+ +
Acrylates	+ +	+ +	+ +	+ +	+ +	+ + +	+ + +	+ + +
Silicones	+	+	+ + +	+	+ + +	+ + +	+ + +	+

4. Conclusion

A very common form of joining glass components to one another or to various primary load-bearing structures made of materials such as steel, concrete or wood is a bolted connection in which holes need to be drilled into the glass panes. These openings negatively affect the load-bearing capacity of the glass components. At present, all-glass constructions are very popular in which the glazing (secondary load-bearing structure) is associated with a primary load-bearing structure which is also made of glass. Nowadays, the glued joint comes to the forefront, which, when used correctly, can be not only stronger, but also more durable than a conventional bolted joint.

5. References

[1] Wurm J 2007 *Glass structures: design and construction of self-supporting skins* (Basel: Birkhäuser) ISBN 978-3-7643-7607-9.

[2] O'Regan Ch and col. 2015 *Structural use of glass in buildings (Second edition)* (London: The Institution of Structural Engineers) ISBN 978-1-906335-25-0.

[3] Eliášová M 2015 *Lectures of Glass structures* (Department of Steel and Timber Structures, Faculty of Civil Engineering, CTU in Prague)

[4] Lorenz J 2016 *Numerical analysis of glass pane* (Master's thesis)

[5] Vencl R 2011 *Analýza chování nepředepnutých šroubovaných spojů konstrukcí ze skla* (Dissetation thesis)

[6] Feldmann M and Kasper R and col. 2014 *Guidance for European structural design of glass components: support to the implementation, harmonization and further development of the Eurocodes* (Luxembourg: Publications Office for the European Union) ISBN 978-92-79-35093-1

[7] Machalická K and Eliášová M 2017 *Adhesive joints in glass structures: effects of various materials in the connection, thickness of the adhesive layer, and ageing* (Int J Adhes)

Acknowledgments

The paper has been supported by The projects: VEGA 1/0172/20" Stress and deformation analysis of load bearing components made out of steel, glass and composite materials" of the Scientific Grant Agency of the Ministry of Education, science, research and sport of the Slovak Republic and the Slovak Academy of Sciences and by the project Slovak Research and Development Agency under the contract No. APVV-15-0486 "Analysis of Shear Connection Influence in Bridges with Encased Beams"

IOP Conf. Series: Materials Science and Engineering **867** (2020) 012028 doi:10.1088/1757-899X/867/1/012028

Blast load analysis of overpass columns with various cross-sections

S Lukić[1] and H Draganić[1]

[1] Josip Juraj Strossmayer University of Osijek, Faculty of Civil Engineering and Architecture Osijek, Department of materials and structures, Croatia

Abstract. The column is one of the most important elements of the bridge because its destruction can cause a collapse of the entire bridge. The aim of the research is to analyse the blast behaviour of the overpass columns on Croatian roads and highways. Based on the archive documents of design plans the most common cross-sectional shapes of the reinforced concrete column were selected. The literature provides formulas for calculation of blast pressure that depends on the distance of the explosive from the target and the amount of explosive. However, the cross-sectional shape of the column plays a significant role in retaining the pressure and impulse on the column and its distribution by column height and width. This cannot be predicted by theoretical expressions, and therefore a numerical simulation that takes into account the column geometry is required. The columns were modelled in the Ansys Autodyn hydrocode software for a supposed scenario in which an auto bomb was placed below the bridge at a distance of 2 m from the column. Based on the numerical simulations it was concluded that the circular column had the lowest pressure on the surface directly exposed to the blast load. This is due to its roundness and the angles at which the pressure was reflected. In contrast, the back surface of the circular column has the highest pressure of all the analysed columns. Generally, rectangular cross-sections have a higher pressure on the directly loaded surface. For more detailed analysis and experimental testing, a rectangular column (P6) with recesses was selected due to its high pressures on both, the front and back surfaces. Selected overpass i.e. its column was designed using the old seismic regulations and as a part of further research, its blast behaviour will be compared to the column designed using modern seismic regulations.

1. Introduction

We are witnessing an increasing frequency of terrorist attacks globally. Following the terrorist threats to bridges in California and New York (2001), greater importance is given to bridges, especially bridge columns, when designing. Damage to the columns puts the entire bridge structure at risk. Bridges are significant because of traffic, connecting certain zones of the city, and apart from their functional role they have great economic value [1].

Davis et al. (2009) state the difference in load on a circular and rectangular column, and the advantages and disadvantages of both. The cross-sectional shape plays a significant role in the magnitude of the pressure and impulse affecting the surface. Compared to the square section, the circular cross-section sustains less pressure and impulse because the pressure wave is reflected at an angle. However, the square cross-section has a larger area in comparison to the circular column of the same dimensions and can more easily resist shear at the base. Therefore, it is not clearly defined which cross-section is better and the advantages and disadvantages of both cross-sections need to be further elaborated [1].

Bridges on Croatian roads and highways were examined and the most common columns were identified and modelled in the Ansys Autodyn software [2]. Based on the archive survey, the most commonly used cross-sections of the columns were listed. A reference column that will be analysed in more detail and experimentally tested was selected based on the blast load maximum pressure.

2. Blast scenario

Due to the complexity of the analysis and the various parameters to take into account when analysing the effect of the explosion on the structure, the scenario of detonation of explosives in the vehicle below the superstructure near the column is considered. According to the research conducted, this is considered as one of the most critical and common scenarios. By collapsing a component such as a column of a bridge, the bridge loses its load-bearing capacity and the entire structure is endangered. In addition to extensive material damage and traffic interruption bridge collapse can cause a large number of casualties [3]. In the past, columns have not been designed to withstand high lateral dynamic loads and are therefore an interesting target for terrorist attacks [4]. The height of the explosive charge is determined by the height and geometry of the vehicle, and in literature [5-7] heights of 0.8, 1.0, and 1.2 m (h), which correspond to the distance of the centre of the charge (located in the vehicle) from the ground level, are most commonly reported. Explosive charge (vehicle) distance from the overpass column (R = 2.0 m) was selected due to existing road barriers around the column. The scenario is shown in figure 1. Table 1 gives the estimated mass of explosives that can be placed in a single-vehicle. Mass of TNT (trinitrotoluene) explosive was selected as 100 kg and based on this value numerical simulations in Ansys Autodyn were performed [8, 9].

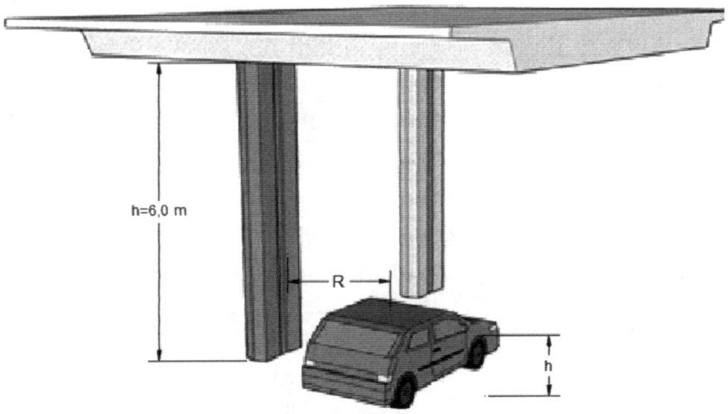

Figure 1. Theoretical blast scenario.

Table 1. Estimated quantities of explosives in various vehicles [8, 9].

Vehicle type	Charge mass [kg]
Compact car trunk	115
Trunk of a large car	230
Closed van	680
Closed truck	2270
Truck with a trailer	13610
Truck with two trailers	27220

3. Columns cross-sections

After reviewing the archive of bridge designs on Croatian roads and highways, the most common cross-sectional shapes of reinforced concrete columns shown in figure 2 have been selected. Circular (O) columns and rectangular columns with different notches (P1-P7) predominate. In the last two decades, rectangular columns with variable cross-sections (V) have been constructed. The cross-section increases linearly from the bottom to the top of the column. Two columns in overpass cross-section dominate as an inter-span supports. The height of the columns can be averaged around 6 m.

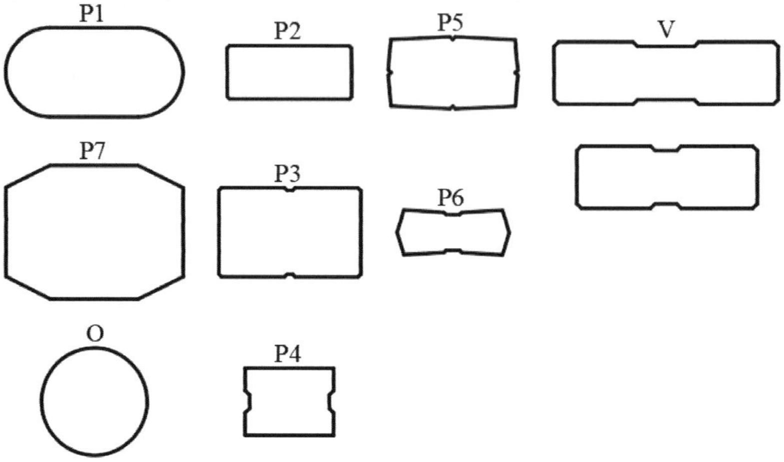

Figure 2. Types of columns cross-sections.

4. Numerical models

The full-scale columns were modelled in Ansys Autodyn. The charge was placed as in the above-described scenario. The air through which the pressure of the explosion wave is propagated is modelled. The blast in air can be modelled using 1D approach. The 1D wedge mesh using axial symmetry in Autodyn is shown in the figure 3(a). Flow out boundary condition is used to eliminate the wave reflection effect. The numerical simulation stops and the data is saved when the blast wave reaches the end of the 1D air mesh. Then the results of blast pressures are remapped to the 3D simulation air surrounding as an initial state of the blast wave. Remap is performed by including symmetry in all directions [10]. Gauge points 1 to 4 are placed on the front surface of the column, where point 1 is placed at the height of 0.8 m what corresponds to the height of the explosive charge, point 2 at a height of 2 m, point 3 in the middle of the column at 3 m, and point 4 at the very top of the column at 6 m. Gauge point locations for column P6 are shown in the figure 3(b) and the locations are the same for all studied columns. The points are set to analyse the pressure distribution across the column. Gauge point 5 is the lateral point, and gauge point 6 is set to the backside, at the height of the explosive charge. The air mesh size plays a significant role in the pressure distribution [11], and in this analysis, a 50 mm element air mesh size was used. The reduction of the mesh size increases the length of the numerical calculation and it is considered that the specified mesh is sufficient to compare obtained pressures. The resulting pressures from Ansys Autodyn may differ due to the size of the air mesh from those obtained analytically using expressions from the literature. Blast pressure rises instantaneously while it takes only a few milliseconds for pressure to attenuate from maximum value to ambient pressure, so the calculation is set to 4 milliseconds, which is sufficient to obtain maximum pressure. If we are interested in the response of the structure to the effect of the explosion, the calculation should last at least 10 times longer. The column is modelled as pure concrete and reinforcement is not considered because the pressure distribution is analysed not structural response. For the sake of faster calculation symmetry is utilized so half of the column is modelled and the blast wave pressure distribution is considered to be symmetrical. The explosive is located at a height of 0.8 m.

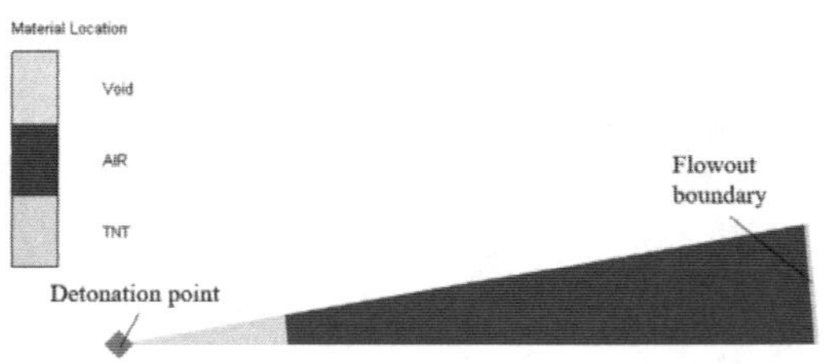

(a)

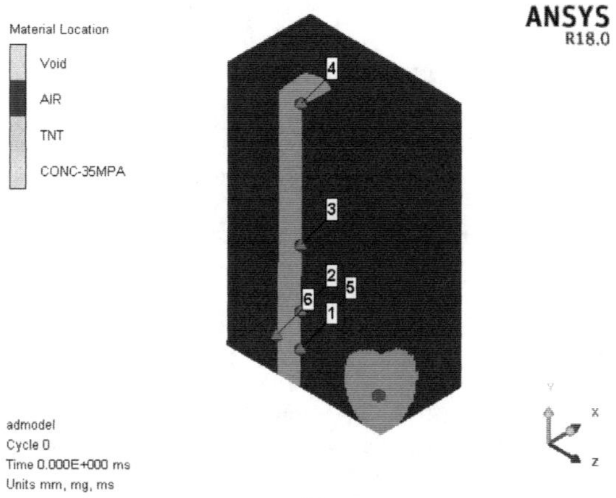

(b)

Figure 3. (a) 1D wedge model of blast wave propagation and (b) 3D column model (P6) in Ansys Autodyn.

5. Results and discussion

At gauge point 1, maximum pressure occurs due to the pressure reflection from the ground, so the reflected pressure interacts with the initial pressure and acts together on the column. The pressure is rapidly expanding upwards and covers gauge points 2, 3 and 4. At these points, the pressure is lower as the distance from the charge increases. The angle of incidence of an explosion wave is directly related to the magnitude of the pressure; a larger angle of incidence means lower pressure. At gauge point 1, the angle of incidence is zero and the pressure is maximum. The lateral gauge point 5 and the back side gauge point 6 are also influenced by the pressure because the pressure rapidly spreads around the column. Figure 4 shows the propagation of the blast pressure wave in time around the column P6.

Figure 4. The blast wave propagation around the column P6.

After the numerical simulation of 9 columns with different cross-sections, shown on figure 2, pressure diagrams and pressure distribution along the columns height was analysed. Figure 5 shows the gauge point maximum pressures for each column. The maximum pressure is at gauge point 1 which is located directly opposite to the detonation point, while in the centre of the column at gauge point 3 there is a significant decrease in blast pressure. The lowest pressure is at gauge point 6 located on the back side of the column. We can also see that column P3 has the highest pressures in comparison to column P6, while the lowest pressures are recorded for circular column O as expected. Figure 6 shows the pressure-time distribution diagrams at gauge point 1 for all column types. Positive phase duration for all recorded pressure-time profiles is around 4 milliseconds.

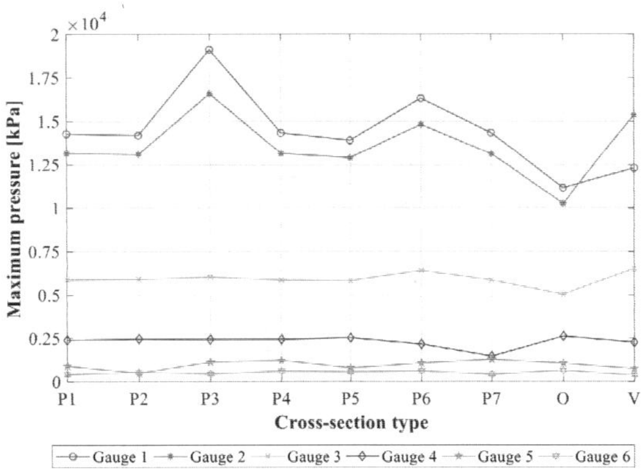

Figure 5. Maximum pressure in gauge points for all types of columns.

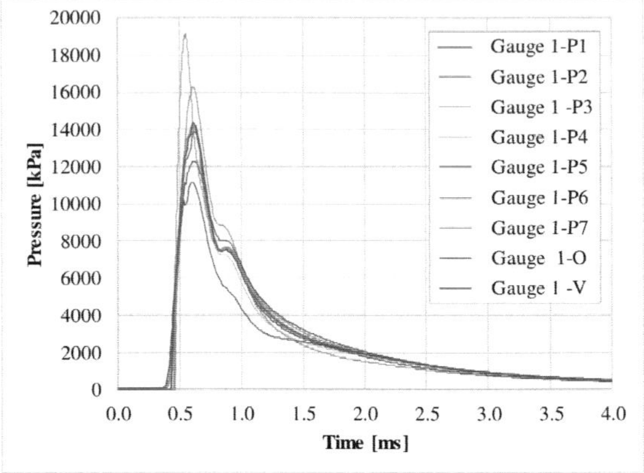

Figure 6. Pressure-time diagrams for gauge point 1.

Of the two columns with maximum pressures, P6 was selected for further analysis because it is more frequent on Croatian highways and is designed according to old seismic regulations. Further research will deal with the blast load behaviour of the columns as they are, studying their blast vulnerability. Furthermore, the column will be designed according to modern seismic regulations, i.e. EN1998 and its blast load behaviour compared to the old column. The bottom part of both columns is exposed to maximum blast pressures. There is a significant increase in the blast pressure due to wave reflection from the ground. Obtained pressure distribution along column height indicates the possible column shear failure. Figure 7 shows the pressure distribution by column height for the selected rectangular column with recesses P6 and for the circular column O. We notice that the largest difference in pressure is in the lower half of the column, below 3 m. Pressures are significantly reduced in the upper half of the column and are almost equal in intensity.

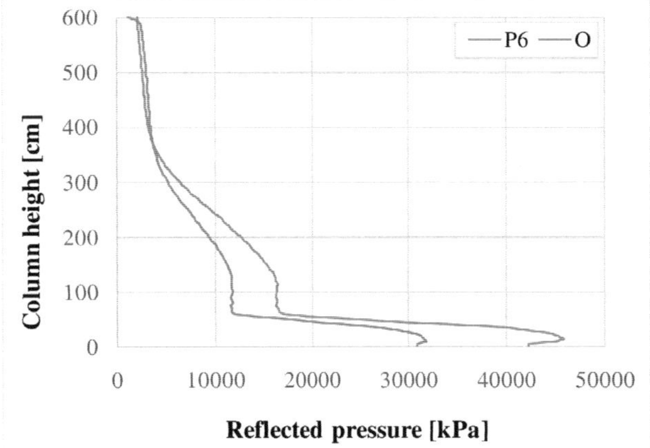

Figure 7. Reflected pressure distribution by column height (P6 and O).

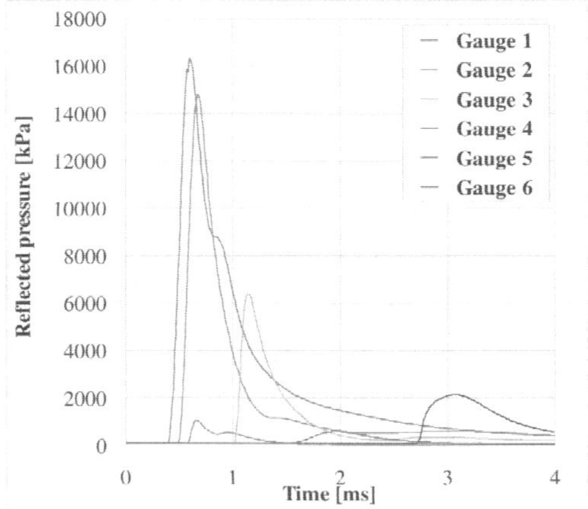

Figure 8. Reflected pressure-time diagrams for all 6 gauge points (P6).

Pressure-time diagrams of all 6 gauge points on the column P6 are shown in figure 8. It is obvious that gauge points 1 and 2 have higher pressures than the rest of the gauge points due to their proximity to detonation point. Gauge points 3 and 4 recorded significantly reduced pressure intensities. Lateral gauge point 5 has its maximum value at the same time instance as gauge point 2 but has a significantly lower pressure values due to its position on the opposite side of the column. The pressure reaches gauge point 6 before reaching the top of the column because the path of the blast wave is shorter than for gauge point 4. Pressure recorded in gauge point 6 indicates that the blast wave envelops and interacts with modelled column.

6. Conclusion

Numerical analysis in Ansys Autodyn confirmed that the column shape plays a significant role in resistance to blast wave propagation. The circular column O has the lowest pressures, while the maximum pressures are observed on the column P3 and P6. The critical zone of the column is the lower half because, in addition to the direct pressure from the explosion, the reflected pressure from the ground also interacts with the column. Due to the high pressure at the bottom, shear failure of the column is expected.

7. References

[1] Davis C E et al. Design and detailing guidelines for bridge columns subjected to blast and other extreme loads 2009 *Structures Congress 2009: Don't Mess with Structural Engineers: Expanding Our Role*
[2] Ansys 2010 *Ansys Autodyn Users`s Manual* Canonsburg PA USA
[3] Cooper J D, Smith M C and Ernst S L, Blue ribbon panel recommendations for bridge and tunnel security
[4] Abedini M et al. 2019 Pressure–impulse (P–I) diagrams for reinforced concrete (RC) structures: a review *Archives of Computational Methods in Engineering* **26**(3) pp 733-767
[5] Winget D G, Marchand K A and Williamson E B 2005 Analysis and design of critical bridges subjected to blast loads *Journal of Structural Engineering* **131**(8) pp 1243-1255
[6] Fujikura S, Bruneau M and Lopez-Garcia D 2008 Experimental investigation of multihazard resistant bridge piers having concrete-filled steel tube under blast loading Journal of Bridge Engineering **13**(6) pp 586-594
[7] Elsanadedy H M et al. 2011 Effect of blast loading on CFRP-Retrofitted RC columns-a numerical study *Latin American journal of solids and structures* **8**(1) pp 55-81
[8] Draganić H and Sigmund V 2012 Blast loading on structures *Technical Gazette* **19**(3) pp 643-65
[9] Hameed A H 2007 Dynamic behavior of reinforced concrete structure subjected to external explosion (Kufa University Jumada Elawal) 1428 p 103
[10] Shi Y, Li Z and Hao H 2008 Mesh size effect in numerical simulation of blast wave propagation and interaction with structures *Transactions of Tianjin University* **14**(6) pp 396-402
[11] Draganić H and Varevac D 2018 Analysis of blast wave parameters depending on air mesh size *Shock and Vibration* **2018**

Acknowledgments

This paper was supported in part by the Croatian Science Foundation (HRZZ) under the project (UIP-2017-05-7041) “Blast Load Capacity of Highway Bridge Columns,” and support for this research is gratefully acknowledged. We thank Croatian Roads (HC) and Croatian Motorways (HAC) for providing the bridge designs documentation and recognizing the importance of this project.

IOP Conf. Series: Materials Science and Engineering **867** (2020) 012029 doi:10.1088/1757-899X/867/1/012029

Recycling of photovoltaic panels - A review of the current trends

K Macalova[1], V Vaclavik[1], T Dvorsky[1], J Svoboda[1], J Charvat[1] and L Gola[1]

[1] VSB - Technical University of Ostrava, Faculty of Mining and Geology, Department of Environmental Engineering, Czech Republic

Abstract. Towards the end of the 20th century there was a huge boom in the development of solar energy, which led to an increase in the production of photovoltaic panels. The producers assume a lifetime of these panels of 30 years. Today, the life of this photovoltaic panels is ending and a large amount of waste is associated with it. This issue is also addressed by the Council of the European Union, which issued Directive 2012/19/EU on waste electrical and electronic equipment. Photovoltaic panels are included in IV. category "Consumer electronics and photovoltaic panels". This Directive mandates that at least 70% of photovoltaic panel materials must be reused or recycled. This paper presents a review of the current trends in recycling of photovoltaic panels and the possibilities of reuse in cement matrix.

1. Introduction

There was a large boom in the field of photovoltaic panels in the early 1990s, when grant programs were launched to support renewable energy sources. Japan and Germany were among the first countries to take advantage of these grant programs. Over time, these grant programs have been implemented in other countries of the world. Photovoltaic energy is generally one of the renewable energy sources with a relatively long life span, which is estimated to be at least 30 years. We are gradually coming to the time of the expected end of life of the first photovoltaic panels and it is necessary to deal with the method of their disposal after the end of their life cycle. Together with the rapid increase in the use of photovoltaic panels, there will be a proportionately increasing production of waste from the area of solar energy production. So far, only panels with mechanical damage caused by poor handling during the installation and transport have been disposed of [1, 2].

The Directive 2012/19/EU of the European Parliament and of the Council on waste electrical and electronic equipment states that at least 70% of photovoltaic panel materials must be reused or recycled [3]. By finding a suitable way of recycling materials from photovoltaic panels we can ensure sustainable production while minimizing the burden on the environment. At present, there are two most widespread methods of recycling photovoltaic cells, namely thermal and mechanical-chemical recycling. As research into the incorporation of waste materials from other industries into the cement matrix [4-7] shows, one of the new potential ways to recycle photovoltaic cells is to incorporate the resulting waste materials into cement composites, as suggested by the studies performed before [8-11].

2. Development of photovoltaic panels

The development of photovoltaic panels can be divided into several stages, when each stage brought a new generation of photovoltaic cells.

IOP Conf. Series: Materials Science and Engineering **867** (2020) 012029 doi:10.1088/1757-899X/867/1/012029

2.1. The first stage of development

The first generation photovoltaic cells were developed during this basic stage. The cells were made of a crystalline structure of silicon. The name "first generation" reveals that it is the oldest technology which is still very popular and commercially the most widespread. This technology is constantly evolving [12-14]. Silicon panels are made of monocrystalline or polycrystalline doped silicon. The structure of the photovoltaic panel is shown in figure 1. The individual cells are connected by means of soldered joints. The bottom solid tedlar plate (a mixture of polyvinylidene fluoride and polyethylene terephthalate) contains a plastic film EVA (ethylene-vinyl acetate), on which the interconnected photovoltaic cells are placed. The cells are covered by a layer of EVA film and hardened glass [13, 15].

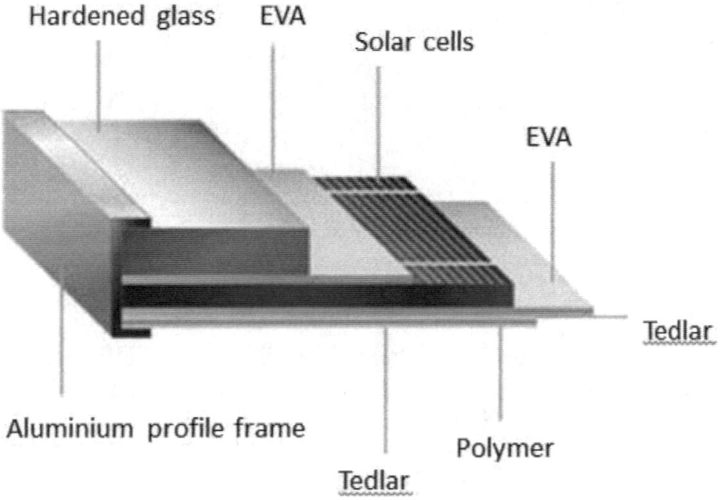

Figure 1. Structure of a photovoltaic panel [15].

2.2. The second stage of development

The second stage of development was focused on the material and manufacturing optimization, with an effort to reduce the production costs and to minimize the use of silicon. This effort resulted in the production of second generation photovoltaic cells, which are cells made from amorphous silicon, microcrystalline silicon and thin-layer technology. The examples of thin-layer technologies used in the production of photovoltaic cells are:

- Technology of Copper Indium Gallium Selenide (CIGS),
- Cadmium Telluride (CdTe),
- Amorphous Silicon (a-Si) [16].

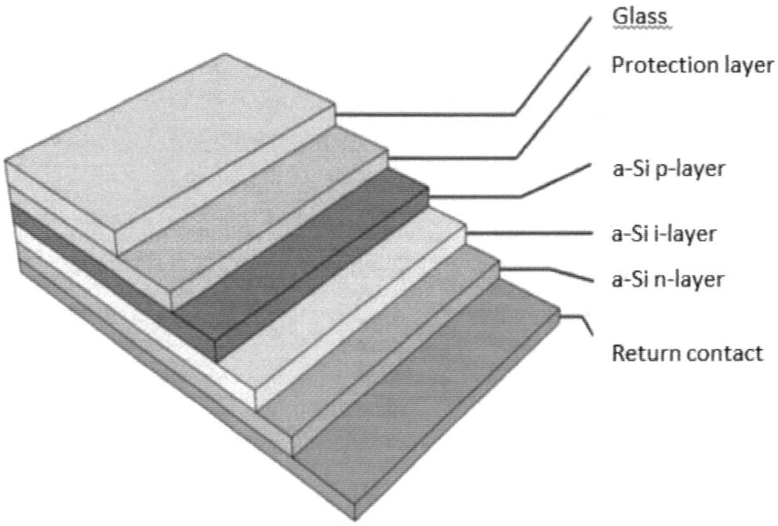

Figure 2. Amorphous silicon panel structure, modified [17].

2.3. The third stage of development

The third stage of development is focused on sustainable production and efficiency improvement using the latest production technologies. The third generation of photovoltaic cells is emerging now. The organic polymer-based prototypes are still in the research and development stage [18, 19].

3. Recyclable photovoltaic panel material

According to German company SolarWorld dealing with the recycling of photovoltaic panels and cells, it is possible to recover up to 84% of the module weight by recycling. Of these, 90% is glass and 95% is semiconductor material [20]. This issue needs to be addressed, because there were 250 000 tonnes of waste from the photovoltaic industry recorded by 2016. By 2050, when the lifetime of most panels will expire, this figure can go up to around 6 million tonnes of waste [2].

The recycled materials include:

- aluminium,
- glass,
- plastic components,
- photovoltaic cells,
- heavy metals (Cd, Pb, Se),
- silver, nickel, tellurium, indium, gallium, molybdenum, ...

The process of recovering crushed recycled material from the photovoltaic panels is shown in figure 3.

IOP Conf. Series: Materials Science and Engineering **867** (2020) 012029 doi:10.1088/1757-899X/867/1/012029

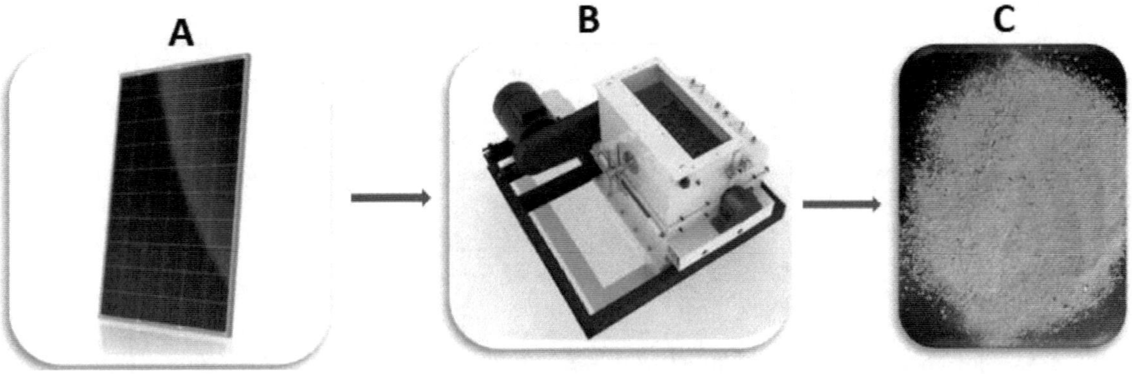

Figure 3. Diagram of obtaining pulp from photovoltaic panels A - photovoltaic panel before recycling, B - hammer crusher, C - recovered pulp - recycled material.

Table 1 shows the percentage of each type of material in photovoltaic panels and their percentage yield during recycling.

Table 1. Yield of the individual material types [21].

Material	Percentage share (%)	Yield (%)
glass	67	>95 %
aluminium (Al)	18	100
plastic materials	11	0
silicon (Si)	3	85
metals	1	80

4. Methods of recycling materials from photovoltaic panels

The basic methods of recycling of photovoltaic panels include thermal recycling and mechanical-chemical recycling. In both methods, it is necessary to first remove the aluminium frame and subsequently recycle the cells [1].

5. Thermal recycling

This method is used to recycle crystalline silicon cells. A cleaned photovoltaic panel is completely placed in an oven heated to a temperature above 500 °C. In this step, the plastic parts (EVA) are evaporated and incinerated in a controlled manner in the following chamber. The remaining material is separated manually. If the recycled material is damaged, up to 85% of the cells can be reused for the repeated production of photovoltaic panels. This technology is considered to be the most advanced method of recycling photovoltaic panels. It is more cost-effective than mechanical-chemical recycling, which requires the use of costly and toxic agents [22].

5.1. Mechanical-chemical recycling

These are separation methods. In the first stage, it is necessary to remove the aluminium frame from the panel. Subsequently, the photovoltaic panels are crushed and the individual components of the panel are sorted according to size fractions. Metals (Cd, Cu, Pb, Ag) are obtained chemically and metallurgically from the separated material. The metals are obtained by leaching or electrolysis. The chemical recycling method is used to clean the first generation photovoltaic cells. The layers deposited on the cells are removed using a solvent. Acidic or basic solutions are used to dissolve the required material. Choosing the right solution with the right concentration and ideal operating temperature is very important with this method [23, 24]. For example, etching can be used to remove the metal coating present on the photovoltaic cell. Older photovoltaic cells contained silver in the back metal coating. In case of this procedure, silver is dissolved in acid and it can be recovered by means of acid electrolysis [25].

6. Conclusions

Nowadays, there has been pressure to reduce the environmental burden and to use primary raw materials in an economic way. This is the reason for the continuous development and search for new materials using mainly recycled secondary raw materials. The issue of recycling photovoltaic cells has so far been addressed only marginally due to the low production of such waste. However, the lifetime of the first types of photovoltaic cells is coming to an end, which will increase this type of waste and put its recycling in the centre of attention. This increase motivates research teams around the world to develop new and sustainable methods of using recycled photovoltaic cells as a secondary raw material.

7. References

[1] Bechnik B 2011 *TZB-info* 7868
[2] Weckend S, Wade A, Heath G. 2016 *End-of-Life Management Solar Photovoltaic Panels.* (Abu Dhabi: IRENA and IEA-PVPS)
[3] European Union. Directive 2012/19/EU of the European Parliament and of the Council of 4 July 2012 on Waste Electrical and Electronic Equipment (WEEE). Official Journal of the European Union. 2012:L 197/38-71
[4] Stevulova N, Schwarzova I, Hospodarova V and Junak J 2016 *Chem. Eng. Trans.* **50** 367–372
[5] Smolakova M et al 2018 *IOP Conf. Ser.: Mater. Sci. Eng.* **385**
[6] Ondova M, Stevulova N and Estokova A 2012 *Procedia Eng.* **42** 1863–1872
[7] Junakova N and Junak J 2017 *Sustain.* **9**
[8] Gavhane A, Sutar D, Soni S and Patil P 2016 *Int. J. Eng. Res.* **51** 594-601
[9] Fernández L J et al 2011 *Sol. Energy Mater. Sol. Cells* **95** 1701–1706
[10] Arliguie G and Grandet J 1990 *Cem. Concr. Res.* **20** 517–524
[11] Appleyard D 2009 *Renew. Energy World Mag.* **12**
[12] Bagher A M, Vahid M M A and Mohsen M 2015 *Am. J. Opt. Photonics* **3** 94
[13] Bechnik B 2011 *TZB-info* 5517
[14] Bruton T M 2002 *Sol. Energy Mater. Sol. Cells* **72** 3–10
[15] RD solar 2020 available from: http://www.rdsolar.cz/sluzby/solarni-panely/
[16] Fraunhofer ISE 2017 available from: www.ise.fraunhofer.de
[17] Alchemie Limited 2013 available from: http://www.solar-facts-and-advice.com/amorphous-silicon.html
[18] Zhou Y et al.2013 *Sci. Rep.* **3** 1–5
[19] Garozzo M 1993 *Ceram. Int.* **19** 287–297
[20] SolarWorld. 2017, available from: https://www.solarworld.de/en/home/
[21] Jahoda J 2013 *Characteristics of recycling methods of used photovoltaic cells* (Ostrava: VSB-Technical University of Ostrava)
[22] Klugmann-Radziemska E and Ostrowski P 2010 *Renew. Energy* **35** 1751–1759
[23] Klugmann-Radziemska E and Ostrowski P 2010 *Ecol. Chem. Eng. S* **17** 387-391
[24] Zeng D, Born M and Wambach K 2004 *J. Environ. Sci.* **16** 889–893
[25] Yi Y K et al 2014 *J. Air Waste Manag. Assoc.* **64** 797–807

Acknowledgments

This research was funded by VŠB–TUO, Faculty of Mining and Geology—grants numbers SP2020/30 and SP2019/29; Moravian-Silesian Region—grant number RRC/10/2018.

IOP Conf. Series: Materials Science and Engineering **867** (2020) 012030 doi:10.1088/1757-899X/867/1/012030

Surface runoff as a potential source of pollution

I Marko[1], R Csicsaiová[1], G Rozsa[1] and Š Stanko[1]

[1] Slovak University of Technology in Bratislava, Faculty of Civil Engineering, Department of Sanitary and Environmental Engineering, Slovakia

Abstract. Due to the accelerated development of urbanized areas, the increase of paved and impermeable surfaces, permanent changes in the environment of the city occur. The spatial structure of the city is changing. There are creating significant artificial landscape elements that are not able to absorb rainwater. As a consequence of this unnatural phenomenon in the city was identified a new area pollution source - surface runoff. The formation of surface runoff is affected by several factors. For us, it is important the character of the pollutants in the surface runoff. The paper aims to point out the impact of surface runoff from urbanized areas on the environment. Compare quality parameters of rainwater runoff from different types of paved areas (different types of roof, roadway, car parking). To get acquainted with the legislation covering the area permitted criteria for the content of undesirable substances and the method of their removal.

1. Introduction

Significant climate change has been increasingly reported in recent years, long periods without rain, extremely high temperatures, especially in summer and frequent heavy rainfall. All of these are natural phenomena that cause many of the problems we have been facing lately. Accelerated land use, elimination of vegetation, and alteration of the landscape's spatial structure by urbanization lead to adverse effects [1], [2]. There is a change in the hydrological system, including infiltration, higher maximum flow and volume, and increased pollutants [1]. One of the reasons for this global problem is the mass migration of the population, resulting from an uncontrolled increase in hard surfaces. Infrastructure development and urbanization have a significant impact on the structure of the city. Instead of conservation vegetation areas and natural conditions of rainwater infiltration across the soil horizon, rainwater from urbanized areas is a diversion into the sewerage network [3], [4]. This currently unacceptable method of draining surface runoff can cause congestion in the sewerage network while increasing the pollution burden by rainwater and the occurrence of urban floods [5].

A significant effort currently is made to develop more resilient and sustainable technologies to address this problem to prevent the spread of impermeable surfaces. Municipal authorities are trying to use nature-friendly measures that support a system of efficient rainwater management [6]. It is a complex of measures and devices with Low - impact development (LID). It is an attractive alternative to the vegetation surface instead of the traditional impermeable surface [3], [7]. These measures can reduce surface runoff in the city, capture maximum rainwater flow and infiltrate rainwater, thereby preventing pollution of highly urbanized areas [7], [8].

There are many possibilities for measures to boost urban green land by applying LID. When properly designing LID, it is essential to know the outflow characteristics resulting from different types of precipitation rainfall events [7], [9]. The result of using LID mainly depends on the quality and quantity of rainwater. At present, great attention is paid to the qualitative and quantitative parameters of rainwater from hard surfaces due to the possible occurrence of hazardous substances. The composition of stormwater can adversely affect water quality and quantity, habitats and biological resources, population health, and the aesthetic appearance of urban waterways [7], [10].

The paper aims to point out the impact of surface runoff from urbanized areas on the environment. Compare quality parameters of rainwater runoff from different types of paved areas (different types of roof, roadway, car parking). To get acquainted with the legislation covering the area permitted criteria for the content of undesirable substances and the method of their removal.

2. Legislative regulation

The European Union has been dealing with the problem of "surface water protection" for decades. The result of which are several standards and recommendations. Among the most important, it is possible to define the Recommendation of the United States Environmental Protection Agency (EPA), the Water Framework Directive (2000/60/EC), which requires the protection and improvement of watercourses [11], as well as other relevant European Parliament Directives on water management.

The management of urban stormwater in the Slovak Republic is not dealt with in detail by regulations and technical standards. Partially, this topic is dealt with in the Water Plan of Slovakia by following Article 4.7 (d), the Water Framework Directive (2000/60/EC) [11], and Government Regulation of the Slovak Republic No. 269/2010 [12]. A shortcoming of the Slovak regulations is the setting of limit values of indicators of rainwater pollution from surface runoff, which infiltrate into the soil horizon. In GR, SR No. 269/2010 in § 9 in points 1–3, only briefly set requirements for discharges of surface runoff water into surface and groundwater [12].

As in Slovakia and Serbia, all standards and recommendations related to this topic are based on EU regulations. Among them, it can define the Water Framework Directive (2000/60/EC), the Water Low of the Republic of Serbia ("Sl. Glasnik" No. 30/10) [13], and the Government Regulation of the Republic Serbia ('Sl. Glasnik' No.24/2014) [14].

Table 1. Comparison of limit values of pollutants in surface waters in Serbia and Slovakia [12], [14].

	MAC for surface water quality ((in parentheses are AA)) - GR SR No.269/2010	MAC for surface water quality ((in parentheses are AA)) – "Sl. Glasnik" RS. No.24/2014
Candium (Cd)[3]	0.45 – 1.5 (0.08 – 0.25)	0.45 – 1.5 (0.08 – 0.25)
Cr (VI)	9.0	-
Copper (Cu)	(1.1 – 8.8)	-
Nickel (Ni)	(20.0)	34.0
Lead (Pb)	(7.2)	(14.0)
Zinc (Zn)	(7.8 – 52.0)	-
Benzo(a)pyrene	0.1 (0.05)	0.27
Fluoranthene	1.0 (0.1)	0.12
Naphtalene	(2.4)	130.0 (2.0)
Phenathrene	2.0 (0.28)	-
Sum of polyaromatic hydrocarbons (PAU)	-	-
Single pestecide	-	-
Sum of pesticicdes	-	-

Notes: AA – annual average
MAC – maximum allowable concentration.

3. Surface runoff

As a result of the urbanization of the city was the areal pollution source. It is a complex element influenced by several factors, such as the duration and intensity of the rainfall event, the length of the previous non-rainfall season, the surface characteristics of land use, and other factors [15]. The size of the surface runoff mainly depends on the percentage of paved surfaces and the total rainfall. During heavy rains, pollutants from impermeable areas (mostly car parks and roads) get in motion and create surface runoff. This type of urban runoff can cause substantial damage in an urbanized area and environmental damage [15].

The nature of the pollutants occurring on the surface of the runoff primarily depends on the pollution of the rain itself and local conditions (industrial zone, countryside, city). When passing through the atmosphere, rain may contain substances such as chlorides, sulphates, nitrates, potassium, calcium, and magnesium in concentrations below 10.0 $mg.l^{-1}$ [17]. As concerns rain from industrial areas, this is an increased share of fossil fuel combustion products. The chemical composition of the surface runoff from urbanized areas mainly affects the material used for the impermeable surface and the utilization of the given area (parking lot, roof, road). Under normal conditions, rainfall from urbanized areas contains substances predominantly from motor vehicles such as heavy metals, oil, fats, organic materials, and others [17]. These usually occur in harmless concentrations. During a long without rainy season, the concentration of these substances can be several times higher.

3.1. Comparison of the quality parameters of rainwater

Currently, many studies deal with the quality of stormwater runoff from different types of impermeable areas. Primarily are monitored the qualitative and quantitative parameters of rainwater from roadways, parking lots, walkways, and roofs. Many authors in their studies focus on the analysis of the quality of urban rainwater runoff from different types of materials from which paved surfaces are made [3], [15], [18], [19]. Angrill (2016), in her research, focused on monitoring rainwater runoff from three types of paved areas (roads, parking, and walkways) and two types of materials (concrete and asphalt) [18]. Another researcher Lee (2012), analysed the rainwater from different types of roof covers (wooden shingles, concrete tiles, clay tile) [19]. Martins and all. (2019) was harvesting rainwater from an urban area in Brazil [20]. In the analysis of captured rainwater samples from paved areas, were monitored chemical, physical, and microbiological parameters. After the analysis, in the samples from the catchment surface, among other specific parameters (ions - Mg^{2+}, NH_4^+, Na^+, etc.) were detect heavy metals like zinc, aluminium, copper, lead, nutrients, toxic compounds, suspended solids [18], [19], [20]. Also, in some samples were found batteries as Escherichia coolly and Enterococci [19]. The pH value, conductivity, and chemical oxygen demand were monitored in each study, as these parameters can significantly affect rainwater quality.

Table 2. Comparison of the qualitative parameters of rainwater from different types of impermeable surface [18], [19].

Roof materials	**pH** [-]	**Al** [$\mu g.l^{-1}$]	**Cu** [$\mu g.l^{-1}$]	**Fe** [$\mu g.l^{-1}$]	**Pb** [$\mu g.l^{-1}$]	**Zn** [$\mu g.l^{-1}$]
Wooden Shingle	6.7	227	58	154	10	135
Concrete Tile	7.1	535	37	160	14	196
Clay Tile	7.1	243	59	155	11	131
Urban surface						
Asphalt road	7.8	-	-	110	-	15
Concrete road	7.5	-	-	35	-	70
Asphalt parking	7.75	-	-	20	-	5
Concrete pedestrian area	8.0	-	-	280	-	20

Table 2 shows the average concentration of some of the quality parameters found in the chemical analysis of rainwater from different surface types. It can be stated that runoff water quality depends on the catchment surface material and types of active development activity. The concentration of the listed parameters in the table 2 above all varied on the duration of rain and the length of the no-rain season.

Increased attention currently is to research into the quality of the first runoff, given that most of the pollutants accumulate at the beginning of the rainfall.

4. Conclusion

Against the current trend in the development of urbanized areas, it is necessary to provide an effective solution to drain and capture rainwater runoff from the impermeable surface. Lack of vegetation causes the formation of surface runoff, and thus increases the maximum flood flow. All this affects the contamination of the pollutants and the degradation of the environment.

As was a state in the previous part of this article, surface runoff can contain substances like heavy metals nutrients, toxic compounds, suspended solids, and bacteria. The occurrence of these substances primarily depends on the type of urban area, material, and the rainy season. Based on the mentioned articles, the most polluted surface runoff is from the roadway and parking; that way is necessary to do something with that.

The new management of rainwater supports the construction of nature-friendly measures. These are measures and equipment with Low – impact development LID. These are mainly infiltration systems such as infiltration ditch, rain gardens, and ponds, and other objects that contribute to the creation of the city's environmental development. Their primary function is to capture, filter, and drain rainwater. The application of these systems would reduce the load on the sewerage network and prevent ground and surface water pollution.

5. References

[1] Chen J, Theller L, Gitua M W, Engel B A and Harbor J M 2017 Urbanization impacts on the summer heavy rainfall climatology over the eastern United States *Journal of Environmental Management* Vol **187** pp 470-481

[2] Zhang J, Hua P, and Krebs P 2017 Influences of land use and antecedent dry-weather period on pollution level and ecological risk of heavy metals in road-deposited sediment *Environmental Pollution* Vol **228** pp 158–168

[3] Liu J, Yan H, Liao Z, Zhang K, Schmidt A R and Tao T 2019 Laboratory analysis on the surface runoff pollution reduction performance of permeable pavements *Science of the Total Environmental* Vol 691 pp 1-8

[4] Brabec E, Schulte S and Richards P L 2002 Impervious surfaces and water quality: a review of current literature and its implications for watershed planning *Journal of Planning Literature* Vol **16 (4)** pp 499-514

[5] Urcikán P and Rusnák D, 2004 *Sewerage and wastewater treatment: Sewerage I. Design of sewerage networks (Bratislava:* Slovak University of Technology in Bratislava) p 324 ISBN 978-80-227-3435-6

[6] Ghofrani Z, Sposito V and Faggani R, 2017 A comprehensive review of blue-green infrastructure concepts *International Journal of Environment and Sustainability* Vol **6** (1) pp 15 - 36

[7] Xu Z, Xiong L, Li H, Xu J, Cai X, Chen K and Wu J 2019 Runoff simulation of two typical urban green land types with the Stormwater Management Model (SWMM): sensitivity analysis and calibration of runoff parameters *Environmental Monitoring and Assessment* Vol 191 pp 1-16

[8] Sansalone J, Kuang X, Ying G and Ranieri V 2012 Filtration and clogging of permeable pavement loaded by urban drainage *Water Research* Vol **46** (20) pp 6763–6774

[9] Davis A P 2005 Green engineering principles promote low-impact development *Environmental Science and Technology* Vol **39** (16) pp 338–344

[10] Zhang J, Hua P and Krebs P 2015 The build-up dynamic and chemical fractionation of Cu, Zn and Cd in road-deposited sediment *Science of the Total Environmental* Vol **532** pp 723–732

[11] The European Parliament and the council of the European Union *Directive 2000/60/EC of the European Parliament and of the Council of 23 October 2000 establishing a framework for Community action in the field of water policy,* European Union 2000

[12] Regulation of the Government of the Slovak Republic *No. 269/2010 Laying down requirements for achieving good water status*, Slovakia 2010

[13] Water Law of the Republic of Serbia *"Sl. Glasnik" No.30/2010,* Serbia 2010

[14] Regulation of the Government of the Republic of Serbia *"Sl. Glasnik" No.24/2014, regulation on the limit values of priorities and priority hazardous substances, which polluting water surfaces and deadlines for reaching them*, Serbia 2014
[15] Li J, Jiang Ch, Lei T and Li Y 2016 Experimental study and simulation of water quality purification of urban surface runoff using non-vegetated bioswales *Ecological Engineering* Vol **95** pp 706-713
[16] Mamoon A A, Jahan S, He, X, Joergensen, N E and Rahman A 2019 First flush analysis using a rainfall simulator on a micro catchment in an arid climate *Science of the Total Environmental* Vol **693** pp 1-12
[17] Pitterm P 2009 *Hydrochemistry 4* Prague, Faculty of Chemical Technology in Prague pp 592 ISBN 978-80-780-701-9
[18] Angrill S, Petit-Boc A, Morales-Pinzón T, Josa A, Rieradevall J and Gabareel X 2016 Urban rainwater runoff quantity and quality – A potential endogenous resource in cities? *Journal of Environmental Management*, Vol **189** pp 14-21
[19] Lee Y, Bak G and Han M 2012 Quality of roof-harvested rainwater – Comparison of different roofing materials *Environmental Pollution* Vol **162** pp 422–429
[20] Martins E H, Nogarotto D C, Mortatti J and Pozza S A 2019 Chemical composition of rainwater in an urban area of the southeast of Brazil *Atmospheric Pollution Research* Vol **10** pp 520–530

Acknowledgments
This work is supported by the Scientific Grant Agency of the Ministry of Education, Youth and Sports of the Slovak Republic and the Slovak Academy of Sciences within the project VEGA 1/0574/19, co-funded by the European Regional Development Fund and by the Slovak Research and Development Agency under the contract No. APVV-18-0203.

IOP Conf. Series: Materials Science and Engineering **867** (2020) 012031 doi:10.1088/1757-899X/867/1/012031

Adsorption of specific chloroacetanilides on granular activated carbon

M Marton[1], J Ilavský[1] and D Barloková[1]

[1] Department of Sanitary and Environmental Engineering, Faculty of Civil Engineering, Slovak Technical University, Radlinského 11, 810 05 Bratislava, Slovak Republic

Abstract. Pesticides, as one of the main monitored micropollutants these days, are monitored and determined even in the drinking water, because of their toxic effect on the living organisms. Presence of these pesticides in the drinking water is undesirable and, therefore, they need to be removed. Chloroacetanilides belong to the group of herbicides, which are commonly used in agriculture. Their complex structure prevents common biological degradation, so more complex processes are needed. One of the most efficient process for their removal is adsorption on activated carbon. In this article adsorption of certain chloroacetanilides from water for two different types of granular activated carbon is presented. Matrix used for dilution of pesticides was drinking water and the efficiency of the removal of pesticides ranged from 5 to 60 % for the specific sorption materials and pesticides. This research has shown that adsorption material Filtrasorb 400 is far more efficient than Norit 1240.

1. Introduction

Chloroacetanilides belong to the group of micropollutants known as pesticides. These pesticides are dangerous for human health, as long-term exposure can cause some serious health problems including breathing problems, irritation of skin and eyes etc [1]. Usage of some pesticides is therefore forbidden, but as they were broadly used, they can still be detected in the sources of water. As the presence of pesticides in sources of drinking water is not desired, they must be removed from drinking water [2]. Pesticides as very complex molecules cannot be removed from water by usual technological processes and more advanced processes are needed. One of the most efficient process of removal pesticides from water is adsorption. Adsorption is a process of attachment or adhesion of atoms, molecules or ions from gaseous, liquid or solution onto the surface of an adsorbent [3]. This process can be carried out by two types of an adsorption. Physical adsorption which is accomplished by the formation of Van der Waals forces (London dispersion forces) between the adsorbent and adsorbate. This type of forces is relatively easy to break by comparison to other types of bonds. Chemisorption is another type of adsorption, where the creation of chemical bonds occurs, which is much more difficult to break than physical bond [4]. A variety of adsorbents is used for the adsorption processes such as zeolites, iron-based adsorbents, different types of clay materials, activated carbon and others [5]. The selection of the right material depends on variety of requirements such as pollution, temperature, chemical structure and the amount of the pollution, which plays a very big role in the selection of adsorbent. Even the economical price of the adsorbent has a high priority [6]. Activated carbon is one of the most used material for adsorption, as it has the best properties out of the mentioned materials. Activated carbon has a very big surface area and it is non-selective, which means it can adsorb almost anything. There are two types of activated carbon used for adsorption processes, powdered and granular activated carbon [7]. Granular activated carbon has some advantages over the powdered. The main reason for usage of granular activated carbon is, that it is possible to reactivate used carbon, which in the case of powder is not possible. Granular activated carbon has particles in the range of 0.2 to 5 mm in diameter and powdered resemble more of the dust-like particles [8].

2. Materials and methods

The pesticide standards were purchased from a company ALS Czech Republic, which also provided us with the sample vials and analysis of pesticides in samples. Pesticides monitored for this study were atrazine, terbuthylazine and their metabolites. Granular activated carbon Norit 1240 W was purchased from Cabot Corporation and Filtrasorb 400 was purchased from Chemviron Carbon. The characteristics of these granular activated carbon are presented in the table I. The selection of these adsorbents was based on previous experiments carried out for heavy metal adsorption [9].

Table 1. Characteristics of Granular activated carbons Norit 1240 W and Filtrasorb 400.

	Norit 1240 W	Filtrasorb 400
Iodine number [-]	975	1000
Particle size > 0.17 [mass%]	10%	5%
Particle size < 0.425 [mass%]	5%	4%
Total surface area B.E.T. [m2/g]	1100	1050
Apparent density [kg/m3]	470	450
Effective size D10 [mm]	0.6 – 0.7	0.6 – 0.7
Uniformity coefficient [-]	1.7	1.7

Stock solution of the pesticides, with concentration of approximately 1 µg/L, was prepared by mixing 50 mL of the pesticide's standards with approximately 5 L of drinking water. This solution was then properly mixed and was used in the experiments. The experiments were performed in the glass bottles with the volume of 200 mL stock pesticide solution. On the analytical scales we weighed out 200 mg granular activated carbon and then it was added to the bottles. Subsequently these bottles were regularly stirred, and the samples were taken out at times 0, 30, 60, 90, 120, 180, 240 and 360 minutes. Samples were taken by pipette into the 40 ml glass vials with added thiosulfate for preservation. After the experiments, these vials were sent to the analytical laboratory which performed the analysis of the concentration of pesticides. Concentration was determined by liquid chromatography using an internal standard method.

3. Results and Discussion

The pesticide detection analysis results after adsorption process in our drinking water spiked with pesticides are shown below. It is important to highlight that these experiments were conducted in the same conditions, such as room temperature of 24°C, neutral range of pH level and equivalent dosage of activated carbon. The goal of this study was to examine and determine the adsorption efficiency of different types of granular activated carbon. The efficiency of granular activated carbon Norit 1240 W and Filtrasorb 400 are shown in tables 2 and 3. In these tables you can see the efficiency of each granular activated carbon and the comparison main pesticides and their metabolites.

Table 2. Adsorption efficiency [%] of granular activated carbon Norit 1240 W.

Time [min]	30	60	90	120	180	240	360
Acetochlor	18.78	20.23	24.19	26.02	29.69	30.48	31.63
Dimethachlor	5.75	7.43	8.94	9.39	10.95	12.37	13.39
Metolachlor	7.82	12.07	16.49	18.37	20.70	22.09	22.29
Alachlor	16.58	19.64	21.76	23.31	25.13	27.08	29.81
Metazachlor	5.38	8.36	10.62	12.79	15.50	16.71	17.18
Propachlor	6.30	8.05	10.13	12.00	14.26	15.68	17.48

Table 3. Adsorption efficiency [%] of granular activated carbon Filtrasorb 400.

Time [min]	30	60	90	120	180	240	360
Acetochlor	44.82	50.56	55.29	58.74	61.01	63.38	63.59
Dimethachlor	18.94	20.72	22.33	25.55	26.94	27.74	30.46
Metolachlor	34.48	39.86	45.93	51.24	53.20	54.36	55.06
Alachlor	46.02	48.28	50.80	53.54	55.69	56.82	58.11
Metazachlor	18.74	20.54	23.23	25.38	28.10	29.66	30.87
Propachlor	12.47	14.68	17.07	19.09	21.84	23.24	25.82

As it is obvious from the values in the tables, Acetochlor proved to be the most efficiently adsorbed from water, where the efficiency was in the range between 18.78 and 31.63% for Norit 1240 W and 44.82 to 63.59% for the Filtrasorb 400. For Norit 1240 W, the least efficient were pesticides Dimethachlor, Propachlor and Metazachlor as it can be observed in the table 1. Filtrasorb 400 proved to be less efficient in the adsorption of these pesticides also, but with a higher adsorption capability.

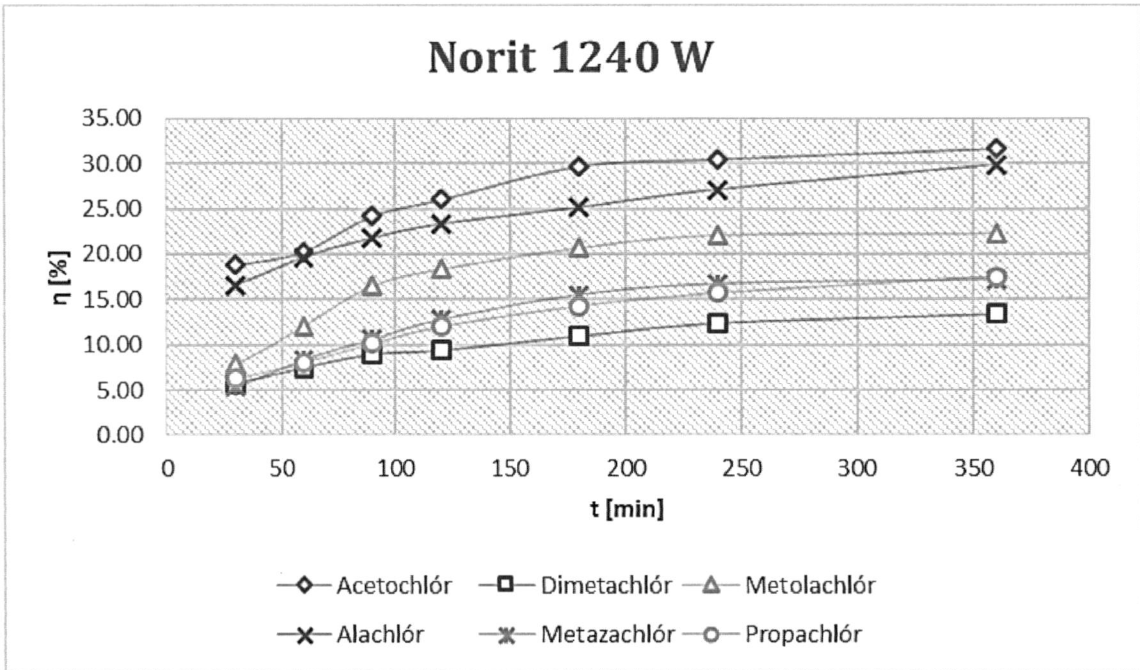

Figure 1. The adsorption efficiency of GAC Norit 1240 W for the time reaction in the range 30 to 360 min.

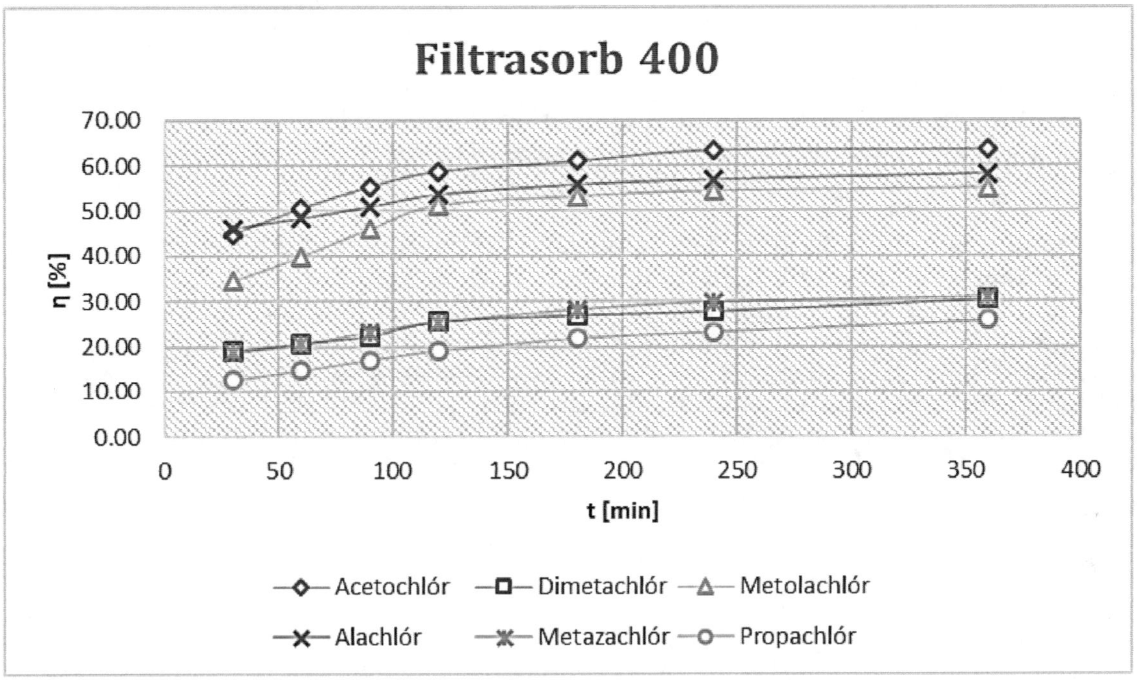

Figure 2. The adsorption efficiency of GAC Filtrasorb 400 for the time reaction in the range 30 to 360 min.

In the presented figures it can be observed that the adsorption efficiency of Alachlor, Acetochlor and Metolachlor was much higher than other pesticides. The adsorption efficiency for Norit 1240 W was in range from 18.78 to 31.63, 7.82 to 22.29 and 16.58 to 29.81 for Acetochlor, Metolachlor and Alachlor respectively. And the adsorption capacity for Filtrasorb 400 was in range from 44.82 to 63.59, 34.48 to 55.06 and 46.02 to 58.11 for Acetochlor, Metolachlor and Alachlor respectively. As it is obvious from these data, Filtrasorb 400 proved to be much more efficient type of granular activated carbon than Norit 1240 W. Each pesticide was adsorbed much more on the surface of Filtrasorb 400 carbon than Norit 1240 W. Adsorption capacity, considering contact time of material with water, for Norit 1240 W and Filtrasorb 400 was considerably different. Adsorption materials showed slightly higher adsorption capacity for Acetochlor, Alachlor and Metolachlor. Adsorption capacity for Norit 1240 W ranges from 0.157 to 0.265 μg/g, 0.125 to 0.225 μg/g and 0.073 to 0.208 μg/g for Acetochlor, Alachlor and Metolachlor respectively. Adsorption capacity for Filtrasorb 400 ranges from 0.375 to 0.532 μg/g, 0.347 to 0.438 μg/g and 0.322 to 0.514 μg/g for Acetochlor, Alachlor and Metolachlor respectively (tables for adsorption capacity is not shown in this article).

4. Conclusions

As presented in this article, the adsorption efficiency of the two studied granular activated carbons, Filtrasorb 400 proved to be more efficient than Norit 1240 W and fulfilled our expectations. Norit 1240 W, on the other hand, showed a bit disappointing results with low adsorption efficiency. The adsorption of the pesticide metabolites showed to be more difficult and less efficient than the main pesticide. This fact is explained by the missing functional groups, which provide better adsorption onto the surface of the activated carbon.

5. References

[1] Heydens W F, Lamb I C and Wilson A G E 2010 *Hayes'Handbook of Pesticide Toxicology (University of California, Riverside)* Third edition vol. 1 Chapter 82 pp 1753 – 1764

[2] Rodriguez-Narvaez O M, Peralta-Hernandez J M, Goonetilleke A and Bandala E R 2017 Treatment technologies for emerging contaminants in water: A review *Chemical Engineering Journal* 323 pp 361-380
[3] Rakic V, Damjankovic L, Rac V, Stosic D and Auroux A 2010 The Adsorption of nicotine from aqueous solutions on different zeolite structures *Water Research* 44 pp 2047-2057
[4] Worch E 2012 Adsorption technology in water treatment Fundamentals, processes, and modeling, de Gruyter.
[5] Marko I, Barloková D ano Ilavský J 2018 Removal of bromates from drinking water with seven types of sorbent materials Pollack Periodica vol. 13 no. 3 pp 231–240
[6] Yang R T 2003 *Adsorbents: Fundamentals and applications* Wiley
[7] Jankovska H 1991 *Active Carbon (Elis Horwood limited, Chichester, England)* p 280
[8] Marsh H, Rodríguez-Reinoso F 2006 *Activated Carbon.* Elsevier Science & Technology Books
[9] Petriláková A and Bálintová M 2011 Evaluation of sorbents efficiency on heavy metals removal from acid mine drainage *Pollack Periodica* vol. 6 no. 2 pp 139–146

Acknowledgments
Authors wishing to acknowledge assistance or encouragement from colleagues, special work by technical staff or financial support from organizations should do so in an unnumbered Acknowledgments section immediately following the last numbered section of the paper.

IOP Conf. Series: Materials Science and Engineering **867** (2020) 012032 doi:10.1088/1757-899X/867/1/012032

Comparative analysis of magnesium oxide boards properties

D Mokrenko[1] **and M Kozlovská**[1]

[1] Technical University of Kosice, Faculty of Civil Engineering, Institute of Construction Technology, Economic and Management, Vysokoskolska 4, 042 00 Kosice, Slovakia

Abstract. The current sustainability trends pose high demands on building materials and products. Those with potential to be labelled as an environmentally-friendly product or technology are getting preferred. One of the natural resources with such potential is magnesite, whose resources are widely found in Slovakia. The article deals with analysis of the magnesium oxide board properties as the most popular MgO-based building product. The comparative analysis of magnesium oxide boards with other available building boards (plasterboards, oriented strand boards) is included in the article and especially from the point of magnesium oxide boards' benefits.

1. Introduction

Currently, buildings account for around 40% of greenhouse gas emissions, with the USA, China and Russia contributing the most [1]. As a result, building ecology is becoming one of the main aspects that can influence the solution of the global climate protection problem and contribute to the sustainable development of the planet as a whole [2]. Most of the research focuses on issues of environmentally sustainable architecture and urbanism, integrated territorial development, the use of energy-efficient equipment, reducing heat loss in buildings and the like. There is less research on the development of environmental building materials and the term 'eco-material' is more of an advertising and marketing meaning. Ecological construction cannot be built without the use of environmentally-friendly materials and products that have a low carbon footprint, emission of hazardous substances, increased biological stability and other properties.

The ideal material for research is magnesium oxide board due to the following factors. It is a 100% environmentally friendly material that can replace almost all available building boards such as plasterboard, cement board, OSB board or plywood [4]; Slovakia is the 6th in MgO mining in the world [3]; MgO is part of various natural building materials that have been used for antiquity and have passed the test of time [5]. What makes a magnesium oxide board different from others is its fire grading, acoustic and heat insulation ability, water and damp resistance, the ability to reflect IR (infrared) rays, elasticity and strength, resistance to pests, fungi, rot, rodents, antiallergic and ecological [4]. Thanks to its properties, the magnesium oxide board is absolutely safe for health and makes the building environment safe and healthy inside. Based on the theoretical analysis and synthesis of acquired knowledge, the aim was to compare the properties of magnesium oxide board with similar plasterboard, cement board, OSB or plywood boards and to give it strengths over others.

2. Description of the present state

Magnesium oxide (burnt magnesium) is produced by the thermal decomposition of magnesium carbonate $MgCO_3$ [8]. High melting points as well as chemical inertness and thermal stability determine the preferred use of magnesia in the manufacture of refractory materials used in the high temperature processes of steel, cement, lime, glass and non-ferrous metals [9].

The world's largest consumer of magnesium is the refractory industry. The metallurgical industry needed high-quality refractory materials that can withstand molten metal. The magnesite has become one of such refractory natural materials because its main quality parameter is high heat resistance. Other sectors - agriculture, chemical, ceramic, construction, pyrotechnic, environmental [10] have their own requirements for specific properties of magnesium utilization.

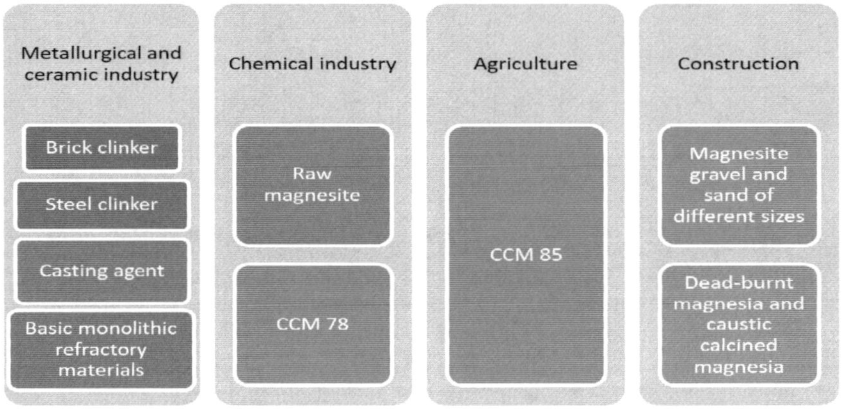

Figure 1. Examples of magnesium-based products across different industries.

Products for the metallurgical and ceramic industry consist of dead-burned magnesite clinkers, which are intended for the production of monolithic materials, products intended for the preparation of new soils, protection of linings of steel aggregates. Products used in the chemical industry are intended for the production of magnesium-based chemicals [11]; the product CCM 78 acts as a binder, the number 78 indicates how much of the magnesium oxide is in the powder [12]. CCM 85 is intended for use in the agricultural industry as a mineral additive for animals, which has a positive effect on their health due to the high magnesium content. In addition, it can be used for the production of fertilizers in the chemical industry, or for eliminating the effects of acid rain [11]. In the construction industry, magnesium is used to the production of building materials and components.

Magnesium oxide board (MgO board) is a large format construction board, generally in white color (Fir.2). MgO board is making from environmentally friendly natural material that is safe for health and does not release toxic substances when used. Important features of the magnesite board are that it is fire resistance, elastic and solid, frost-resistant, waterproof, ecological, antiallergic, heat-reflective, healthy, resistant to mold and rodent [7]. A unique feature of the magnesium oxide board is that the board actively consumes CO_2 from the air throughout its life cycle, thereby continuously cleaning it and improving the quality of the indoor environment [4]. The magnesium plate can be used practically everywhere. This product is used for ceilings, partitions, fireplaces, all types of interior and exterior wall cladding, ventilated facades, creative ceilings and wherever a building needs to be protected from fire, rot and mold. They can be easily tiled, plastered or painted.

According to the market analysis of MgO boards for years 2018-2023 [19] the global MgO board market is segmented into five major regions, namely, North America (U.S., Canada), Europe (U.K., France, Germany, Russia), Asia-Pacific (China, Japan, Australia), South America (Brazil, Argentina), and Middle East & Africa (South Africa, Saudi Arabia). There are many producers and even more suppliers in these regions. Each producer of MgO boards has its own recipe for his product, which results in some different properties. In general, magnesium oxide board is made up of "Sorel cement" - a mixture of magnesium oxide, provided with additional elements. Zhangjiagang Oriental Construction Material Co., Ltd. states [21], that magnesium oxide board is made of high purity magnesium oxide, MGO sulfate ($MgSO_4$), replace of magnesium chloride ($MgCl_2$), high strength fiber glass mesh, perlite, non-woven fabrics. According to North American MgO LLC [20] MgO board consists of 67% MgO, 18% SiO, 5% Fly Ash Filler, 9% additives and 1% Fiberglass gauze. Exclusive importer of MgO boards in Slovakia shows again a different structure of boards [6]: 50% MgO, 25% MgCl, 15% Cement-based sealants, 8% SiO, 1% Fiberglass gauze, 1% other.

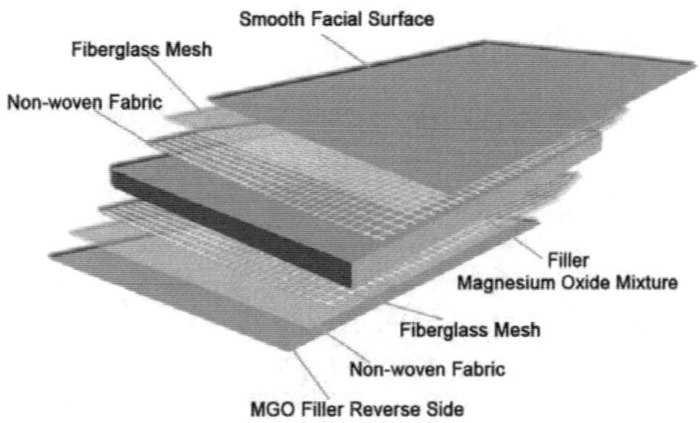

Figure 2. Examples of magnesium oxide board structure [22].

From the study of several sources such as Product Catalog by MGO Slovakia [4], Magnesium Oxide by G Swanson [7] and Technical data sheet by RETRAST TRADE [13] it is possible to consider the strengths and weaknesses of MgO boards (table 1).

Table 1. Strengths and weaknesses of MgO boards.

	Strengths
Fire resistance	The material is resistant to temperatures of 800 ° C and can withstand the effects of an open flame for up to 240 minutes when it is part of prefabricated structures.
Resistant to moisture	It does not deform, softens and rot under the influence of moisture or steam.
Frost resistance	The material can withstand more than 50 freeze-thaw cycles, which does not destroy it.
Strength	It has good impact resistance and increased strength. Material resistant to impacts greater than 5 KJ / m2. The flexural strength is from 15 MPa to 22 MPa.
Soundproofing	The material has good sound absorption and provides high sound insulation. Noise protection board 8 mm - 29 dB.
Thermal insulation	It has low thermal conductivity properties - the thermal conductivity coefficient is 0.216 W / mK.
Constant dimensional stability	It is not affected by moisture, heat or cold.
Thermal accumulated effect	This is a newly revealed property of the magnesite board, which is 50% higher for plasterboard.
	Weaknesses
Price	In today's building materials market, the price for this product is more expensive when compared to similar materials.
Demand	Demand for these materials is relatively low due to competition in the market cheaper materials and lack of consumer knowledge.
Availability	Relatively few countries are involved in MgO mining due to a lack of awareness of its properties.
Quality	Occurrence of defects due to not always correct material composition.

This comparison shows that MgO boards have more strengths than weaknesses and give the possibility to recognize that magnesium board has great potential in the construction market as such. Approximately, since 2012 MgO boards have used more intensively also in Slovakia. The exclusive importer is MGO Slovakia s.r.o., which sold it on the market under the name SUPERDOSKA. Importers themselves declare that there is no uniform recipe for the production of MgO boards and boards of individual producers differ significantly in their properties. In short, what applies to one board may not apply to another manufacturer's boards at all. From the references of many who used MgO boards in construction in Slovakia, the biggest discussed problem was the moisture of boards used in exteriors. Paradoxically, moisture resistance is reported as one of the benefits of MgO boards. There are two possible ways of explaining this "quality". Some manufacturers in the material composition do not adequately apply the necessary additives, or on the other hand, users do not apply all the technological procedures for the use of MgO boards outdoors - especially in terms of their subsequent surface treatment.

Despite the fact, the biggest barrier in the expansion of the use of MgO boards appears to be their price. Further comparisons should show whether the disadvantage in terms of price has its justification in relation to better material and construction parameters of MgO boards compared to other built boards.

3. Materials and methodology

To compare the parameters, building boards on different material bases were chosen, which are most often used for dry construction in Slovakia (table 2). The most used are plasterboards, produced on the basis of plaster. Plasterboard is a panel made of calcium sulfate dehydrate (gypsum), with or without additives, typically extruded between thick sheets of facer and backer paper. The most commonly used wood-based boards are OSB and Plywood. OSB (Oriented strand board) is a type of engineered wood similar to particle board, formed by adding adhesives and then compressing layers of wood strands (flakes) in specific orientations. Plywood is a material manufactured from thin layers or "plies" of wood veneer that are glued together with adjacent layers having their wood grain rotated up to 90 degrees to one another. These boards are traditional and long-used building boards (for example OSB board was invented by Armin Elmendorf in California in 1963). From modern innovative board was chosen board Cetris. Cetris is a cement-bonded particleboard with smooth naturally cement-grey surface. In order to make the parameters more objectively comparable, boards were selected with the same thickness of 12 mm.

Table 2. Selected products for comparison of building board parameters.

Material base	Product	Producer/Supplier
Magnesium	SUPERdoska Basic (MgO board), 12x1220x2440mm	MGO SLOVAKIA s.r.o.
Cement bonded particle	CETRIS-BASIC, 12x1250x3350mm	Bauintegra, s.r.o.
Gypsum	Plasterboard RB (A), 12x1200x2500mm	Saint-Gobain Construction Products, s.r.o.
Wood veneer	Plywood joinery Beech/Spruce B/CP, 12x1250x2500mm	JAF HOLZ Slovakia spol. s.r.o.
Wood fiber	OSB 3 4P+D uncut Kronospan, 12x625x2500mm	JAF HOLZ Slovakia spol. s.r.o.

Based on available data from technical literature, data sheet and catalogs of MGO Slovakia [4], Retrast Trade [13], OBI [14], Relend [15], GOST 3916.1-96 [16], STN 73 0862 [18] and Cetris [17] for comparison were selected following parameters:

- market prices per 1 m^2;
- mass weight – the higher density makes material stronger and more resistant to change its shape;
- elastic modulus - the more flexible the material, for example, allows to create curves in the construction design;
- thermal resistance – the higher the thermal resistance of the less heat is lost through the construction;
- acoustic resistance – the higher the value, the more resistant the material is to the sound transmission;
- beam strength, tensile strength, compressive strength, impact strength - the higher the value, the more durable the material;
- content of formaldehyde – the lower the formaldehyde content, the healthier the material;
- fire reaction class - ranging from the best class (A - nonflammable) to the worst class (C – flammable);
- resistant to moisture – possibility to use in the wet environment;
- mold and pest resistance – contributes to a healthier indoor environment.

4. Results

Data on individual boards found from the above sources were processed into a table 3. A gray background highlights the best result in a given property.

Table 3. Comparative analysis of MgO board, Cetris, Plasterboard, Plywood and OSB board properties.

Properties		MgO board	Cetris	Plasterboard	Plywood	OSB board
Market price	**[EUR/m²]**	**9,28**	**9,5**	**4,05**	**5,49**	**4,68**
Mass weight	[kg/m3]	1000	1350	680	600	620
Elastic modul	[MPa]	6045	6800	2200	3500	3500
Beam strength	[MPa]	18	11,5	2,2	16	16
Tensile strength	[MPa]	11	0,63	1,8-2,5	0,3	0,3
Compressive strength	[MPa]	18	15	6,8	25	20
Impact strength	[kg/m2]	5	-*	3,6	3,4	-*
Thermal resistance	[m²K/W]	1,14	0,037	0,1042	0,12	0,12
Acoustic resistance	[dB]	48	30	25	23	25
Content of formaldehyde	[g/100g]	0	0	0	max.8	max.8
Fire reaction class		A1	A2	A2	C2	C1
Resistant to moisture		Yes	Yes	No	No	No
Mold and pest resistance		Yes	Yes	No	No	No

* producers have not included this value in the product catalog;
- gray color means a **better result**.

In terms of prices follows that MgO board has a comparable (even lower) price than a board Cetris. Prices of these two modern boards (MgO board, Cetris) are about 50% higher than the prices of drywall or wood-based boards. The significant difference between boards that are "more expensive" and "cheaper" boards is in almost twice value of mass weight. Further more or less different strength parameters are deriving from this parameter. In terms of the modulus of elasticity, the values of MgO board and Cetris board are comparable, but in other strength parameters the MgO board appears to be better. Higher than all boards, MgO board has beam strength, significantly better is tensile strength and impact strength. In terms of compressive strength, it has somewhat lower values than wood-based panels, but better than Cetris and of course plasterboard. The MgO board has a nearly twice as better parameter as all other boards in terms of acoustic resistance, and even more significant are differences in thermal resistance. The MgO board also shows the best values in fire reaction, while of course the worst reaction to fire is wood-based boards. In terms of health safety, the worst are wood-based boards again, since they are glued and contain a small (although allowed) amount of formaldehyde. Only MgO board and Cetris board have resistance to moisture and mold and pest resistance. Wood-based boards and plasterboard do not have these properties.

It follows that the higher price of MgO board is clearly justified, as it shows significantly better values in almost all parameters. The Cetris board, which has even higher price, although only some of the parameters have comparable values to the MgO board.

5. Conclusion

Building boards are an integral part of all buildings. They have a wide range of applications (walls, partitions, floors, roofs, soffits) as well as a relatively broad material base (gypsum, wood fiber, particle, plywood, fiber cement). The paper analyzed the properties of a relatively new - innovative building board, made on a magnesium base. Many sources list a number of their benefits that improve the performance of traditional gypsum and wood-based building boards (fire, mold and pest, acoustic or thermal resistance, strength parameters - especially tensile and impact strength, health safety, resistant to moisture, mass weight and elastic modul). On the other hand, it appears that their wider use is hindered by a price that is higher than traditional building boards. The aim of the paper was to prove that a relatively higher price is justified in relation to better properties. For comparison were selected three traditional gypsum or wood-based boards and one modern board - based on cement bonded particle. In relation to the price were examined 12 physic-mechanical and environmental parameters. The results confirmed that the price of traditional boards is about 50% lower than for innovative boards, on the other hand, all properties showed worse parameters, some even more than 50% (except for compressive strength wood-based boards - whose value was comparable). Another finding points to the fact that, from the perspective of modern innovative boards, the price of MgO board is even lower than that of Cetris, with the MgO board showing significantly better values in six parameters (beam, tensile and compressive strength; thermal, acoustic and fire resistance). Based on the results, it can be concluded that MgO boards have an unambiguous justification and potential for their wider use, both in terms of price and in terms of their better parameters.

6. References

[1] World Business Council for Sustainable Development 2009 *Building Energy Efficiency: Market Transformation* (Switzerland)
[2] Benuzh A and Kolchigin M 2012 *Analysis of the concept of green construction as a mechanism for ensuring the environmental safety of construction activities* (Russia)
[3] Komelin I and Lysenko A 2019 *Interaction of magnesium production salt melts with atmospheric air. Izvestiya Vuzov. Tsvetnaya Metallurgiya* (Russia)
[4] MGO Slovakia s.r.o. 2011 *Product Catalog 2011* (Slovakia) [Online] Available at: http://www.benat.sk/files/katalog_superdoska.pdf
[5] Voroncov V and other 2008 *Natural materials in architecture: training aid* (Belgorod, Russia)
[6] STEELSOFT s.r.o. 2012 *Safety data sheet* (Kosice, Slovakia) [Online] Available at: http://www.mgodosky.sk/download/MgO_Dosky_KARTA_BEZP_UDAJOV_TSUS_SK.pdf
[7] Swanson G 2010 *Magnesium Oxide, Magnesium Chloride and Phosphate-based Cements* (Austin, Texas, United States)) [Online] Available at: http://www.greenhomebuilding.com/pdf/MgO-GENERAL.pdf
[8] Durisin J and other 2004 *Chemistry I examples and tasks* (Kosice, Slovakia)
[9] Knunanc I and other 1990 *Chemical Encyclopedia. Volume 2* (USSR)
[10] Shipcov V 2011 *Magnesian RAW: history, world capacity and resources of Karelia* (Russia)
[11] SMZ, a.s. Jelsava 2015 *CCM 85* (Slovakia) [Online] Available at: https://www.smzjelsava.sk/sites/default/files/ccm85-en-v4.pdf
[12] Selmo 2018 *Periclase powder PPE-88* (Dnipro, Ukraine) [Online] Available at: https://selmo.com.ua/poroshok_periklazovyy_ppe-88/
[13] RETRAST TRADE, s.r.o. 2012 *Technical data sheet* (Prague, Czech Republic) [Online] Available at: http://www.retrast.cz/downloads/TECHNICKY_LIST_RT_desky.pdf
[14] OBI [Online] Available at: https://www.obi.sk/
[15] Relend 2015 *Glass-magnesite sheet: application, disadvantages of the board* (Russia)
[16] GOST 3916.1-96 *General purpose plywood with outer layers of hardwood veneer. Technical specifications*(Russia)
[17] CETRIS® *CETRIS BASIC Technický list* (Slovakia) [Online] Available at: https://www.cetris.sk/dokumenty/technicke-listy/basic.pdf
[18] STN 73 0862 *Determination of flammability degree of building materials*
[19] Magnesium Oxide Board (MgO Board) Market 2018-2023 *Analysis, Segmentation And Forecast appeared first on America News Hour*
[20] North American MgO LLC 2018 *MagTech Ultra Magnesium Oxide Board - Environmental Product Declaration* [Online] Available at: https://assets.northamericanmgo.com/documents/MagTech-Ultra-EPD.pdf
[21] Zhangjiagang Oriental Construction Material Co., Ltd. *Innovative home building system OCM boards and Housing Panel*
[22] ZJG LEADER *MgO Board* (China) [Online] Available at: http://www.zjgleader.com/wp-content/uploads/2016/11/Mgo-BoardMagnesium-Oxide-Board.pdf

Acknowledgments
The activity presented in the paper is part of the research grant VEGA 1/0557/18 “Research and development of process and product innovations of modern methods of construction in the context of the Industry 4.0 principles”.

IOP Conf. Series: Materials Science and Engineering **867** (2020) 012033 doi:10.1088/1757-899X/867/1/012033

Trend analysis of drought indices in the eastern Slovakia

P Nagy[1], P Purcz[2], S Galas[3], M M Portela[4], Helena Hlavatá[5] and Dorota Simonová[5]

[1] Institute of Environmental Engineering, Department of Environmental Engineering, Faculty of Civil Engineering Technical University of Kosice, Vysokoskolska 4, Kosice 04200, Slovakia
[2] Institute of Construction Technology, Economics and Management, Faculty of Civil Engineering Technical University of Kosice, Vysokoskolska 4, Kosice 04200, Slovakia
[3] Department of Environmental Analysis, Geological Mapping and Economic Geology, AGH Krakow, Adama Mickiewicza 30, 30-059 Kraków, Poland
[4] Instituto Superior Tecnico, Universidade de Lisboa, Av. Rovisco Pais 1, 1049-001 Lisboa, Portugal
[5] Slovak Hydrometeorological Institute, Branch Office Košice Ďumbierska 26, 041 17 Košice, Slovakia

Abstract. Extreme weather changes have been increasing in the recent times also in the eastern Slovakia. Extremely dry periods alternate with above-average precipitation. In the paper the drought indices were evaluated: Standardized Evapotranspiration Index (SPEI) and Standardized Precipitation Index (SPI). Using the non-parametric statistical Mann-Kendall test. The evaluated period was 1960 – 2015 at selected river and climatological stations in the eastern Slovakia. The results showed that no clean significant trends occurred in all evaluated indices and stations.

1. Introduction

Climate change is a global phenomenon of the scientific community and the public. Recently, global air temperature has risen, potentially increasing global temperature [1-2]. The increase in temperature is associated with a basic hydrological cycle that modifies precipitation changes, evapotranspiration changes, and thus increases the risk of extreme events such as drought and tidal floods from extreme precipitation [2]. The characteristics of the drought may vary with respect to the area where the drought occurs, the time in which climate area and various other aspects. Drought events are also occurring in the Central Europe, which is in a temperate climate zone and where are located the selected weather stations [3].

Drought classification is a prerequisite for research on dry and wet seasons. Many different drought indices are currently being developed and are used according to the area we want to explore [4-6]. Indices can be classified mainly into four basic areas: meteorology, soil moisture, hydrology and remote sensing. I dealt with meteorological indices, which are Standardized Precipitation index (SPI) and Standardized Evapotranspiration Index (SPEI). The World Meteorological Organization recommends the non-parametric Mann-Kendall trend test to detect trends in climate variables [7-8].

2. Materials and Methods

2.1. Methods

The Standardized Precipitation Index (SPI) is a simple drought index that characterizes the meteorological drought in time periods ranging from 1 month to 36 months. In the paper 12 months SPI was used. The SPI can calculate them on the basis of a minimum time series of 20 years. I had available time series data of 30 or more years. The Drinc software was used for SPI calculation [9-11].

Standardized evapotranspiration index (SPEI), is based on water balance. The input data for this index calculation are precipitation and air temperature. This index combines potential evapotranspiration (PET) and precipitation. The PET calculated using the Thornthwaite method and the SPEI is calculated be using the Rgui statistical program [13-14].

The Mann-Kendall trend test is a statistical non-parametric test method and is considered as a good tool for trend analysis. The Mann-Kendal test is independent of data distribution and less sensitive to missing data and is therefore most commonly used to assess trends for various climatological and hydrological variables [12].

2.2. Datasets and study area

For calculation of SPI and SPEI indices I used monthly average temperatures and precipitation from selected climatological stations in eastern Slovakia. Table 1 describes selected climatological stations. It presents the evaluated period for which we had available data and average, minimum, maximum values of climatological variables. The were provided by the Slovak Hydrometeorological Institute branch office Košice. Seven climatological stations were assessed. The climatological stations Červený Kláštor, Poprad and Mníšek nad Hnilcom are located in the north of the eastern Slovakia, the stations Streda nad Bodrogom and Košice are located in the south of Slovakia. Humenné, Kamenica nad Cirochou and Krásny Brod are located in the eastern part of eastern Slovakia and the Mníšek nad Hnilcom and Švedlár are located in the western part of the eastern Slovakia.

Table 1. Input data.

Station	Value variables	Evaluated period	Min monthly	Max monthly	Average monthly
Poprad	Precipitation [mm]	1960-2014	0.4	220.4	50.1
	Air temperature[°C]	1960-2014	-24.7	25.65	6.15
Chmeľnica	Precipitation [mm]	1980-2010	4.6	259.3	64.7
Červený Kláštor	Air temperature [°C]	1961-2014	-29.9	26.7	6.34
Mníšek nad Hnilcom	Precipitation [mm]	1980-2010	1.1	261.2	59.8
Švedlár	Air temperature [°C]	1961-2014	-22.4	25.4	6.56
Humenné	Precipitation [mm]	1980-2014	4.8	271.7	61.4
Kamenica nad Cirochou	Air temperature [°C]	1960-2014	-21.1	29.6	8.7
Krásny brod	Precipitation [mm]	1986-2010	1.4	245.5	69.2
Tisinec-Stropkov	Air temperature [°C]	1963-2010	-22.3	26.8	8.14
Streda nad Bodrogom	Precipitation [mm]	1980-2014	0.6	260	53.1
Košice	Air temperature [°C]	1960-2014	-18.4	29.75	9.1

As you can see from table 1 precipitation varies from 0.4 o 271.7 mm per month. Air temperature varies from -29.9 to 29.75.

IOP Conf. Series: Materials Science and Engineering **867** (2020) 012033 doi:10.1088/1757-899X/867/1/012033

3. Results

Using the nonparametric Mann Kendall test, is possible to identify trend in indices. In table 2 12-month Standardized Precipitation Index in the evaluated climatological stations are interpreted. There is no trend in Poprad station resulting from the alternation of dry and wet seasons, and in other stations there is a negative trend, with dry periods prevail.

Table 2. Mann-Kendall trend test of 12 months SPI.

Climatological station seasons	Mann-Kendall trend test		Test interpretation
Poprad 1960-2014	Kendall's tau	0.154	H0: There is no trend H1: There is a trend **There is no trend**
	S	221.000	
	Var(S)	17967.000	
	p-value (Two-tailed)	0.101	
	alpha	0.05	
Mníšek nad Hnilcom 1980-2010	Kendall's tau	0.631	H0: There is no trend H1: There is a trend **There is a trend**
	S	773.000	
	Var(S)	14291.667	
	p-value (Two-tailed)	< 0.0001	
	alpha	0.05	
Humenné Humenné 1980-2010	Kendall's tau	0.384	H0: There is no trend H1: There is a trend **There is a trend**
	S	471.000	
	Var(S)	14291.667	
	p-value (Two-tailed)	< 0.0001	
	alpha	0.05	
Krásny Brod 1986-2010	Kendall's tau	0.552	H0: There is no trend H1: There is a trend **There is a trend**
	S	676.000	
	Var(S)	14290.667	
	p-value (Two-tailed)	< 0.0001	
	alpha	0.05	
Chmeľnica 1980-2010	Kendall's tau	0.520	H0: There is no trend H1: There is a trend **There is a trend**
	S	637.000	
	Var(S)	14291.667	
	p-value (Two-tailed)	< 0.0001	
	alpha	0.05	
Streda nad Bodrogom 1980-2010	Kendall's tau	0.411	H0: There is no trend H1: There is a trend **There is a trend**
	S	501.000	
	Var(S)	14275.000	
	p-value (Two-tailed)	< 0.0001	
	alpha	0.05	

In table 3 is presented in the evaluated climatological stations. There is a trend in the stations Poprad, Mníšek nad Hnilcom and Chmeľnica in the SPEI index and it is decreasing, so there are more dry periods. In Humenné, Krasny Brod and Chmeľnica there is no trend characterized by alternating periods.

Table 3. Mann-Kendall trend test of 12 months SPEI.

Climatology station season	Mann-Kendall trend test		Test interpretation
Poprad 1960-2014	Kendall's tau	-0.201	H0: There is no trend H1: There is a trend **There is a trend**
	S	-393.000	
	Var(S)	28427.000	
	p-value (Two-tailed)	0.020	
	alpha	0.05	
Mníšek nad Hnilcom and Švedlár 1980-2010	Kendall's tau	0.365	H0: There is no trend H1: There is a trend **There is a trend**
	S	148.000	
	Var(S)	2842.000	
	p-value (Two-tailed)	0.006	
	alpha	0.05	
Humenné and Kamenica nad Cirochou 1989-2010	Kendall's tau	-0.067	H0: There is no trend H1: There is a trend **There is no trend**
	S	-29.000	
	Var(S)	3141.667	
	p-value (Two-tailed)	0.617	
	alpha	0.05	
Krásny Brod and Tisinec-Stropkov 1980-2010	Kendall's tau	0.225	H0: There is no trend H1: There is a trend **There is no trend**
	S	62.000	
	Var(S)	1625.333	
	p-value (Two-tailed)	0.130	
	alpha	0.05	
Chmeľnica and Červený Kláštor 1986-2010	Kendall's tau	0.448	H0: There is no trend H1: There is a trend **There is a trend**
	S	182.000	
	Var(S)	2842.000	
	p-value (Two-tailed)	0.001	
	alpha	0.05	
Streda nad Bodrogom and Košice 1980-2014	Kendall's tau	-0.200	H0: There is no trend H1: There is a trend **There is no trend**
	S	-87.000	
	Var(S)	3141.667	
	p-value (Two-tailed)	0.125	
	alpha	0.05	

4. Conclusions

The work was focused on occurrence of trend in climatological indices Standardized Precipitation Index (SPI) and Standardized Evapotranspiration Index (SPEI). The results show that in the SPI no trend was only in the station Poprad, in other stations evaluated the trend has a decreasing character. In the SPEI, no trend is in three stations: Humenné/Kamenica nad Cirochou, Krásny Brod/Tisinec, Streda nad Bodrogom/Košice and the other three: Poprad, Mníšek nad Hnilcom/Švedlár, Chmeľnica/Červený Kláštor. Where the trend does not occur is the alternation of dry and wet periods but is not possible to determine the frequency of the alternation of periods. The results would be useful for further research into climate and hydrological models, which would be relevant for strategic water resource planning and drought prevention.

The main factors of climate change are decreasing precipitation and rising air temperature, which contribute to drought formation. As the temperature rises, the surface water vapor increases, which implies that precipitation should increase but this is not the case. Drought occurrence continuous research and should be given more attention. Reducing the length and intensity of droughts would be possible if the drought could be more accurately predicted and water management could be better managed. During the dry season it is necessary to increase the soil moisture, which would be beneficial for agriculture, overestimating water levels in the river basins and in wet periods it was necessary to retain water in the country.

5. References

[1] Mishra S, Ashok K and Singh V P 2010 *J. Hydrol.* **391** 202-216
[2] Mesaros P, Smetankova J and Krajnikova K 2020 Lecture Notes in Civil Engineering 47 307-314
[2] Mesaros P, Mandicak T, Behun M and Smetankova J 2018 *ICETA 2018 - 16th IEEE International Conference on Emerging eLearning Technologies and Applications*, Proceedings **10** 367-372
[2] Fendekova, M, Poorová J and Slivova V 2018 *Hydrologické sucho na Slovensku a prognóza jeho vývoja* (Bratislava: Comenius University)
[3] Didovets I, Krysanova V, Bürger G, Snizhko S, Balabukh V, and Bronstert A 2019 *J. Hydrol.: Reg. Stud.* **22** 100590
[4] Milanovic M, Gocic M, and Trajkovic, S 2015 *Agric. Agric. Sci. Procedia* **4** 167-174
[5] Zeleňáková M, Purcz P, Ondrejka Harbuľáková V, and Oravcová A 2016 *Desalin. Water Treat.* **57** 2693-2701
[6] Harbulakova V O, Zelenakova M, and Sugarekova M 2018 *International Multidisciplinary Scientific GeoConference: SGEM: Surveying Geology & mining Ecology Management* **18** 391-398
[7] West H, Quinn N, and Horswell M 2019 *Remote Sens. Environ.* **232** 111291
[8] Tirivarombo S, Osupile D, and Eliasson, P 2018 *Phys. Chem. Earth. Parts A/B/C* **106** 1-10
[9] Svoboda M, Hayes M, and Wood D 2012 *Standardized precipitation index user guide* (World Meteorological Organization Geneva, Switzerland)
[10] Tigkas D, Vangelis H, and Tsakiris, G 2015 *Earth Sci. Inf.* *8* 697-709
[11] Svoboda M, and Fuchs B 2016 *Handbook of drought indicators and indices*
[12] Mann H B 1945 *Econometrica: Journal of the Econometric Society* 245-259
[13] Thornthwaite C W 1948 *Geogr. rev.* **38** 55-94
[14] Spinoni J, Barbosa P, De Jager A, McCormick N, Naumann G, Vogt J V, and Mazzeschi M 2019 *J. Hydrol.: Reg. Stud.* **22** 100593

Acknowledgments

This work has been supported by the Slovak Research and Development Agency by supporting the project SK-PT-18-0008 and project SL-PL-18-0033.

IOP Conf. Series: Materials Science and Engineering **867** (2020) 012034 doi:10.1088/1757-899X/867/1/012034

The conformity of the tools of selected software programs for 4D building modeling

Y Nechyporchuk[1] **and R Bašková**[1]

[1]Technical university of Kosice, Faculty of Civil Engineering, Department of Construction Technology, Economy and Management, Slovakia

Abstract. 4D modeling is a part of the BIM (Building Information Modeling) platform. It represents the design approach, when the building is designed not only in space (3D) but also in time (3D + time). The aim of the article is to compare Autodesk Navisworks and Synchro Pro in terms of the conformity with the requirements for creating and using 4D building models. The introduction represents the resource search where the advantages and ways of using 4D modeling for construction and the situation of its applying in Slovakia and abroad are summarized. Besides, the 4D model of the building Water World, created in Autodesk Navisworks and Synchro Pro, is introduced there. In the conclusion, there is a comparison of the tools of this software in user's several points of view.

1. Introduction

Today is not the first year that the era of the fourth industrial revolution (Industry 4.0). "Construction 4.0" is a branch of Industry 4.0. This term is used to refer to the digitalisation of the construction industry. BIM is one of the elements of Construction 4.0. The AEC (architecture, engineering, and construction) industry is in a constant search for new methods and technologies that will reduce costs and project implementation time, increase productivity and quality. Building information modeling (BIM) offers ways to achieve these objectives [1]. BIM provides a technology information of all elements of the building that can be used in designing, planning, and managing the project in collaboration and coordination environment. Building information modeling is a digital representation of all kinds of building data, and is used as a shared knowledge resource to make resolutions on a facility throughout its lifecycle [2].

Today BIM is the most popular and promising technology in AEC industry. The usage of BIM is growing in most countries. UK has become one of the world's leaders in the implementation of BIM thanks to the government support, more than two thirds of the country's firms use BIM. Since 2016, all government construction projects in the UK are required to achieve BIM Level 2 [3]. In Singapore 80% of the construction industry uses BIM (compulsory for practitioners to submit architectural or engineering plans in the BIM format for regulatory approval). In Spain BIM used to be mandatory in public construction projects from December 2018 and in infrastructure projects by July 2019. More and more governments support the BIM initiative. Since 2014, according to EU Directive 2014/24/UE, the member states are invited to encourage and require the usage of BIM in construction projects financed by EU public funds as of 2016. Year by year the usage of BIM increases [3].

The basis of the BIM model is a geometric 3D model; in addition, information on time (4D BIM model), finances (5D BIM model) or other project resources depending on need can be provided.

2. 4D modeling in construction

The four-dimensional model was recognized as an approach that improves the methods of planning and construction management [4]. The usage of 4D adoption is increasing within the construction sector and upper management starts realizing the potential benefits of using it as a viable planning tool [5]. 4D models can be used to understand the relationships between the various stages and processes of construction work [4], identify and eliminate errors in the project before they occur at the construction site, coordinate work and improve the management of the work of teams and subcontractors, provide information on the current status of the project for hour plan adjustment and other important decisions, in addition, it can be used to create simulations and animations of the construction process for presentation her purposes.

2.1. The level of detail for 4D modeling

The level of detail (LOD) shows with what accuracy and how detailed the data provided in the project is developed. The level of detail of the information in a building model depends on the functions it will be used for. For 4D CAD schedule analysis model must contain temporary works (scaffolding, excavation) and show how the construction will be phased (how deck pours will be made, the sequence of wall erection, the reach of tower cranes, and so forth) [7].

Since the 4D model presents graphical and temporal data, it has the graphical level of detail and the temporal level of detail (or temporal resolution).

The graphical level of detail of a simulation is entirely reliant on the detail contained within the preliminary 3D CAD model. If a 3D model contained little graphical detail some tasks were overlooked in the simulation, while if the 3D model had too much detail the project planner was not interested in many aspects, but the work got complicated due to the large file sizes [6].

The temporal level of detail relates to the interval of time that is shown between state changes in the 3D model and how often the 3D model is updated to show the progress of construction operations. The construction process is highly dynamic and that's why the state of the construction 'product' can change hourly. Conversely, some construction operations are slow to evolve and in this case it may take a week for any significant state change to occur in relation to the 3D model [6].

Both graphical and temporal detailing of 4D modeling should be such that it can contribute to the adoption of effective management decisions and, consequently, the construction processes during the project. As such, the LOD for 4D is still not well defined, and there is no way to develop a simulation with a specific approach to LOD management [8].

2.2. Current 4D software applications

Today there are enough software for 4D modeling that provide different levels of capabilities and toolkits. Consider some of them.

Vico Schedule Planner and 4D Player. That's a part of the platform Vico office from company Trimble. The building model is imported from adequate modeling application. Here, the 3D model can be formed in multiple formats such as Revit, Tekla, ArchiCAD, CAD-Duct, IFC files, SketchUp and 3D DWG files [9]. Scheduling can be created in Vico Schedule Planner or schedule dates can be imported from Primavera or MS Project. Changes in the scheduling system are automatically reflected in the 4D visualization [7].

Innovaya Visual 4D Simulation. Innovaya Visual Simulation associates Building Information Models (BIM) objects with scheduling activities, performing 4D construction planning and constructability analysis. The building model can be imported from any 3D design program. It doesn't have tools for scheduling, but schedule dates can be imported from Primavera or MS Project [10] [11].

Navisworks (Autodesk). Navisworks technologies enable us to assemble all project data into a single model that enables us to make design decisions more effectively, manages the project, reduces errors. The building model can be imported from any 3D design program thanks to the support of the IFS format. Navisworks allows you to import a schedule from Primavera, MS Project and Microsoft Excel. It has a module that allows collision detection in real time [11].

Synchro Pro. Software created by Synchro Software, Ltd., purchased in 2018 by Bentley Systems Incorporated. This is a well-developed, complicated stand-alone 4D BIM tool [7]. Synchro has tools for creating schedule (uses the CPM method) or you can import data from specialized software (Primavera or MS Project). Import of 3D models is possible using the IFC format or special plugins.

Powerproject BIM (Elecosoft). Powerproject BIM is an extension for Powerproject which allows you to import IFC data into the project plan and to associate 3D model elements with a schedule. The construction schedule is created in Powerproject, the 3D model is imported in IFC format.

3. Comparison of selected software programs for 4D modeling

The article will compare two programs for 4D modeling - Autodesk Navisworks and Synchro PRO. Most often these programs can be found in the literature over the past ten years. The main characteristics that the requirements for 4D modeling software should have will be considered. Comparisons are made based on the compliance with the above requirements and specifications.

3.1. Characteristics of the software for 4D modeling

For comparison, we will use characteristics that are important for representatives of programs in this area, affect the construction process or simplify the use of these products and / or their functions.

3.1.1. Level of Scheduling. Planning and scheduling the project is one of the most important part of construction phase. Because it depends on this plan and schedule how quickly and efficiently the project will be completed, as well as the financial part. In general, planning is the organization and distribution of work, depending on the availability of resources. Gantt charts are great for managing various kinds of task relationships. For example, the task sequence, the dependency of tasks, the completion time of tasks, the duration of every single task and more. In this case, a construction Gantt chart offers a great network based on a logical progression toward the final success of the project.

From the point of view of Scheduling, it is possible to divide programs for 4D modeling into two categories: programs that have functions for creating a schedule and programs that do not have such functions, but use graphics created in other programs. Programs of the first category also often use graphics from third-party programs, but even in this case they can edit data without resorting to the programs from which the import was performed. Representatives of the second category are deprived of these advantages. An important role is played by what formats the program supports for importing schedule and whether it's possible to update the data if it was changed in the program from which the import was made. The level of scheduling in programs for 4D modeling is usually lower than that which is needed to draw up a high-quality construction schedule.

3.1.2. Work with 3D models. Since 4D models consist of graphic and temporal data, the ability of the program to work with graphic data of the project is important. It is also important which formats can be imported. Of course, you can import models using the IFC format, but with this approach quite often errors occur (incorrect display of objects, bindings, etc.). The number of files that can be downloaded and which you can work with simultaneously, updating the model when making changes, the ability to adjust the model directly in the program, and the user interface are of great importance.

3.1.3. Linking of 3D model/s with schedule. Linking 3D objects to the construction plan can be done manually or automatically. Automatic linking is possible when names of the object match the items in the construction plan. Automatic linking is only relevant for simple projects, and even in this case the result of such a link needs to be modified. However, this feature sometimes allows you to reduce the time to create a 4D model.

3.1.4. Clash detection. One of the functions of 4D modeling is collision detection. Standard clash detection identifies clashes between static objects such as beams, columns, pipes, and ducts. 4D clash detection can detect clashes between permanent and temporary objects, whether static or moving (such as tower cranes and trucks), clashes of equipment and material delivery [7]. A systematic step-by-step phasing in 4D BIM eliminates the possibility of encountering any workflow clash and avoids site downtimes to keep up the man-efficiency of the task force.

3.1.5. Simulation and animation. To search for 4D collisions, we need information about all objects and their movement (location and time of temporary objects, the movement of machines and mechanisms, for example, the movement of cranes and others). To do this, the software must distinguish different types of objects (temporary, permanent, demolition objects, and so on) and be able to simulate movement at a construction site. Also, one of the reconciliations of 4D modeling is the use of materials as presentation materials. And the possibility of creating animation and its export plays a role here.

3.2. Comparison of Navisworks and Synchro Pro based on the characteristics of the software for 4D modeling

Based on the categories presented above, comparisons were made between the Synchro Pro and Navisworks software. The comparison results are shown in table 1.

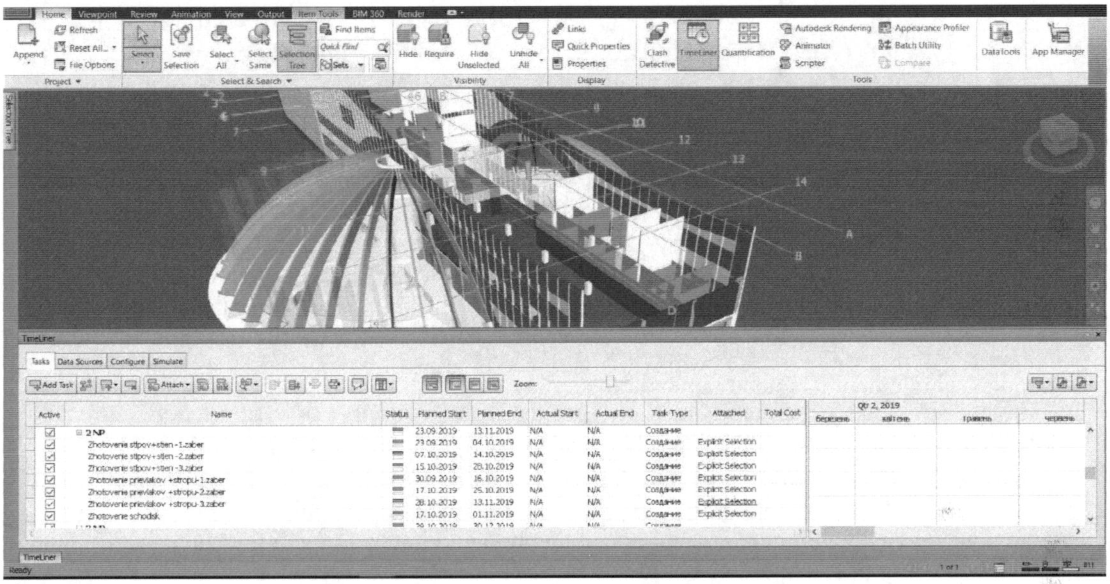

Figure 1. Autodesk Navisworks workspace.

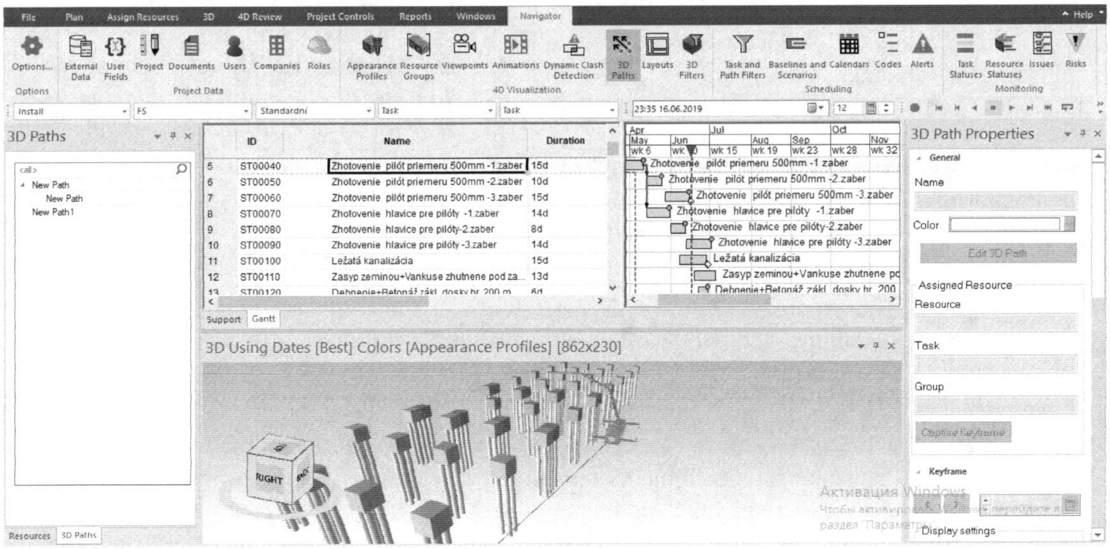

Figure 2. Synchro Pro workspace.

Table 1. Comparison of Navisworks and Synchro Pro based on the characteristics of the software for 4D modeling.

Characteristics of the software	Navisworks	Synchro Pro
Level of Scheduling	It has no tools for network process analysis (Gantt chart). It allows you to create a hierarchically structured list of construction processes and assign the start and the end date to the process, displaying a line schedule of processes. The construction schedule can be imported from MS or Primavera or downloaded in CSV format. You can easily update the data if it has been changed in MS Project or Primavera. You can load several charts at the same time and work with them.	He has tools for creating a Gantt chart. It has tools for summarizing resources (histograms). It does not have tools for analysing labor and material resources. It uses a network analysis method with only simple links for time scheduling of processes that does not meet the requirements for quality tools for planning construction processes and resources in time and space. Synchro allows you to import a schedule from Primavera, MS Project, Safran Project and Microsoft Excel. It has more options for editing schedule in its software environment than Navisworks.
Work with 3D models	Models created in any product from the Autodesk family can be imported. Simple update of the 3D model. It is possible to work with several 3D models at once. Supports import using IFC format.	To import a 3D model into the Synchro software environment, IFC format or plugins can be used, for example, for import from Revit (There is an API interface that allows you to write your plugins). Data can be updated when a change is made in the source file of the 3D model. It is possible to work with several 3D models at once. You can also attach equipment and mechanisms to tasks from the schedule.
Linking of 3D model/s with schedule.	Automatic or manual linking of 3D objects and schedule is possible. Convenient filter system for manual connection.	Only manual connection of 3D model objects and construction schedules is possible.
Clash detection	The program is one of the best for searching for 3D collisions and preventing them. Automatic check for 4D collisions is not available.	Automatic search for spatial collisions is possible. Automatic check for 4D collisions is not available
Simulation and animation	It is possible to assign a certain type of task to an object (creation, temporary objects), you can also create your own types of tasks, for example, to designate different subcontractors. The movement of machines and mechanisms cannot be	It is possible to assign a certain type of task to an object (creation, temporary objects), you can also create your own types of tasks, for example, to designate different subcontractors. The movement of machines and mechanisms can be tied to tasks, for example, moving a

	tied to tasks. A wide range of animations for presentation purposes.	crane before starting work on the next part of the object. A wide range of animations for presentation purposes.

4. Conclusion

In this article, the state of 4D modeling, possibilities and capacities of software tools for 4D modeling were analysed. A comparison was made between two Synchro Pro and Navisworks products. For comparison, the possibilities and tools that are important in creating and working with 4D models have been analyzed. Each of the programs has its weak and strong points, and the choice between them should be based on the analysis of the tasks that are being implemented. For example, if the source data were developed in a product from the Autodesk family or you do not need to display the operation of machines and mechanisms, but only the construction process itself, then it may be better to use Navisworks. If you need to model the construction site, the operation of machines and mechanisms and the like in more detail, then Synchro Pro is probably the best option. The choice of program should be based on the analysis of the source data, available resources and desired results. It is also impossible not to notice that both of these programs do not have the function of automatic detection of 4D collisions. This function is very important and would significantly increase the efficiency and level of demand for this software. These programs have a sufficient number of useful functions and methods of application, but in addition to this there is a great potential for development.

5. References

[1] Azhar S 2011 Building Information Modeling (BIM): Trends, Benefits, Risks, and Challenges for the AEC Industry *J. Leadership Manage. Eng.* **11** 241

[2] Amjed N Hasan and Sawsan M Rasheed 2019 The Benefits of and Challenges to Implement 5D BIM in Construction Industry *Civ. Eng. J.* **5** 412

[3] Shimonti P 2018 BIM Adoption Around the World: How Good are We? *J. Geospatial World*

[4] Butkovic B and Heesom D 2017 Towards A Framework For Multi-LOD 4D BIM Simulations *Proc. Int. Research Conf. 2017: Shaping Tomorrow's Built Environment* (University of Salford, Salford, UK) p 578

[5] Gledson B 2015 Investigating the diffusion of 4D BIM Innovation *Proc. 31st Annual ARCOM Conf.* vol.1 (Lincoln, UK) p 641

[6] Heesom D 2006 Interactive Generation of 'Multi-Level of Detail' 4D CAD Simulation *6th Int. Conf. on Construction. Application of Virtual Reality* (Orlando, Florida,USA)

[7] Sacks R Eastman C Lee G and Teicholz P 2018 *BIM Handbook: A Guide to Building Information Modeling for Owners, Designers, Engineers, Contractors, and Facility Managers* (John Wiley & Sons, Inc.: Hoboken, New Jersey) p 229,253,247

[8] Guevremont M Hammad A 2018 Multi-LOD 4D Simulation in Phased Rehabilitation Projects *17th Int. Conf. on Computing in Civil and Building Engineering* (Tampere, Finland)

[9] Pucko Z Suman N and Klansek U 2014 Building Information Modeling Based Time And Cost Planning In Construction Projects *OTMC: An Int. J.* **6** 958

[10] Aouad G Lee A Wu S and Onyenobi T 2012 *Computer Aided Design Guide for Architecture, Engineering and Construction* (SPON Press: NY, USA) p 100

[11] Martynets I 2010 Clash detection in three-dimensional modeling of industrial plants and objects *J. CAD and Graphics (САПР и Графика)* **3** 50

Acknowledgments

The paper presents a partial research results of project VEGA 1/0557/18 "Research and development of process and product innovations of modern methods of construction in the context of the Industry 4.0 principles" and project VEGA 1/0828/17 "Research and application of knowledge-based systems for modelling cost and economic parameters in Building Information Modelling".

Spatial-temporal statistical analysis of daily rainfall for Košice station in eastern Slovakia

A. Repel[1], V Jothiprakash[2], Helena Hlavatá[3] S Galas[4], and M M Portela[5]

[1] Institute of Environmental Engineering, Department of Environmental Engineering, Faculty of Civil Engineering Technical University of Kosice, Vysokoskolska 4, Kosice 04200, Slovakia
[2] Department of Civil Engineering, Indian Institute of Technology, Bombay, Mumbai 400076, India
[3] Slovak Hydrometeorological Institute, Branch Office Košice Ďumbierska 26, 041 17 Košice, Slovakia
[4] Department of Environmental Analysis, Geological Mapping and Economic Geology, AGH Krakow, Adama Mickiewicza 30, 30-059 Kraków, Poland
[5] Instituto Superior Tecnico, Universidade de Lisboa, Av. Rovisco Pais 1, 1049-001 Lisboa, Portugal

Abstract. Hydrologic processes, especially rainfall is characterized for their variability in time and space. These processes are known as stochastic processes. For engineering practice is very important to know the behavior of time series for long period. This paper is focused on spatial-temporal analysis of daily rainfall for Košice station in eastern Slovakia. Time series of daily rainfall from period 1951 to 2018 was analyzed by basic statistical analysis and also by descriptive statistical analysis. Results of basic statistical analysis showed the basic parameters of time series, such as average daily rainfall, maximum daily rainfall or total annual daily rainfall and other basic parameters. Descriptive analysis is used to describe behavior of a time series. Results showed, that time series of daily rainfall in Košice station is stationary, homogeneous and there is no noise in time series, which is good for using this time series in further research

1. Introduction

Distribution of rainfall and its changes during the time is one of the most important aspects of climate change and it needs detailed investigation is the time. One of the most important climatic variables is rainfall and it is because it affects important hydrological events such as floods and droughts. Rainfall and other hydrologic processes are known as stochastic processes. It means, that they evolve in space and time in a way that is partly predictable or deterministic and partly random [1].

Rainfall and other hydrological processes are characterized for their variability in space and in time. There are many ways how to analyzed the variability, for example by examining trend, stationarity, homogeneity, periodicity or noise [2]. Long term rainfall data presents time series, which is a series of data points indexed in time order.

Many authors in the world analyzed historical time series of rainfall in many countries in the world. In Asia, between Nepal and Tibet in the Himalayan region was showed that the annual precipitation amount is increasing in the period between 1943 and 1993 [3]. Similar results were presented also in other studies in many Canadian series [4] or in many places in South America, where authors analyze the number of rainy days per year and the daily rainfall amount [5]. In Europe, studies on Ireland series collected from 1940 to 1990 confirm the increasing of the total annual precipitation, in particular on the west coast [6]. Study from Ireland also shows how the intensity of rainfall is growing up and the return period of the extreme events is reducing. Results of this study is, that 30-year return period events are becoming 10-year events nowadays [7]. In Italy, it was proved that dry periods tend to be longer with increasing variability of the length of the dry spells and this is also associated with shorter duration of rain episodes with an evident effect on rainfall extremes [8]. In Slovakia, there are not many studies of rainfall time series in last years. Daily rainfall data were analyzed by Dub in 1950 and second, very complex analyze was done in 1973 by Šamaj and Valovič. Since these studies, not many authors have dealt with this issue. Statistical analysis of daily rainfall is very important for further use of data, for example for engineering practice.

2. Study area and data

Slovakia country with area of 49 035 km² is located in central Europe. There are very diverse topographic patterns throughout Slovakia. The relief declines from the Greater Carpathian range in the North, with an elevation 2655m at Gerlachovsky peak to lowlands in western and eastern part of Slovakia with lowest altitude around 100m. The city of Košice is located in the eastern part of Slovakia in the Košice region, near the border with Hungary (20km), Ukraine (80km) and Poland (90km) at the crossroads of historic trade routes. It is about 400 km from the capital Bratislava. City of Košice lies in the river Hornad basin in the altitude 202 m above the sea level and GPS coordinates of the city are 48°43'15.994"N, 21°15'28.447"E. Territory, where Kosice city is situated belongs to the northern temperate climatic zone with a regular change of four seasons of the year and variable weather with a relatively smooth distribution of precipitation during the year. Average annually temperature in Košice is around 9°C, average annually rainfall is around 650mm [9].

For the statistical analysis of rainfall data was used manually recorded daily data from period 1951 to 2018, the observation period is 68 years. Time series of manually recorded daily rainfall data is shown in Figure 1. These data were provided by Slovak hydrometeorological institute (SHMI). Basic and descriptive statistical analysis were done for the time series. Statistical analysis includes trend, homogeneity, periodicity and noise for the time series.

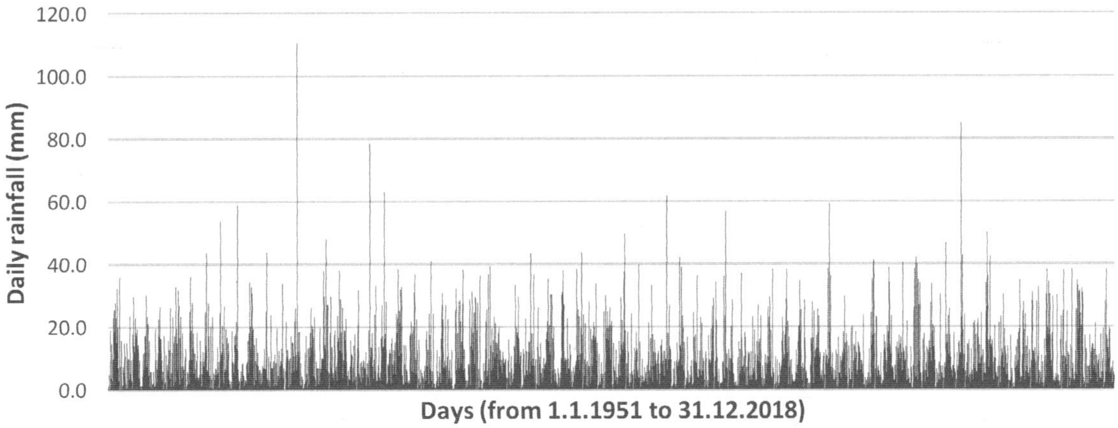

Figure 1. Time series of daily rainfall in Košice station for the period from 1951 to 2018

3. Methods and methodology

The temporal variation and behavior in the rainfall data time series was analyzed using several statistical tests. Two types of statistical analysis were done and it was basic statistics and descriptive statistics. Basic statistics consist of mean (average), standard deviation, skewness, kurtosis, maximum, percentage of 0 and number of 0 what presents number of days without rainfall (dry days). Descriptive statistics consist of trend analysis, stationarity, homogeneity, periodicity and noise tests. Trend in the time series is estimated using Mann Kendall (MK) test, Sen´s slope and Spearman´s rank correlation. Stationarity of time series was tested by ADF test, Phillips-Perron test and KPSS test. Stationarity tests serves to analysis the behavior of the mean and variance in time series. Homogeneity tests indicates the variations in the time series (like stationarity tests) bud homogeneity analysis also gives the break point where changes happened in time series (if there is break point in time series). Homogeneity tests used in present study are Pettit test, SNH test, Buishand´s test, Von Neumann ratio. For determining white noise in time series, these tests are used: Box Pierce, Ljung-Box, McLeod-Li, Fisher´s kappa, Bartlett´s Kolmogorov-Smirnov. For determining the periodicity of time series, spectral analysis was used. All the tests were carried out at 5% significance level. In all the tests, p value is calculated, and if the p value is less than 0.05 (5% significance level), then the null hypothesis is rejected and the alternative hypothesis is accepted. The methodology and types of tests used in the present study are shown in Figure 2.

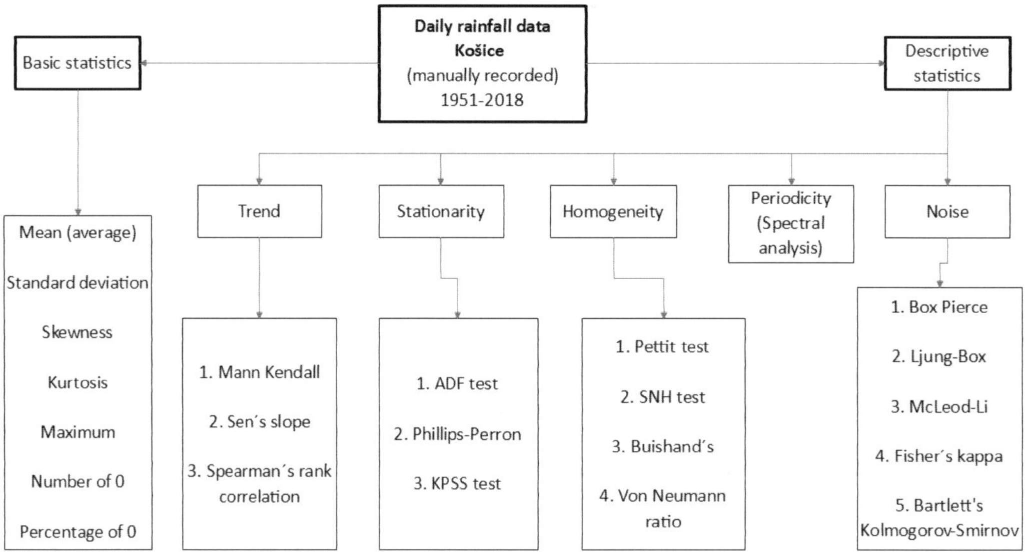

Figure 2. Flow chart of methodology used in the study

4. Results and discussion

4.1. Basic statistics

The basic characteristics of the time series, such as maximum or average rainfall are shown in Table 1. Average annual rainfall is 614.8 mm per one year. Maximum of annual rainfall was in 2010 and minimum was in 1956. Maximum daily rainfall was observed in august 1963 and it was 110.5 mm.

Table 1. Basic characteristics of the daily rainfall time series in Košice

Total of rainfall (in mm) (1.1.1951-31.12.2018)	41 806.10
Average daily rainfall (in mm)	1.684
Average annual rainfall (in mm)	614.8
Maximum and minimum annual rainfall (in mm)	958.9 (in 2010), 418.9 (in 1986)
Maximum daily rainfall (in mm)	110.5 (13.8.1963)
Average number of dry days in year	220 (60.3%)

IOP Conf. Series: Materials Science and Engineering **867** (2020) 012035 doi:10.1088/1757-899X/867/1/012035

For each day in the year, the maximum of daily rainfall was found in 68 years of observing. Time series of maximum daily rainfall for each day is shown in Figure 3.

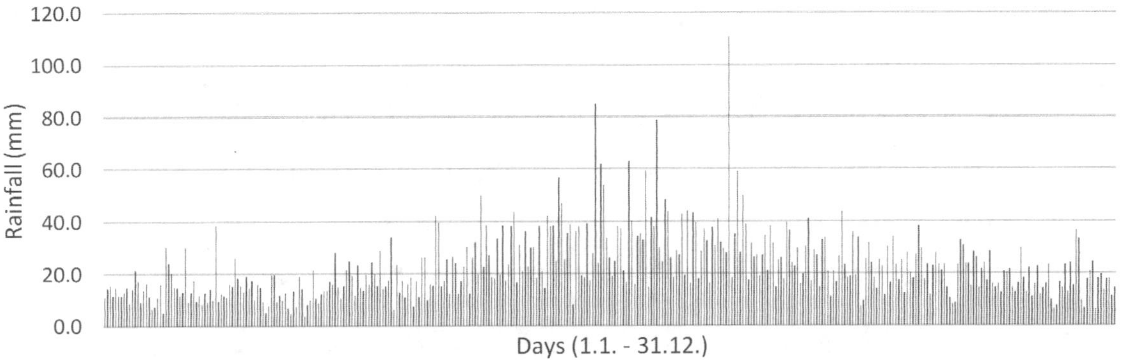

Figure 3. Maximum of daily rainfall for each day for the period from 1951 to 2018

From the Figure 3 it is clear, that maxims of the daily rainfall occurred especially in summer period from Jun to August and these rainfalls are presented as storms from the heat. Months with the lowest maxims of daily rainfall are from January to April.

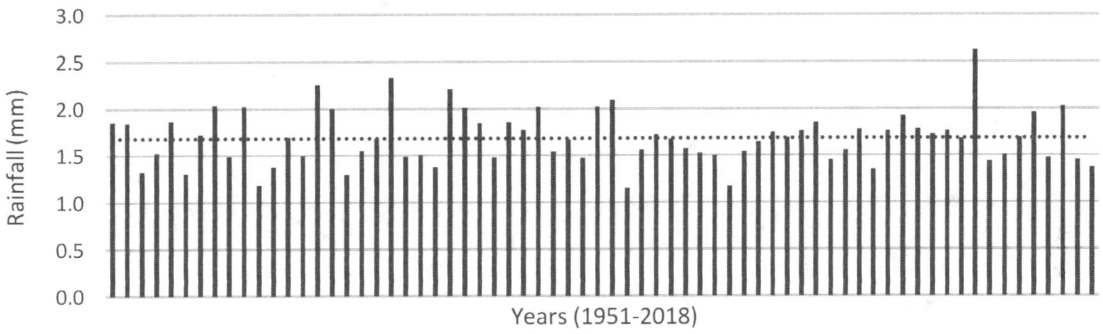

Figure 4. Average daily rainfall in the period from 1951 to 2018

Average daily rainfall remains constant over the observation period, what is shown in Figure 4. There is also no trend in average daily rainfall during last 68 years (from 1951 to 2018). Average daily rainfall was maxim in 2010 and it was 2.627mm per day.

Maximum of daily rainfall in each year is shown in Figure 5. It can be said that there is also no change in these values. Highest maximum daily rainfall was observed in 1963 and in 2010.

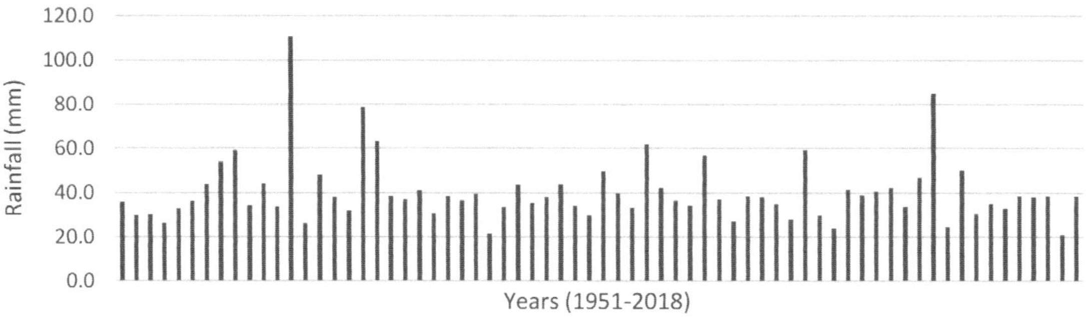

Figure 5. Maximum daily rainfall in the period from 1951 to 2018

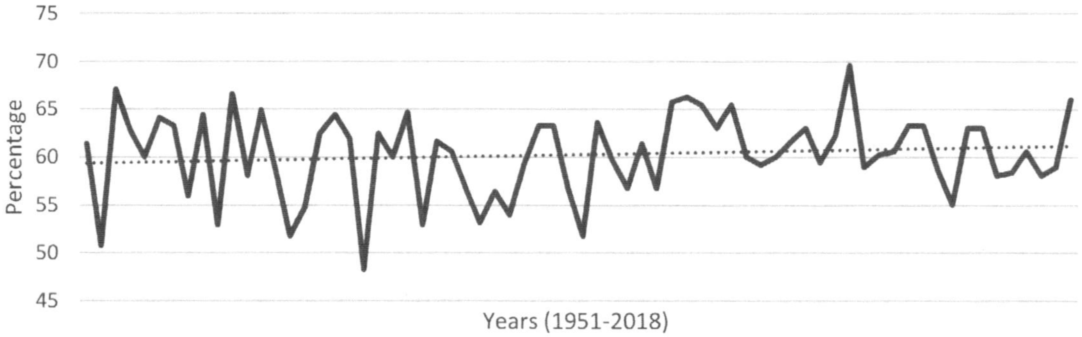

Figure 6. Percentage of dry days (days without rainfall)

In the Figure 6, there is percentage of dry days for each year in the period from 1951 to 2018. Percentage of dry day means the quantity of days without rainfall in each year. Average percentage is 60% and from the Figure we can said, that there is positive trend in percentage of dry days. It means, that there are more dry days in one year than in the past.

4.2. Descriptive statistics

Descriptive statistical analysis consists of trend, stationarity, homogeneity, periodicity and noise tests and was done on time series of daily rainfall. A description of each method is given in the section Methods and the methodology. The results of descriptive statistical analysis are in Table 2.

TREND	**STATIONARITY**	**HOMOGENEITY**	**PERIODICITY**	**NOISE**
Mann Kendall: Kendall´s tau = -0.0066 p-value = 0.16 alpha= 0.05 **NO TREND**	ADF test p-value= 0.00001 alpha= 0.05 **STATIONARY**	Pettit test: p-value= 0.278 alpha= 0.05 **HOMOGENEOUS**		Box pierce: p-value= 0.0001 alpha= 0.05 **NO NOISE**
Sen´s slope: Slope= $6.1x10^{-5}$ **NO TREND**	Phillips-Perron p-value= 0.0015 alpha= 0,05 **STATIONARY**	SNH test p-value= 0.550 alpha= 0.05 **HOMOGENEOUS**	365.25 days	Ljung-Box: p-value= 0.0001 alpha= 0.05 **NO NOISE**
Spearman´s rank: Coeff.= 0.35 **WEAK TREND**	KPSS test p-value= 0.972 alpha= 0.05 **STATIONARY**	Buishand´s : p-value= 0.6992 alpha= 0,05 **HOMOGENEOUS**		McLeod-Li: p-value= 0.0001 alpha= 0.05 **NO NOISE**
		Von neumann ratio:		Fisher´s kappa:

		p-value= 0.3915 alpha= 0.05 **HOMOGENEOUS**		p-value= 0.0001 alpha= 0.05 **NO NOISE**
				Bartlett´s Kolmogorov-Smirnov: p-value= 0.0001 alpha= 0.05 **NO NOISE**

The trend in the daily rainfall time series was estimated using MK test, Sen´s slope test and Spearman´s rank correlation coefficient methods. Globally we can say that there is no trend in the time series. The stationarity test was carried out by ADF test, Phillips-Perron test and KPSS test. The result of all the three methods concludes that the daily rainfall series in Košice station is significantly stationary at 5% significance level as shown Table 2. It means, that any hydrologic model can be used to model daily rainfall. It can be concluded that the daily rainfall series is not showing much variation in mean and standard deviation. The homogeneity was tested using Pettitt's test, standard normal homogeneity (SNH) test, Buishand test and von Neumann's test. Time series of daily rainfall in Košice station is homogeneous, what was proven by all four tests at 5% significance level. It means, that there is no the break point in the time series. From the table 2 it is clear, that there is no white noise in the time series and periodicity of the series is 365 days.

5. Conclusion

Climate change is a frequently mentioned term nowadays. This paper is focused on the evaluation of daily rainfall and statistical analysis of daily rainfall in Košice station in the period from 1951 to 2018. Statistical analysis was divided into two parts, basic statistical analysis and descriptive statistical analysis. Basic statistical analysis consists of basic parameters, such as mean, maximum, minimum, total, standard deviation, number and percentage of zeros and other. Basic statistical analysis is done in tabular and also in graphic form. Average daily rainfall in Košice station is 1.684 mm per day, average annual rainfall is 614.8 mm per year. The rainiest year was the 2010 with total annual rainfall 958.9 mm. Average daily rainfall remains constant over the last 68 years. There has been a slight increase in the number of days without rainfall in recent years, while the annual total rainfalls remain approximately the same. Descriptive statistical analysis shows the characteristics of the time series of daily rainfall. Descriptive statistical analysis consists of trend, stationarity, homogeneity, periodicity and noise tests. These parameters were tested by multiple tests. The results show that the time series of daily rainfall does not show positive or negative trend and the time series is stationary, homogeneous with periodicity 365.25 days and there is no noise in time series. These results show that the data in time series are reliable and should be used for further analysis or more complex statistical methods.

6. References

[1] V. Te Chow, *Applied hydrology* (Tata McGraw-Hill Education, 2010)
[2] Drissia T K, Jothiprakash V and Anitha A B 2019 Statistical classification of streamflow based on flow variability in west flowing rivers of Kerala, India *Theor. Appl. Climatol.* **137** 1643-1658
[3] Cislaghi M, De Michele C, Ghezzi A and Rosso R 2005 Statistical assessment of trends and oscillations in rainfall dynamics: Analysis of long daily Italian series *Atmos. Res.* **77** 188-202
[3] Sharma K P, Moore B and Vorosmarty C J 2000 Anthropogenic, climatic, and hydrologic trends in the Kosi Basin, Himalaya *Clim. Change* **47** 141-165
[4] Hamilton J P, Whitelaw G S and Fenech A 2001 Mean annual temperature and total annual precipitation trends at Canadian biosphere reserves *Environ. Monit. Assess* **67** 239-275
[5] Lucero O A and Rozas D 2002 Characteristics of aggregation of daily rainfall in a middle-latitudes region during a climate variability in annual rainfall amount *Atmos. Res.* **61** 35-48

[6] Hoppe H and Kiely G 1999 Precipitation over Ireland—Observed change since 1940 *Physics and Chemistry of the Earth, Part B: Hydrology, Oceans and Atmosphere* **24** 91-96
[7] Kiely G 1999 Climate change in Ireland from precipitation and streamflow observations *Adv. Water Resour.* **23** 141-151
[8] Cislaghi M, De Michele C, Ghezzi A and Rosso R 2005 Statistical assessment of trends and oscillations in rainfall dynamics: Analysis of long daily Italian series *Atmos. Res.* **77** 188-202
[9] L. Miklós 2002 *Landscape atlas of the Slovak Republic*

Acknowledgments
This work has been supported by the Slovak Research and Development Agency by supporting the project SK-PT-18-0008 and project SL-PL-18-0033.

IOP Conf. Series: Materials Science and Engineering **867** (2020) 012036 doi:10.1088/1757-899X/867/1/012036

Possibilities of usability of sediments flowing into a combined sewer network

G Rózsa[1], R Csicsaiová[1], M Šutús[1] and I Škultétyová[1]

[1] Slovak University of Technology in Bratislava, Faculty of Civil Engineering, Department of Sanitary and Environmental Engineering, Slovakia

Abstract. Primarily there are gravel and sand particles contained, which are, together with the surface runoff, discharged through the rainwater inlets into the sewer system. The concentration of such sediments varies depending on the urbanization density and built infrastructure in the site. In the part of cities with increased occurrence of trees and greenery are the sediments mostly of organic origin (leaves, grass, branches, etc.). With fully functional urban stormwater drainage are these sediments collected directly in the drain inlets and during the dry weather conditions discharged by the authorized persons into the maintain truck. In our research, we focus on the quantity of the rainwater runoff sediments in general view while analysing the possibility and methods of utilization of these sediments.

1. Introduction

During rainfall, rainwater runoff is created on the impermeable surfaces, which is drained to the sewerage system through rainwater gullies. Stormwater drain inlets (also known as gullies) are an essential part of the sewer system responsible for collecting and draining excess water from infrastructure. These objects are commonly designed as gully pots, referring to the presence of a sand trap [1]. However, it is well known that wastewater and urban runoff transport a variety of solid particles that can create obstructions in the sewer system, thus deteriorate its operation. Sediment related blockages may result in the release of raw sewage into the environment due to a reduced hydraulic capacity that increases the probability of local sewer congestion and relating events, e.g. sewer system overflow [2]. By capturing suspended particles from the runoff, silting, and wear of downstream sewer components are reduced [3]. The gully pots contribute to reducing the pollutions of receiving water bodies, especially during storm events. Nowadays, it is widely recognized that the accumulation of sediments in the urban sewer system can lead to its congestion, flooding, premature overflow, and water pollution [4].

The composition of sediments is varied and depends on the urbanization of the site. From the areas with an increased occurrence of vegetation, sediments are predominantly of organic origin (leaves, grass, branches, etc.) on the contrary to the sites that are mostly paved with concrete and asphalt. Rainwater runoff collects these sediments into the specialized baskets situated in gully pots. These baskets are emptied during the dry weather season by road maintenance personnel and are further processed either for recovery (composting) or disposal, depending on the level of contamination. The frequency of cleaning gully pots depends on the maintenance budget, location, or vulnerability of the urban environment. The effectiveness of this type of management depends on the number of blockages in the system within the specified interval [5]. If data on the operational condition of gully pots are utilized to determine the maintenance interval, it is possible to balance the effectiveness of strategies and the associated resources to provide cost-effective service provision [1].

This study focused on solving the current state of use of sediments flowing into the combined sewer network and their future effective method for recycling, in order to improve the current state of waste management in cities and municipalities.

Figure 1. Gully pot with sediments by flowing rainwater (Authors, 2019).

2. Methodology

At present, there is still no exact definition of biodegradable waste. Literature reviews and legislation define biodegradable waste in different ways.

Biodegradable waste is waste that is able to be decomposed by an anaerobic or aerobic way, in particular: food waste, paper, and paperboard waste, garden and park waste [6]. This general definition is taken from the European Directive 1999/31/EC on the landfill of waste.

Bio-waste is a biodegradable garden and park waste, food waste, and kitchen waste from households, restaurants, catering and retail establishments, and comparable food business waste [6]. This definition of bio-waste does not include, for example, forestry, agricultural, manure, sludge, and other biodegradable waste such as natural textiles, paper or processed wood [7].

The Waste Management Program and the Landfill Restriction Strategy mainly focus on the sorting of components of this type of waste that need to be reduced in landfills. By joining the European Union, each member state has committed to meet the objectives set in the European Union's directives. Similar acting is expected from the countries outside the European Union. The main aim of the European Union is to reduce the landfilling of biodegradable municipal waste. Bio-waste should be managed by the waste hierarchy established by the European Directive, which gives priority to the prevention and recovery of waste over its disposal based on the strategy for the circular economy. The most suitable forms of recovery of biodegradable waste include composting and anaerobic digestion.

European legislation on the management of biodegradable waste gives the Member States the freedom to choose the most appropriate form of treatment and recovery of this type of waste, but the conditions and requirements laid down in the Directives must be respected.

Based on expert estimations, bio-waste represents from 40 to 45% of the total amount of municipal waste. Early, the production of municipal waste is expected to increase in the European Union, in connection with increasing living standards and a higher rate of urbanization. Therefore, it will mainly depend on the residents, whether they will separate waste and increase the level of separation of municipal waste.

Biodegradable municipal waste has nowadays an excellent potential for reuse since this waste does not dispose of harmful properties. When choosing the method and technology of biowaste treatment, it is necessary to know the amounts and types of bio-waste, as they have different physical properties and chemical composition and, therefore, may not be suitable for the chosen treatment method. The problem arises in the improperly chosen way of the handling and disposal of waste, such as landfilling and incineration. These methods of waste disposal produce substances that adversely affect the environment. From a microbiological and the water eutrophication point of the view, the sludge from sewage treatment plants, fertilizer wastes from agricultural production, and kitchen and restaurant wastes are hazardous biowaste. Risks can be eliminated by the correct method and technology of biowaste treatment [8].

Before selecting a suitable method for processing or disposal, it is necessary to monitor the qualitative and quantitative parameters of the selected site. It is also required to follow the ratios between paved and pervious areas, the function of buildings, and the quantity of the greenery around. During the monitoring of the samples of sediments from gully pots, parameters such as mass, humidity, pH, heavy metal concentration, volatile substances, and sediment composition should be mainly controlled. Based on the information obtained from the monitoring of sediments, the most effective method for further waste management - its disposal or recovery, will be proposed.

3. Possibilities of usability of sediments

All the produced wastes must be treated, disposed of or recovered in an appropriate way. The choice of method is mainly based on the characteristics of the waste. Regarding biodegradable waste, it can be disposed of or recovered by various processes having different environmental impacts. Landfilling and incineration are the most harmful ways for disposal of biowaste. By using other treatments, we can more effectively use the potential of this type of waste while reducing negative environmental impacts among such methods of processing, respectively. Such techniques include composting, anaerobic digestion, mechanical-biological treatment, energy-related processing, and biofuel production (such as chips, pellets, briquettes, alcoholic fermentation, and re-esterification). When choosing the method and technology of biowaste treatment, it is necessary to know the amounts and types of bio-waste, as they have different physical properties and chemical composition and, therefore, the chosen treatment method may not be the most suitable one [8].

3.1. Landfilling

Biodegradable waste that is disposed of by landfilling is harmful to the environment since it produces landfill gas and leachate. Landfill gas contains methane and carbon dioxide, which are greenhouse gases. If they are not trapped and escape freely into the atmosphere, they contribute significantly to the greenhouse effect. Methane is particularly problematic because it is able to capture approximately 20 times more infrared energy than carbon dioxide [9]. In addition to landfill gas, landfill leachate may also harm the environment, as it may contaminate groundwater and soil if landfill sealing and containment systems are not provided or are broken. Since landfilling generally has many adverse effects on the environment, there is a tendency to limit this type of disposal. Nor this disposal method is considered to be a sustainable solution in waste management [10].

3.2. Combustion

Incineration of biodegradable municipal waste in waste incineration plants is also not considered as an environmentally friendly way of disposal because it generates toxic gases (e.g. dioxins) due to its high humidity and salt (chlorine) content. In addition to dioxins in the presence of chlorine, the combustion of this type of waste also produces highly aggressive hydrochloric acid, hazardous chlorinated hydrocarbons, and furans. Burning bio-waste from gardens is also a burdensome element for the environment. The smoke produced by this imperfect combustion process contains harmful substances for human health, such as carbon monoxide, various hydrocarbons, tars, but also the dioxins mentioned [8].

3.3. Composting and Anaerobic digestion

Composting and fermentation processes are among the methods of processing various types of bio-waste. The aim of the biological processing of organic material is its highest possible microbial degradation (stabilization). Deodorization (odor removal) is also achieved in this way. Biodegradable municipal waste from urbanized areas often needs to be further cleaned (metal, plastic, glass) and pasteurized. Therefore, a sorting and pasteurization line is included before the mechanical treatment step of the substrate.

3.3.1. Composting. Composting is a natural biochemical process in which microorganisms degrade organic material in the presence of oxygen. The resulting products of the process are carbon dioxide (CO_2), water (H_2O), heat, and compost. Composting is an adequate way of recovering organic waste, since in terms of ecosystem and sustainable development, composting by the natural cycle of biogenic elements is associated with natural energy transfer and biological matter degradation [11].

The composted material must contain both a sufficient amount of degradable organic components and nutrients without the presence of harmful substances. The most commonly used materials for composting include bio-waste such as green garden waste, garden and park waste, industry and municipal waste with a similar composition of household (municipal) waste [11].

The final quality of compost is influenced by several major technological factors, such as the composting method chosen and the physical, chemical, and microbiological properties of the composted raw materials. Other factors include composting time, formula recipe, preparation of raw materials before composting and storage, monitoring of the composting process, and the final maturity and stability of the compost [12].

3.3.2. Anaerobic digestion. Anaerobic digestion (methane fermentation, methanization or anaerobic composting) is a process in which the biodegradable organic matter gradually decomposes without access of air. Several groups of anaerobic microorganisms are involved in this process, where the product of one group becomes the substrate of the other, so the failure of one group can disrupt the entire process system. The final product is biogas and non-degradable residues of organic matter (so-called digestate), which is stabilized from the hygienic and sensory point of view [13].

The main product of anaerobic digestion is biogas. Biogas is a colorless gas whose main components are methane (55-75%), carbon dioxide (23-43%), and hydrogen (around 2%). The advantage of fermentation as opposed to composting is the possibility of processing and recovering liquid organic wastes such as canteen liquid wastes, cooking oils, greases and fat-trapping contents or bio-waste contaminated with no toxic components such as heavy metals. Furthermore, organic municipal and industrial waste - kitchen waste, garden waste, and various organic waste from the garden - is processed; organic special agricultural wastes (possible manure treatment), municipal sewage sludge produced in aerobic wastewater treatment stage and specific industrial wastes. From the environmental and economic point of view, mixing household waste with sludge or waste contaminated with toxic components is necessary [11].

4. Conclusion

Sediments from drain gully pots can be recovered or disposed of by various methods. According to the analysis of these techniques, it can be concluded that the most optimal process for the treatment of sediments from gully pots is anaerobic digestion. The advantage of this method is that contaminated heavy metal materials can also be further processed. In the presence of higher oil concentrations in sediments, it is necessary to mix with contamination-free materials to optimize the treatment process. Treated waste in a biogas plant can be used as an organic fertilizer if it complies with the limits of foreign substances (especially in the case of heavy metals), or it can be composted or treated for agricultural substrates.

5. References

[1] Post J A B, Pothof I W M, Dirksen J, Baars E J, Langeveld J G and Clemens F H L R 2016 *Water Research* Vol. **88** pp 245-56

[2] Xie Q, Bharat C, Nazim Khan R, Best A and Hodkiewicz M 2017 *Urban Water Journal* Vol. **14** pp 669-75

[3] Ashley R, Bertrand-Krajewski J L, Hvitved-Jacobsen T and Verbanck M 2004 *Solids in Sewers* (London - IWA Publishing Company) ISBN 1 900222 91 4

[4] Morrison G M, Revitt D M and Ellis J B 1995 *Water Science and Technology* Vol. **31** I 7 pp 229-36
[5] Swanson L 2001 *Int. J. Production Economics* **70** pp 237-44
[6] The Act of The National Council of the Slovak Republic *No.79/2015 on waste and on amendments to certain acts*, Slovakia 2015
[7] Ministry of environment of the SR, *Strategy to limit the disposal of biodegradable waste in landfills* 2010
[8] Moňok B *Disposal with biodegradable waste* 2010 Košice, Priatelia Zeme – SPZ. ISBN 978-80-967972-8-8
[9] Eleazer W E, Odle W S, Wang Yu-Sheng, and Barlaz M A 1997 *Environmental Science & Technology* Vol. **31** I 3 pp. 911-17
[10] Promoting a European framework for corporate social responsibility, *Green Book* 2001 Luxembourg: Office for Official Publications of the European Communities
[11] Čermák O and Kebísek M 2008 *Waste Management. Composting* Bratislava, Slovak University of Technology in Bratislava ISBN 8022729208
[12] Pliva P and team 2006 *Creation, process and management of composting process* Praha Research Institute of Agricultural Technology ISBN 80-86884-11-2
[13] Židek M 2004 Anaerobic digestion of selected substrates on laboratory fermenter Brno Energy from biomass III - seminar

Acknowledgments
This work was supported by the Scientific Grant Agency of the Ministry of Education, Youth and Sports of the Slovak Republic and the Slovak Academy of Sciences within the project VEGA 1/0574/19, co-funded by the European Regional Development Fund and by the Slovak Research and Development Agency under the contract No. APVV-18-0203.

IOP Conf. Series: Materials Science and Engineering **867** (2020) 012037 doi:10.1088/1757-899X/867/1/012037

Database of building information models - identification and design of parameter structure

J Smetanková[1]

[1] Technical University of Kosice, Faculty of Civil Engineering, Institute of Construction Technology, Economic and Management, Vysokoskolska 4, 042 00 Kosice, Slovakia

Abstract. Evolution in the construction industry is constantly progressing. Many applications are currently being developed to track the building's life cycle. Increasing emphasis is placed on managing the entire life cycle of buildings. New technologies are evolving, for example building information modelling. The output of this technology is a building information model. Building information model processes digital content related to the whole life cycle of a building. Developers create information models. These models are then archived to mass repositories such as databases and libraries. The databases contain models that contain graphical information and selected non-graphical information. The main aim of the article is to design and identify the structure of a database, that will be containing graphic models with selected non-graphical information based on the acquired knowledge.

1. Introduction

Lifestyle change in modern society is reflected in increased levels of flexibility, productivity and technological development. The construction industry supports process automation. Automation in construction industry brings achieve the following advantages:

- uniform quality with higher accuracy,
- replacing of human labor in tasks involving heavy physical work or monotonous work,
- replacement of human work in dangerous environments,
- making tasks that are beyond human capabilities easier.
- increasing productivity and efficiency of work at reduced costs,
- economic improvement,
- improvement of working environment [1].

Automation in the building industry is manifested, for example, by the use of drones, the automation of prefabricated construction of houses, the application of robotics, the application of information modelling technology and support autonomous equipment for example stiles machinery, hundegger, Trimble, 3D printing concrete and other [2].

2. Building information models

Construction of buildings is a time-consuming and costly process. The following steps are important to improve the construction and use management process:

- improve cooperation between information suppliers and users,
- provide more accurate background information to support decision-making,
- provide a standard way to store information,
- integrate data collection and information retention into business processes [3].

To improve the construction process, has been developed building information modelling technology Building information modelling is an important technology in the building industry. BIM is a process that involves generating and managing physical and functional project information. The output of the process is building information models. Building information models are digital files that describe every aspect of a project and support decision making throughout the project cycle [4].

Objects are also defined as parameters and relations to other objects, so that if there are changes in a related object, dependent or adjacent ones will automatically change or adjust, as well [4].

Building Information Modelling is a digital representation of the physical and functional characteristics of an element. The model represents a shared source of knowledge that forms a reliable basis for decision making during the life cycle of an element [3].

The traditional building design relied heavily on two-dimensional technical drawings. BIM extends traditional technical drawings to 3D dimension. 3D models to give primary spatial dimensions (width, height, and depth).In addition to the three-dimensional model, BIM also provides additional information such as for example the fourth dimension - time parameter, the fifth dimension - cost parameters, the sixth dimension - environmental and sustainability analysis of buildings and the seventh dimension taking into account facility management across the life cycle [5][6][7][8].

The use of information models goes beyond the planning and design phase of a project to its use and demolition. Within the building lifecycle, it includes cost management, construction management, project management and operations management processes.

3. Database design

The database is an organized collection of data. The data is stored and accessed electronically from a computer system. The database is usually managed by a database management system (DBMS). The database management system consists of an integrated set of computer software that allows the user to interact with one or more databases and provides access to all the data contained in the database [9].

The created database provides four main functional groups:

- data definition –create, change, delete definitions that define,
- update –input, change and delete actual data,
- retrieval–providing information - directly applicable or suitable for further processing by other applications,
- administration– user registration and monitoring, data security, performance monitoring, data integrity maintenance, concurrency control and information retrieval [10][11][12].

Data within the most common types of databases in operation today is typically modelled in rows and columns in a series of tables to make processing and data querying efficient. The data can then be easily accessed, managed, modified, updated, controlled, and organized. Most databases use structured query language (SQL) for writing and querying data [9].

3.1. Types of Databases

Since its establishment in the early 1960s, databases have been constantly evolving. One of the primary types of databases was the creation of navigation databases as a system of hierarchic database (tree model- one relationship between items) and network database (more flexible model that allowed multiple relationships between items). These systems were characterized by simplicity but were inflexible. The development of database types is constantly evolving. We know the following types:

- relational databases – the items are organized as a group of tables with columns and rows. Relational database technology provides the most efficient and flexible way to access structured information,
- object-oriented databases – information is represented in the form of objects,
- distributed databases – a database consists of two or more files located at different sites; the database can be stored on multiple computers,
- data warehouses – central data archive, a type of database that is specially designed for fast search and analysis,
- NoSQL database or a non-relational database allows the storage and manipulation of unstructured and semi-structured data (as opposed to a relational database that defines how all data entered in the database must be stored),

IOP Conf. Series: Materials Science and Engineering **867** (2020) 012037 doi:10.1088/1757-899X/867/1/012037

- graph databases – store data in terms of entities and relationships between entities,
- OLTP databases – a fast, analytical database designed for a large number of multi-user transactions [9].

In addition to these types of databases, specific scientific, financial and other functions are complementary and adapted to the development and automation effects. The latest databases include:

- Open source databases – database system with open source,
- Cloud databases – a set of structured or unstructured data on a private, public or hybrid cloud computing platform. There are two types of cloud databases: traditional and cloud-based databases with database services (DBaaS) - also include administrative and maintenance services,
- Multimodel database – combined different types of database models into one integrated system,
- Document / JSON database – the database is designed to save, retrieve and manage information focused on documents representing a modern way of storing data in JSON format instead of rows and columns,
- Self-driving databases – the latest and most innovative type of self-managed database (automated management) is cloud systems that uses machine learning to automate database tuning, security, backup, updates, and other routine management tasks traditionally performed by database administrators [9].

3.2. Database structure

The main aim is to design a database of elements respectively information models, which will be contain graphical and non-graphical information. It is important to design in the database individual links between elements, apply machine learning to optimize building solutions, support multi-criteria evaluation in the selection of elements and support the possibility of combining elements based on user-entered values.

3.3. Database output – building information model

The database structure was created on the basis of a secondary survey. The survey showed that there is a lack of a united environment on the market that has elements containing graphical, technical, economic and other selected parameters. The elements in the database are divided based on the functional parts of the building. The individual functional parts or elements of the structure will contain, besides graphic and technical parameters, economic and time information, durability parameters and environmental parameters.

3.3.1. Division of the construction into functional parts. When dividing the construction into functional parts, it was important to take into account the function of the element, the way of making and taking into account the project documentation in terms of the calculation formula. In addition to the division into functional parts, it was important to take into account the technological aspect. The technological aspect of the construction process represents the division of the object, element into phase and partial processes. Individual function components are divided into types and kinds. Each type and kind of construction work will have specific information for material, technical and technological aspect and so on, see the example in figure 1.

IOP Conf. Series: Materials Science and Engineering **867** (2020) 012037 doi:10.1088/1757-899X/867/1/012037

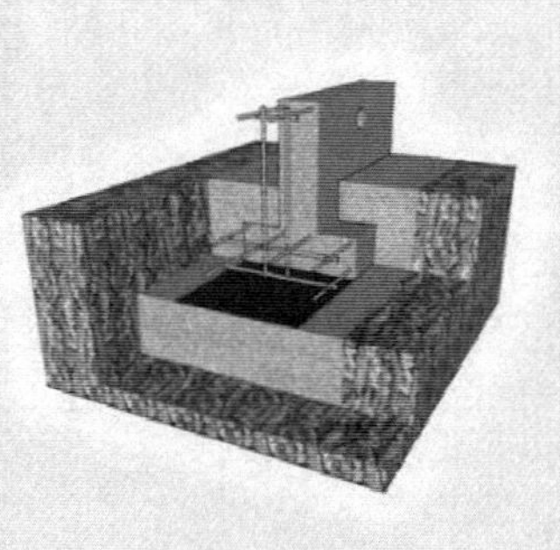

Technical parameters
- Foundations
- Horizontal foundation constructions
- Belt foundations
- Graduated foundations of reinforced concrete
- Placed on gravel bed (150 mm)

Dimensions
- width of foundation's belt - 900 mm
- extension of foundation's belt - 200 mm
- height of foundation's belt - 1000 mm

Figure 1. Example of a structural element - Belt foundation - description and specification.

3.3.2. Economic and time aspect. Individual construction elements disponible cost information across the building's life cycle. Procurement costs will be divided according to the calculation formula into direct material, direct wages, machinery and equipment costs, other direct costs, production and administrative overheads, profit and risk. In addition to the above economic parameters, the element will also have information about time, ie unit labor, total labor, technological breaks and the like (see figure 2).

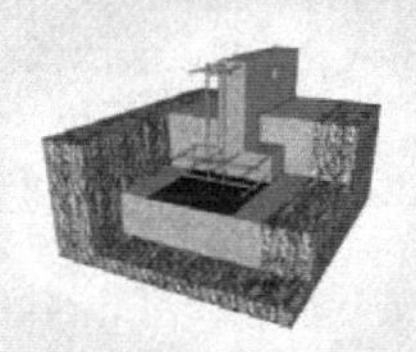

Economic parameters			
Process	Unit of measure (UM)	Unit price (UM/€)
Gravel bed under belt foundation with compacting from gravel 0-32 mm fraction	m3	49,078
Concreting of reinforced concrete belt foundation (reinforcement 10505))	m3	123,485
Formwork of foundation strips - components - performance	m2	11,548
Formwork of foundation strips - components - removal	m2	2,475
	Total price	**172,563**

Time parameters		
Process	Unit of measure (UM)	Unit elaboratenes (Nh)
Placement of gravel bed with compacting from gravel 0-32 mm fraction	m3	1,60
Reinforcement process of foundation strips - reinforcement 10505	100 kg	1,16
Concreting of reinforced concrete belt foundation	m3	1,27
Formwork of foundation strips - components - performance	m2	0,23
Formwork of foundation strips - components - removal	m2	0,12
Total eleboratenes- hour		4,38
Total eleboratenes - days		1
Technological breaks -days		2
Total eleboratenes (work+breaks) - days		**3**

Figure 2. Example of a structural element – Belt foundation – economic and time parameters.

3.3.3. Element durability. The lifetime of an element is the period of time from the actuation of the element to the state when the desired properties fall below an acceptable minimum.

There are two types of durability:
- technical durability – the period during which the construction element performs its function, meets the essential safety requirements and does not endanger the occupants of the building,
- economic durability – time from the beginning of use of the building to its economic disappearance - time of replacement of the original element with a new one. After this time, the construction element becomes unprofitable, so the cost exceeds profit. The economic durability is shorter than the technical durability [13].

Individual building elements have information about the technical and economic durability of the element. Construction elements dispends information also about possible failures, their causes and ways of eliminating the resulting failures. An example of the information included in the information model is shown in figure 3.

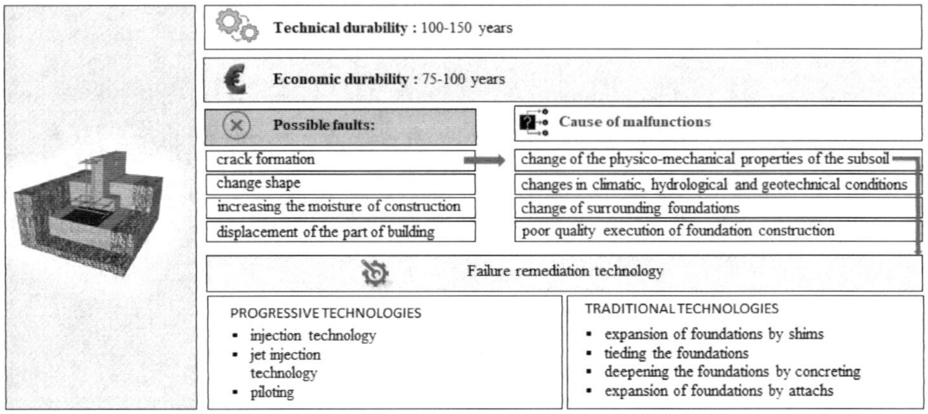

Figure 3. Example of a structural element – Belt foundation – information of durability.

3.3.4. Environmental aspect, concrete sustainability feature. The priority of the created database is to consider and integrate selected environmental aspects of individual elements. Construction elements can have different impacts on the environment, so it is important to consider, for example, resource use, environmental pollution [13][13], the quality of the indoor environment and the functional, aesthetic and social context of the element.

4. Conclusion

Many of the challenges faced by the industry stem from the working environment, which is often unstructured and faces unpredictable and uncontrollable elements. Individual industries are ahead of the construction industry in terms of technological development. Building production is a costly and demanding process. When managing construction production, it is important to have a many of information that will help you manage effectively process. Many innovative technologies and methodologies are used for this purpose. The article deals with database design and identification of parameter structure of information models of individual construction elements. The individual models in the database contain, in addition to the graphical information - 3D model of the element, also information about acquisition costs, information taking into account the time demands realization of the selected structural element and information on the service life of the element and its environmental burden.

5. References

[1] Tejankar A S 2019 *The Use of Automation in Construction Industry* [Online] Available at: https://www.engineeringcivil.com/the-use-of-automation-in-construction-industry.html [accessed 2. Feb. 2020]

[2] Novotny R 2019 *How Automation in Construction is Reshaping the Construction Industry* [Online] Available at: https://esub.com/reshaping-the-construction-industry-automation-tools-and-technology/ [accessed 2. Feb. 2020]

[3] National BIM Standard – United States ™ Version 2 *Frequently Asked Questions About the National BIM Standard-United States™* [Online] Available at: https://web.archive.org/web/20141016190503/http://www.nationalbimstandard.org/faq.php#faq1[accessed 2. Feb. 2020]

[4] Cherkaoui H 2019 *What is BIM? What are its benefits to the construction industry?* [Online] Available at: https://www.letsbuild.com/blog/what-is-bim-what-are-its-benefits-to-the-construction-industry [accessed 2. Feb. 2020]

[5] Jacobi J P E 2011 *4D BIM or Simulation-Based Modeling* [Online] Available at: https://web.archive.org/web/20120528051608/http://structuremag.org/article.aspx?articleID=1234 [accessed 2. Feb. 2020]

[6] Web Archive *Introduction to BIM, 4D and 5D* [Online] Available at: https://web.archive.org/web/20130403091827/http://www.cadsoft-consult.com/blogs/architecture/2009/09/ashrae-introduction-to-bim-4d-and-5d/ [accessed 2. Feb. 2020]
[7] BIMestiMATE *The theory of evolution. BIM 3D...7D* [Online] Available at: https://bimestimate.eu/en/the-theory-of-evolution-bim-3d-7d/ [accessed 3. Feb. 2020]
[8] Bimpanzee.COM *Im 3D, 4D, 5D, 6D and 7D* [Online] Available at: http://www.bimpanzee.com/bim-3d-4d--5d--6d---7d.html [accessed 3. Feb. 2020]
[9] ORACLE *What is database* [Online] Available at: https://www.oracle.com/database/what-is-database.html [accessed 3. Feb. 2020]
[10] Merriam Webster *What is update* [Online] Available at: https://www.merriam-webster.com/dictionary/update [accessed 3. Feb. 2020]
[11] Merriam Webster *What is retrieval* [Online] Available at: https://www.merriam-webster.com/dictionary/retrieval [accessed 3. Feb. 2020]
[12] Merriam Webster *What is administration* [Online] Available at: https://www.merriam-webster.com/dictionary/administration [accessed 3. Feb. 2020]
[13] Vyparina M 2011 *Specifics of determining the general value of real estate and buildings (Špecifiká stanovenia všeobecnej hodnoty nehnuteľností a stavieb)*, Proceedings of seminar lectures University of Tilin at EDIS-2011 ISBN 978-80-554-0334-2 p 44-45
[14] Zeleňáková M, Purcz P, Ondrejka Harbuľáková V and Oravcová A 2016 *Desalin. Water Treat.* **57** 2693-2701
[15] Harbulakova V O, Zelenakova M and Sugarekova M 2018 *International Multidisciplinary Scientific GeoConference: SGEM: Surveying Geology & mining Ecology Management* **18** 391-398

Acknowledgments
The paper presents a partial research results of project APVV-17-0549 “Research of knowledge-based and virtual technologies for intelligent designing and realization of building projects with emphasis on economic efficiency and sustainability” and project VEGA 1/0828/17 “Research and application of knowledge-based systems for modelling cost and economic parameters in Building Information Modelling”.

IOP Conf. Series: Materials Science and Engineering **867** (2020) 012038 doi:10.1088/1757-899X/867/1/012038

Experimental investigation on bond-slip behavior of selfconsolidating rubberized concrete

K Strukar[1], M Jeleč[2], T Kalman Šipoš[1], M Grubišić[1] and S Lukić[2],

[1] University of Josip Juraj Strossmayer of Osijek, Faculty of Civil Engineering and Architecture, Department for technical mechanics, Croatia
[1] University of Josip Juraj Strossmayer of Osijek, Faculty of Civil Engineering and Architecture, Department for structures and materials, Croatia

Abstract. Reducing the industry's impact on the environment is the goal of sustainable construction which can be achieved by using renewable and recyclable resources such as scrap tire waste. Great potential of this waste is found in construction mostly as a partial replacement of mineral aggregate in concrete and mortars, asphalt mixtures and geotechnical works. Previous studies on the selfconsolidating rubberized concrete (SCRC) are mostly concerned with mechanical properties of rubberized concrete. However, in order to allow structural application, it is also necessary to investigate bond-slip behavior of SCRC and reinforcement bars. Thus, in this paper pull-out tests are performed on 24 cubic samples with reinforcement confinement. Three different SCRC mixtures are made, first with concrete without rubber particles as reference mixture, second with 10%, and third with 15% rubber particles and 5% silica fume. The other parameter varied is diameter of steel bars (Ø12, Ø16, and Ø20 mm). From the results obtained in experimental investigation, key bond parameters and failure modes of SCRC are compared with bond-slip behavior of self-consolidating concrete (SCC).

1. Introduction

In order to reduce waste and use recyclable materials, great potential is found in scrap tire waste due to their enormous amounts produced each year which is resulting in landfilling, burning, burying or grinding. Least harmful solution, as it is for the environment, so it is for people, is grinding scrap tires and reusing them. Potential use of tire waste is found in construction mostly as a partial replacement of mineral aggregate in concretes and mortars, asphalt mixtures and geotechnical works [1].

Previous studies have shown both positive and negative impact on concrete's properties by replacing mineral aggregate with rubber particles [2], [3]. Increased hardness and ductility, enhanced dynamic properties, increased resistance to freezing and thawing, and resistance to HCl are some of improved properties. On the other hand, reduced resistance to carbonation, increased resistance to abrasion, water absorption and permeability, reduced compressive strength and modulus of elasticity of the concrete are results of negative impact of rubber addition.

Another very important property is bond-slip behavior of self-consolidating rubberized concrete (SCRC) and steel bars in reinforced concrete constructions. Main assumption in case of these structures is strong connection between concrete and rebar, and compatibility of deformations of between these two materials, $\varepsilon_c = \varepsilon_s$. Thus bond strength between them enables stress transition from rebar to concrete without slip. In order to ensure these compatibility conditions, it is necessary to follow directions regarding adhesion and anchoring of reinforcement bars given in guidelines [4].

Some of the investigations on this property of rubberized concrete were carried out by Hall and Najim [5]. They conducted pull-out test in order to compare bond-slip behavior for four different mixes, plain rubberized concrete (PRC) and self-consolidating rubberized concrete (SCRC) with plain concrete (PC) and self-consolidating concrete (SCC), respectively. From the obtained results it was observed that the maximum bond strength τ_{max} was decreased while the bond coefficient values γ were increased in PRC and SCRC mixes compared to PC and SCC mixes. Also, representing slip displacement that occurred at the maximum bond slip stress τ_{max} decreased significantly for mixes with rubber. Bompa and Elghazouli [6] examined the complete bond-slip behavior between deformed rebar and rubberized concrete. Their test results offered direct interpretation of the bond behavior under low levels of confinement as well as its influence on the failure modes. Anu Bala et al. [7] investigated how will replacement of mineral aggregate by rubber particles and of cement by fly ash (FA) up to 40% and 30%, respectively, influence bond strength. Results shown that the addition of rubber decreased bond strength, while the addition of FA improves it, however for replacement levels up to 20% of FA. Gesoglu et al. [8] replaced total aggregate volume in concrete with up to 30% of rubber particles. Results of their experiment have shown that bond strength of those concretes with rubber was lower than that of plain concrete. They found the reason is because rubber particles are much softer than the natural aggregate and that causes bond strength loss as the friction between reinforcement bar and the concrete decreases with the use of rubber aggregates.

In this paper pull-out tests are performed on 24 cubic samples of self-consolidating concrete with reinforcement confinement in order to investigate key bond parameters and failure modes of bond-slip behavior of SCRC concrete.

2. Case study

In order to assess the bond-slip (τ-s) behavior between steel bars and concrete, 24 cubic specimens with 0reinforcement confinement were prepared in order to perform pull-out test. The parameters investigated in this research were diameter of the rebar (Ø12, Ø16, Ø20 mm) with a rubber amount in self-consolidating concrete. Two mixtures of SCRC were made, one as referent, without rubber, and the other one with 10% rubber particles replaced by volume of both fine and coarse mineral aggregate. Compressive strength, and modulus of elasticity of concrete were also tested to obtain general mechanical properties.

2.1. Materials

Materials used for production of concrete mixes include Portland cement 42.5 R, which conforms to HRN EN 197-1:2005, mineral (MA) and recycled aggregates (RA), silica fume (SLF), dolomite powder (DP), water and admixtures. Mineral aggregates consisted of fine aggregates (FA) of 0–2 mm and 2–4 mm fraction, and coarse aggregates (CA) which included gravel of 4–8 mm and 8–16 mm fraction. Recycled aggregate crumb rubber (CR) with particles of 0.5–4 mm size and density of 1050 kg/m3 was used in concrete mixes as substitute for 10%, and 15% ratio of the volume of both mineral and recycled aggregate. Dolomite powder was also used in order to fill pores in concrete mixture. For bars were used hot rolled steel bars with diameter of 12, 16 and 20 mm. Steel grade was B500B.

2.2. Mix design

Self-consolidating concrete (SCC) mixtures are labelled with SCC-xCR-ySLF where CR and SLF represents crumb rubber and silica fume, respectively. Labels x and y refers to the percentage of crumb rubber (0-15%) and Silica fume (0-5%) in mixtures, respectively.

From table 1 can be seen that in each concrete mixture, water to binder (w/b) factor kept constant. Referent SCC had a composition of 450 kg/m^3 of cement, 938.45 kg/m^3 of sand, and 814.90 kg/m^3 of gravel of different fractions, 180 kg/m^3 of tap water, 6.75 kg/m^3 of super-plasticizer (SP) Energy FM 500 to obtain better workability. Additional water was necessary due to the dry surface of natural aggregates. Other rubberized concrete mixtures in this group contained rubber which replaced 10– 15% of volume of sand. Quantities of rubber used were 66.0 kg/m^3 and 98,75 kg/m^3 for ratio of 10% and 15% of crumb rubber, respectively. Super-plasticizer Glenium ACE and viscosity modifying agent

RheoMATRIX (RM) were used in order to enable the right balance between fluidity, passing ability, and resistance to segregation.

Table 1. Concrete mixtures design.

Mixture ID	w/b	Cement 42.5 R [kg/m³]	SLF [kg/m³]	Water [kg/m³]	SP [kg/m³]	VMA [kg/m³]	DP [kg/m³]	CR 0–4 mm [kg/m³]	FA 0–2 mm [kg/m³]	FA 0–4 mm [kg/m³]	CA 4–8 mm [kg/m³]	CA 8–16 mm [kg/m³]
SCC-0CR-0SLF	0.40	450.00	0.00	180.00	6.75	1.35	80.00	0.00	324.45	614.00	362.18	452.72
SCC-10CR-0SLF	0.40	450.00	0.00	180.00	6.75	1.13	80.00	66.00	324.34	438.42	362.05	452.56

The results of fresh properties of concrete are presented in table 2 and it can be seen that mass and density are decreased with the addition of rubber particles. On the other hand, porosity increased with the addition of rubber while addition of silica fume decreased it. Results regarding viscosity, flowability and passing ability met criteria for selfconsolidating concrete.

Table 2. Results of fresh concrete's properties.

Mixture ID	Mass [kg]	Density [kg/m³]	Porosity [%]	Viscosity	Flowability	Passing abbility	
				T500 [s]	Slump flow [mm]	PJ [mm] ≤ 10 ≤	H_2/H_1
SCC-0CR-0SLF	8.06	2388.15	1.50	2.79	800	7	0.95
SCC-10CR-0SLF	7.854	2327.11	1.90	3.26	735	10	0.90

2.3. Specimen preparation and testing

According to HRN EN 12390-1:2012 [9] guidelines for shape, dimensions of specimens and moulds, and HRN EN 12390-2 [10] guidelines for making and curing specimens, both cubic (150×150×150 mm³) and cylindrical (∅150 mm×300 mm) specimens were casted for each mixture. In figure 1 is shown the process of making cubic specimens for pull-out test, while in table 3 are given numbers of specimens per mixture and testing. The cylinders and cubes were demoulded 24 h after casting and then continuously moist-cured for 28 days. After this period, test specimens were measured and prepared for testing.

(a) (b) (c)

Figure 1. (a) confinement of pull-out test specimens; (b) concrete poured in moulds; (c) specimens after 28 days ready for testing.

Compressive strength and modulus of elasticity were tested under uniaxial compression in a four column high stiffness welded frame Controls Automatic Compression machine, while tensile strength of steel bars and pull-out tests were conducted on tensile testing machine Shimadzu AGXplus. Experimental setups are shown in figure 2.

Table 3. Number of specimens per mixture and testing.

Mixture ID	Diameter [mm]	Cubic specimens	Cylindrical specimens	
		Pull-out test	Compressive strength f_c [MPa]	Modulus of elasticity E_{cm} [MPa]
SCC-0CR-0SLF (M1)	ϕ12	4		
	ϕ16	4	3	3
	ϕ20	4		
SCC-10CR-0SLF (M2)	ϕ12	4		
	ϕ16	4	3	3
	ϕ20	4		
Σ		24	6	6

(a) (b) (c)

Figure 2. Experimental setup for (a) compressive strength; (b) modulus of elasticity; (c) pull-out test.

3. Test results and data analysis

3.1. Compressive strength and modulus of elasticity of self-consolidating rubberized concrete

Results obtained from tests that include mean values (μ) and coefficient of variation (*CoV*), are given in table 4. It can be seen that the ultimate compressive strength reduces with increase of rubber particles amount. Results of modulus of elasticity also show that it is decreased with the addition of rubber, while silica fume did not show any positive impact.

Table 4. Test results of compressive strength and modulus of elasticity.

Mixture ID	*Compressive strength f_c* [MPa]			*Modulus of elasticity E_{cm}* [MPa]		
	μ		*CV*	*μ*		*CoV*
SCC-0CR-0SLF (M1)	43.7	REF	0.06	38576.62	REF	0.08
SCC-10CR-0SLF (M2)	31.25	-28.49%	0.09	35256.04	-8,57%	0.03

3.2. Bond-slip behavior of self-consolidating rubberized concrete

Results of pull-out tests describe bond-slip behavior (τ-s) of SCRC which is presented in figure 3 as comparison of τ-s curve obtained for three different bar diameters for each concrete mixture. In figure 4 is presented comparison of τ-s curves for the same bar diameter but different concrete mixture. From these figures can be seen that bond strength (τ_{slip}) is reduced both with the addition of rubber and increase in bar diameter. Similar to the compressive strength, difference in ultimate bond strength (τ_{max}) between M1 and M2 is not big as that between M2 and M1, again probably due to the addition of silica fume. Another observation is made regarding relative slip (s). From the results given in table 5 it can be seen that relative slip s at τ_{max} reduces with increase in bar diameter.

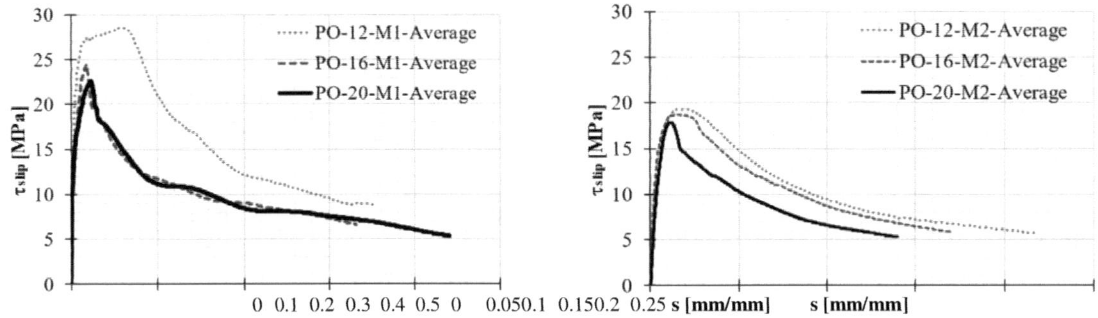

(a) (b)

Figure 3. Comparison of average τ-s curves of different bar diameter for each concrete mixture: (a) SCC-0CR-0SLF (M1); (b) SCC-10CR-0SLF (M2).

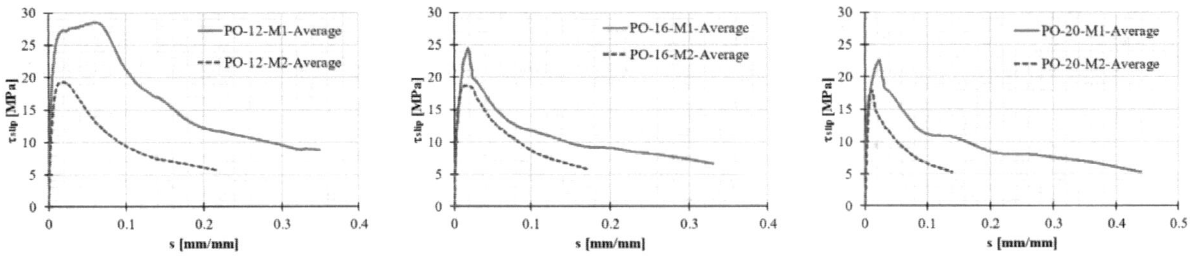

(a) (b) (b)

Figure 4. Comparison of average τ-s curves of different concrete mixture and the same bar diameter: (a) ϕ12; (b) ϕ16; (c) ϕ20.

Table 5. Comparison of pull-out test results.

Diameter [mm]	Mixture ID	Bond strength τ_{max} [MPa]			Bond strength τ_{max} [MPa]		
		μ μ		*CV*	μ		*CV*
ϕ12	SCC-0CR-0SLF	28.71	REF	0.01	0.064	REF	0.08
	SCC-10CR-0SLF	19.54	-31.94%	0.03	0.018	-71,87%	0.27
ϕ16	SCC-0CR-0SLF	24.86	REF	0.09	0.019	REF	0.10
	SCC-10CR-0SLF	18.88	-24.05%	0.25	0.015	-21.05%	0.27
ϕ20	SCC-0CR-0SLF	20.42	REF	0.01	0.014	REF	0.66
	SCC-10CR-0SLF	17.88	-12.44%	0.08	0.011	-21.43%	0.05

From figure 4 also can be observed two types of specimen's failure modes First possible failure mode is due to the concrete splitting and second due to the bar pullout. From presented τ-s curves is concluded that in specimens named with PO-16-M1-Average, PO-20-M1-Average, and PO-20-M2-Average failure occurred due to the concrete splitting. In figure 5 are shown specimens after testing of both concrete mixture in order to see relation between shape of the curve and failure modes. From these observations can be concluded that addition of rubber particles in concrete mixture contributes to more ductile failure, without concrete splitting. Similar conclusion was observed in [11] regarding compressive stress-strain curves of rubberized concrete.

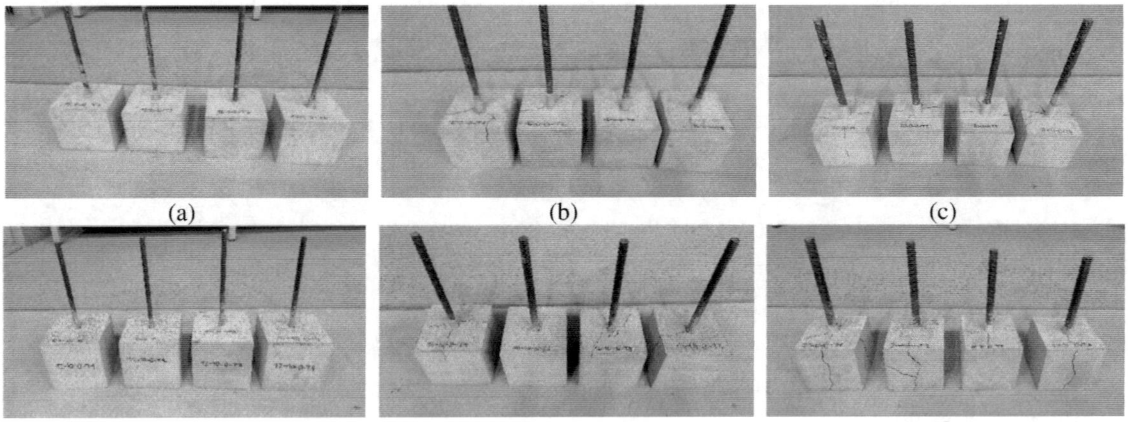

(a) (b) (c) (d) (e) (f)

Figure 5. Failure modes of SCC-0CR-0SLF specimens with (a) ϕ12; (b) ϕ16; and (c) ϕ20 rebar diameter; and of SCC-10CR-0SLF specimens with (d) ϕ12; (e) ϕ16; (f) ϕ20 rebar diameter.

4. Conclusions

Based on experimental results of compressive strength, modulus of elasticity and bond-slip behavior of self-consolidating rubberized concrete (SCRC), it can be concluded that the addition of rubber has significant influence compared to the referent self-consolidating concrete (SCC). Thus, compressive strength was reduced by 28.49% for 10% rubber particles, respectively, while modulus of elasticity was reduced by 8.57%.

On bond-slip behavior of SCC and SCRC influenced both rubber replacement and steel bar diameter. Results have shown that increasing both rubber particles amount and diameter decreased bond strength and slip displacement. Bond strength was decreased by 31.94% (ϕ12), 24.05% (ϕ16), and 12.44% (ϕ20) for 10% rubber replacement, respectively. From the failure modes that occurred on specimens, it was concluded that steel bar diameter has great influence on the type of failure. Reduction of bond strength and concrete splitting due to the greater bar diameter occurred probably due to three reasons: (1) lower ratio of concrete and steel bar of grater diameter; (2) ribs on steel bar cause earlier concrete splitting; (3) greater surface of steel bar gets in touch with concrete matrix and in case of higher amounts of rubber particles friction between them is much lower because of rubber's softness and much lower modulus of elasticity. However, in order to derive more accurate conclusions, additional investigations need to be carried out.

5. References

[1] Oikonomou N and Mavridou S 2009 The use of waste tyre rubber in civil engineering works *Sustainability of Construction Materials*

[2] Bušić R, Miličević I, Šipoš T K, and Strukar K 2018 Recycled rubber as an aggregate replacement in self-compacting concrete-literature overview *Materials* (Basel) vol. 11 no. 9

[3] Strukar K, Kalman Šipoš T, Miličević I, and Bušić R 2019 Potential use of rubber as aggregate in structural reinforced concrete element – A review *Engineering Structures*

[4] Narayanan R S 2001 *EN1992 Eurocode 2: Design of concrete structures* Proc. Inst. Civ. Eng. - Civ. Eng

[5] Hall M R and Najim K B 2014 Structural behaviour and durability of steel-reinforced structural Plain/Self-Compacting Rubberised Concrete (PRC/SCRC) *Constr. Build. Mater* vol. 73 pp 490–497

[6] Bompa D V and Elghazouli, A Y 2017 Bond-slip response of deformed bars in rubberised concrete *Constr. Build. Mater*

[7] Bala A, Sehgal V K and Saini B 2014 *Effect of Fly ash and Waste Rubber on Properties of Concrete composite* (Www.Crl. Issres.Net)

[8] Gesoglu M, Güneyisi E, Hansu O, Ipek S, and Asaad D S 2015 Influence of waste rubber utilization on the fracture and steel-concrete bond strength properties of concrete *Constr. Build. Mater*

[9] HRN EN 12390-1:2012 2012 *Testing Hardened Concrete—Part 1: Shape, Dimensions and Other Requirements for Specimens and Moulds* (Croatian Standards Institute: Zagreb, Croatia)

[10] HRN EN 12390-2:2009 20019 *Testing Hardened Concrete—Part 2: Making and Curing Specimens for Strength Tests* (Croatian Standards Institute: Zagreb, Croatia)

[11] Strukar K, Kalman Šipoš T, Dokšanović T, and Rodrigues H 2018 Experimental Study of Rubberized Concrete Stress-Strain Behavior for Improving Constitutive Models *Materials* vol. 11 no. 11 (Basel) p 2245

Acknowledgement
This paper was supported by Croatian Science Foundation under the project UIP-2017-05-7113 Development of Reinforced Concrete Elements and Systems with Waste Tire Powder—ReCoTiP.

IOP Conf. Series: Materials Science and Engineering **867** (2020) 012039 doi:10.1088/1757-899X/867/1/012039

Identification and risk assessment of historic and listed buildings

J Stuchlík[1]

[1] Czech Technical University in Prague, Faculty of Civil Engineering, Prague, Czech Republic

Abstract. The article deals with the identification and evaluation of risks of publicly funded construction projects. The study presents a risk analysis, incl. draft measures to eliminate some of the most serious risks that may affect the completion date and contractual budget. Risks are qualified as the product of the probability of the occurrence of an adverse event and the potential damage caused by the event. Risk identification was based on a list of 30 publicly funded national heritage constructions. The most serious risks from the contractor's point of view were identified as a poor-quality preparation and formulation of the contract agreement, climatic effects and shortcomings in the project documentation. Possible implementation of the proposed solutions to these risks in the construction process can eliminate the contractor's failure to meet the deadline for completion of the construction and exceeding the contractual budget.

1. Introduction

By awarding public contracts, the state provides significant opportunities for the public to use public funds in the economy. In the Czech Republic (CR), 31,428 public contracts were awarded in 2018, with a total volume of CZK 495.50 billion (excluding VAT). Of this amount, roughly 2/5 of the construction is allocated. The value of public contracts increased by 43.50% and the number of contracts by 15.60% compared to 2017 [1]. Given the amount of public funding provided to construction, the use of funds should be considered and directed towards meeting the needs of the general public. In spite of all the efforts of the state to check compliance with the agreed procedures during the construction of the work, serious deficiencies are repeatedly identified in some projects leading to failure to meet the deadline for completion or failure to comply with the contractual budget of the construction. E.g. in the Czech Republic (CR) four construction projects out of ten have not been met and six out of ten construction projects have not been respected [2]. Problems such as budget failures and building cost increases occur both in developing and developed countries. Bent Flyvbjerg, a professor and scientist at Oxford's Saïd Business School, specializing in the management and financing of world construction projects, said that nine out of ten major construction projects will be more expensive in the world. The planned costs for the analysed buildings are exceeded in the range of 50-100% [3]. Another study states that construction projects have an average cost increase of around 33% [4]. Risk analysis is an irreplaceable tool for decision-making and legislative activities in the construction industry and is important for the success of the project. In developed countries, very high amounts are spent on safety, risk reduction and health protection and their volume is constantly increasing [5]. Project risk management is recognized as a major contributor to the success of a project, but its implementation in practice is far from easy as the process requires a thorough understanding of the 'components' that influence project development [6]. The aim of the risk identification phase is to identify as many potential project risks as possible, to understand them and then to describe them correctly, thus increasing the project's preparedness for possible threats. Risk is a factor that is relatively difficult to assess. In principle, public procurement must take into account all possible risks: political, economic (e.g. market risk, exchange rate risk, credit

IOP Conf. Series: Materials Science and Engineering **867** (2020) 012039 doi:10.1088/1757-899X/867/1/012039

risk), security, environmental, etc., since public procurement is linked to the Public Procurement Act and they are strictly codified [7]. Many domestic and foreign publications have addressed the subject of risk identification and risk analysis / management, where researchers analysed various aspects of risk management through the implementation of different research methodologies [5], [8]. Considerable less attention has been paid to the remaining risk management processes such as risk response and risk monitoring and control over the last few years. Many of these studies dealt with risk management at the construction project level, while research into the impact of risks on the construction company and the construction sector as such has been neglected. The most common way to obtain data for carrying out risk assessments was to use a questionnaire or a case study, with output in the form of statistical data [8]. Expert publications are being published and updated to provide a generalized approach to risk management based on modern knowledge [6]. This article aims to identify and qualify the severity of risks that may adversely affect the work of the contractor, thereby leading to a failure to meet the deadline for completion of the construction or to fail to meet the contractual budget of the project. Analysis of construction risks in the Czech Republic is needed to improve the process of implementation of public buildings that will be economically advantageous for the state, as defined by the Public Procurement Act, No. 134/2016 Coll.

2. Methods

The risk assessment process includes risk identification, risk analysis and risk assessment. Risks can be identified using various methods, which are largely universal [9]. In this risk analysis, the world-renowned risk management approach of the US Project Management Institute, specified by the PMBOK Guide (2004), was chosen. The risk identification is based on the definition of risk (R) given by prof. M. Tichý as the product of the probability of realizing an adverse event (P) and the potential damage / impact caused by the event (D) according to the formula: $R = P \times D$ [9]. The components of risk identification were: 1) study of the project, tender documentation, consideration of circumstances, 2) identification of all hazards and possible hazard scenarios, and 3) list of possible hazards, hazard scenarios, trigger mechanisms. One of the possible methods [10] was chosen for risk identification: brainstorming and consultation with three construction professionals from 10 to 30 years, from the position of site manager to construction company management. Furthermore, historical data and previous experience with own projects were used. Risks were identified and analysed on the basis of an author-created list of 30 public construction projects implemented in the Czech Republic between 2008 and 2019, in the financial range from CZK 700,000 to CZK 30 million. These were listed buildings (churches, historic facades, etc.). The database of implemented projects was compiled from in-house data sources and from publicly available data sources, i.e. portals that publish information on public procurement. The period from the takeover of the construction site to the handover of the completed construction work was included in the risk analysis. Individual risks were classified according to the subject matter into: construction - technical - technological; manufacturing; economic and financial; market; credit; legislative; political; environmental; force majeure; strategic; information and risk with human factor. The second stage of risk assessment is risk analysis, a concept whose interpretation is understood and interpreted differently within risk management. Risk analysis can be quantitative and qualitative [5]. The aim of risk analysis is to make a sound estimate or to directly calculate the probability of occurrence of a certain hazard and the amount of expected adverse impact. In practice, the most frequently encountered are methods of expert estimation, which is the case of this study, ie evaluation by description and scale. Therefore, for the purposes of this study, the probability of risk occurrence and the risk impact rate were defined by the experts as follows (table 1):

Table 1. Probability Scale and Risk Impact Scale.

Probability of risk - scale	Numerically	Impact of risk on the construction process	Numerically
Almost excluded	*1*	*Unimportant*	*1*
Unlikely	*2*	*Low*	*2*
Possible	*3*	*Medium*	*3*
Very likely	*4*	*High*	*4*
Almost certain	*5*	*Catastrophic*	*5*

A risk matrix is designed to give a likelihood of occurrence and a measure of impact to each identified risk that graphically assesses the significance of the risks. Quantitative risk analysis, using a mathematical logical risk model and software to simulate risk impact, was not used in this study because of the lack of necessary numerical data on the risks examined, as a result of human error or sociological linkage and uncertainty for which theoretical models are lacking. The third stage was risk assessment and selection of those that could have the most serious negative impact on the project, and subsequently possible measures to minimize the impact of risks from the contractor's point of view were proposed.

3. Results

As part of the risk analysis, a list of risks to the construction process in its various phases was created (table 2). Only high and very high risks are included in the list of risks, i.e. there are no acceptable and increased risks, therefore the market and environmental risks are missing from the list of risks.

Table 2. List of the most serious risks of the construction process.

Risk area (code)	Risk scenario	Consequences of risk implementation	Likelihood of risk	Risk impact	Product
1. CONSTRUCTION TECHNICAL AND TECHNOLOGICAL RISKS					
Project documentation (1.1)	*Incompleteness and non-fulfillment of claims on PD.*	*Changes in PD. Financial loss and failure to meet the deadline.*	4	3	12
Cultural and archeological heritage (1.2)	*Insufficient archaeological survey. Possible damage to cultural and archaeological monuments located near the implemented project.*	*PD change required. Performing archaeological research. Possible financial loss and completion date.*	3	4	12
2. PRODUCTION RISKS					
Building budget (3.1)	*Incorrectly processed BOQ. Underestimation of work in the budget according to PD.*	*Failure to meet the deadline for completion of the construction. Possible financial loss.*	3	4	12
3. LEGISLATIVE RISKS					
Additional authorizations (5.1)	*Necessity to obtain additional permits related to the realization of the work (restoration plans; binding opinions; etc.).*	*Possible financial loss and failure to meet the deadline.*	3	3	9
4. INFORMATION RISKS AND HUMAN ERRORS					
Qualification of workers and subcontractors (6.1)	*Untrained and unskilled workers / subcontractors carrying out professional construction work.*	*Additional costs related to removal of defects. Failure to meet the deadline.*	3	5	15
5. FORCE MAJEURE					

Climatic effects (7.1)	*Negative influence of the building process by climatic influences (floods, landslides, heavy rains, earthquakes, etc.). Suspension of construction.*	*Failure to comply with the contract price of the construction and the date of completion of the construction.*	3	4	12
		6. STRATEGIC			
Contracts (8.1)	*Poorly compiled contracts. Contract changes caused by any party involved in the construction process.*	*Failure to comply with the contract price of the construction and the date of completion of the construction.*	4	5	20

A risk matrix (table 3) was drawn up for expert graphical assessments of the significance of risks, where the intensity of risk impact intensity is plotted on the vertical axis and the risk probability value is plotted on the horizontal axis. The risk assessed is the more significant the higher the probability of its occurrence in combination with the intensity of its possible impact on the project.

Table 3. Risk matrix.

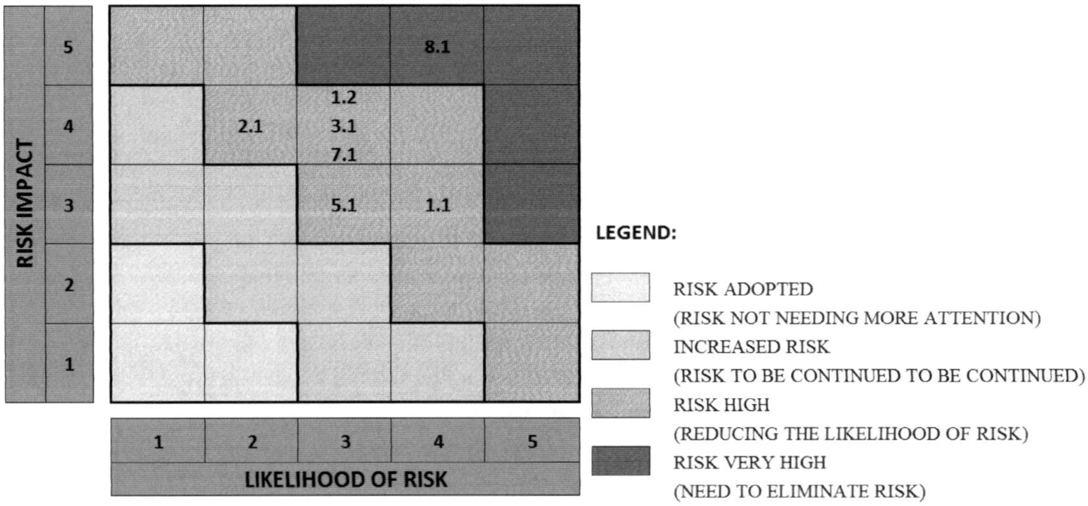

The following paragraphs present options for managing selected risks, with the highest probability of occurrence:

- *Risk CONTRACTS (Risk Severity Rate: 20)*

The contract for work implies to the contractor obligations that must be fulfilled. The draft contract must be properly studied and commented before it is signed. Where possible, it is good to quantify the contractor's obligations, such as the extent of the salinity survey (number of probes), the extent of the mycological survey (description of the structures and / or location where it is to be carried out) elements of the building, etc.

- *Risk: QUALIFICATION OF WORKERS AND SUB-CONTRACTORS (Risk Severity Rate: 15)*

It is advisable that the contractor invests time and money to train the staff and not to select subcontractors only at the lowest bid price, but also on the basis of past experience (e.g. reference buildings, client satisfaction, financial analysis, etc.). Consequently, the Contractor may eliminate the aforementioned risk of financial loss and failure to comply with the contractual completion of the works.

- *Risk: PROJECT DOCUMENTATION (Risk Severity Rate: 12)*

The contractor confirms the accuracy and completeness of PD by signing the Contract for Work. It may be the case that the client does not have to acknowledge to the contractor future work resulting from the discovered new facts during construction, which were not mentioned in the original PD. The transferred

IOP Conf. Series: Materials Science and Engineering **867** (2020) 012039 doi:10.1088/1757-899X/867/1/012039

risk from the PD processor to the contractor must be properly contracted in advance. One way is, for example, to ensure that the processor of the PD will be obliged to submit to the contractor a new technical solution by a certain date, on the basis of which the contractor shall prepare a change sheet. Should this not be the case, the contractor shall not be obliged to commence the aforementioned works and shall not be responsible for any failure to meet the deadline for completion of the construction.

- *Risk: CULTURAL AND ARCHEOLOGICAL HERITAGE (Risk Severity Rate: 12)*

The contract for work implies to the contractor obligations that must be fulfilled by the contractor. The draft contract must be properly studied and commented before it is signed. It is necessary to pay attention, for example, to the delegated responsibility from the client to the contractor in the framework of archaeological research in case of positive findings. Furthermore, it is necessary to ensure that if a positive finding of archaeological elements is found, the contractor is not responsible for any failure to meet the deadline for completion of the construction.

- *Risk: BUILDING BUDGET (Risk Severity Rate: 12)*

The Contractor confirms within the conclusion of the Contract for Work that the budget together with the PD has been studied and it is correct and complete. It may be the case that the client does not have to accept the contractor´s future works resulting from a poorly prepared budget (BOQ). A risk delegated from the designer of project to the contractor must be properly contracted. One way is to ensure that the designer will be obliged to submit to the contractor a properly processed BOQ by a certain date, on the basis of which the contractor prepares a change sheet, which the client is obliged to confirm and accept. Should this not be the case, the Contractor shall not be obliged to commence the aforementioned work and shall not be responsible for any failure to meet the deadline for completion of the construction.

- *Risk: CLIMATIC EFFECTS (Risk Severity Rate: 12)*

The contractor draws up a construction schedule, which is usually a necessary annex to the contract for work. It is advisable for the contractor, when drawing up the construction schedule, to include a time reserve to cover suspended construction work due to adverse climatic effects (long-term frosts, heavy rains, floods, etc.). If the construction schedule is correctly designed so that the individual construction phases and professions are smoothly connected, the late completion date due to adverse climatic effects is eliminated. Furthermore, the contractor may be insured against adverse weather conditions.

4. Conclusion

Today, public procurement authorities have a number of tools and procedures at their disposal to increase the success of the project. An example would be a way to at least mitigate the effect of a decrease in the number of contractors' bids per contract from the contracting authority (i.e. competition), which has a negative impact on rising prices, declining quality and sometimes the need to cancel orders for lack of interest. Public procurement competition, which is a key tool for achieving maximum price / quality in public procurement, is experiencing the weakest period in the construction sector (and other sectors) from 2006 to 2018, with an average of 6 bids (2016) falling below 4 bids. (2018). This leads to more expensive public contracts. Competition for public procurement therefore appears to be one of the most important indicators of good practice in public procurement. From the point of view of the contracting authority, the precise preparation of the tender documentation and proper market research can increase the chance of success of the contract. In addition, other important factors are: the time of the contract (between April and June); type of procedure (open procedure); receiving electronic offers (not paying so far; paid only in the long run); contracting authorities' profiles (e.g. EZAK and e-contracts show better competition compared to NEN and Tender arena, especially in the small-scale segment); size of the contract (division of a large contract into smaller units, under the law). For each construction project it is necessary to identify the required risks, determine their importance, size, evaluate them and take steps to eliminate them. The most serious risks are the omission of essential moments for the successful start of the work, such as failure to estimate the archaeological and geological conditions (within the project documentation), poorly compiled contract with the client or subcontractors, use of bad materials or neglect of safety measures in individual intermediate construction works. Sometimes there are shortcomings in the provision of a sufficient number of qualified experts who are able and willing to participate in the project and with their professional profile to guarantee the quality and credible result of the project. Risks can be eliminated to some extent, managed. However, there may be risks that cannot be predicted and eliminated, such as terrorism. From the perspective of the contractor, 11 possible causes of deviations from the construction work schedule and cost overruns were defined in this study. If sufficient time is not foreseen in the schedule, the whole project may be delayed in addition to the delay of one activity. Each project is unique and during the implementation of the project there may be a situation where an unexpected cause or a combination of causes that will delay the project may arise, especially if the performance of the public contract covers a longer period of time. The main reasons for not respecting the budget of the public contract in the Czech Republic are: changes and errors in project documentation (32%), insufficient geological survey (14%), non-compliance or change in construction technology (12%), climatic effects (9%) (6%), unprofessional project management (4%), and archaeological exploration (4%) [2]. According to the above analysis, the greatest risks for contractors are unclear contracts, insufficient project documentation and insufficient qualification of workers and subcontractors. The last is closely related to poor quality of work, e.g. failure to follow the technological procedure, failure to observe the material specified in the project documentation, speed of construction (time pressure usually reduces the quality of work). Therefore, it is necessary to pay maximum attention to these areas before the start of the project and also to regularly inspect the riskiest areas during construction. Of course, the magnitude of the financial impact on a building depends directly on the type and extent of the cause. Late contracting of subcontractors, poor technological continuity of work in the schedule and poor timing of the duration of activities are significant risks that can affect the work schedule. The prolongation of the works can be very likely, and is often caused by either the contractor or additional investor requirements. Therefore, it is important that extra work is adequately treated in the contract of work. Thanks to the qualitative analysis carried out, it is possible to identify the risks to be further focused and the risks to which particular attention should be paid. The weakness of qualitative risk analysis, where the scale defined by the expert team is used to assess the probability of occurrence and intensity of the negative, is the risk of subjective assessment of the problem. The uncertainty of expert estimates can be reduced, for example, by breaking down the unfavourable situation into smaller, sub-parts, where it is easier to estimate the probability and the amount of negative impact.

5. References

[1] ČTK *Objem veřejných zakázek v ČR loni vzrostl o 44 % na 495,5 miliardy* [online] [cit. 2020-01-10] Available from: https://www.ceskenoviny.cz/zpravy/objem-verejnych-zakazek v-cr-loni-vzrostl-o-44-na-495-5-miliardy/1715755

[2] Stuchlík J and Frková J 2016 *Analysis of leading causes od overpriced public building contracts* Construction Maeconomics Conference (2016) [online] [cit. 2019-11-12] Available from: http://www.conferencecm.com/podklady/history7/Prispevky/paper_Stuchl%C3%ADk_FrkovC3%A1.docx

[3] Tramba D *Svět v opojení megaprojektů* [online] [cit. 2019-11-12] Available from: https://ekonom.ihned.cz/c1-64326010-svet-v-opojeni-megaprojekty

[4] Hartley J R 1992 *Concurrent Engineering: Shortening Lead Times, Raising Quality, and Lowering Costs* Productivity Press ISBN 1563270064

[5] Teplý B *Je analýza rizik ve stavebnictví užitečná?* Časopis Stavebnictví 02/10

[6] Chapman R J 2013 *Rules of Project Risk Management: Implementation Guidelines for Major Projects* 28.12.2013 (London: Taylor & Francis) ISBN 1472411951

[7] Dziadosz A, Tomczykb A and Kapliński O 2015 *Financial risk estimation in construction contracts.* Procedia Engineering pp 120 – 128

[8] Cakmak P I and Tezel E 2019 *A Guide for Risk Management in Construction Projects: Present Knowledge and Future Directions* DOI: 10.5772/intechopen.84361

[9] Tichý M 2006 *OVLÁDÁNÍ RIZIKA: Analýza a management 1.* Praha: C. H. Beck p 396 ISBN 80-7179-415-5

[10] ČSN EN 62198 (010694) 2014 *Management rizika projektu – Směrnice pro použití* Vyd. 1. Praha: UNMZ p 20 katalog. číslo 95697

IOP Conf. Series: Materials Science and Engineering **867** (2020) 012040 doi:10.1088/1757-899X/867/1/012040

Numerical nonlinear analysis of a membrane structure subjected to dynamic load

L Stulerova[1], L Kapolka[1] and S Kmet[1]

[1] Technical university of Kosice, Faculty of Civil Engineering, Institute of Structural Engineering, Vysokoskolska 4, 042 00 Kosice, Slovakia

Abstract. The aim of this paper is the numerical analysis of a thin technical textile in shape of hyperbolic paraboloid. It is the membrane structure, which can function only in tension and it is geometrically nonlinear, therefore nonlinear analysis has to be carried out. In this case, the structure was subjected to the dynamic loads effects. Natural vibration was applied as a fundamental dynamic characteristic. All of this results in natural frequencies and mode shapes and deflections. Acquired results describe the dynamic behaviour of nonlinear system. This part of the analysis could represent a first step in the dynamic analysis. To complete the dynamic analysis of the membrane structure, the forced vibrations as nonlinear time history analysis will have to be performed in a future.

1. Introduction

Membrane structures can be defined as a material of 6. generation along the traditional ones: brick, wood, steel, stone and concrete. Membrane are made of the technical textiles. This type of material directly relates with their positive characteristics. They are very light-weighted and can be used for the wide-span structures without a need of intermediate supports. With this fact coheres a material saving and therefore a lower initial costs. And the economic aspect is completed with the fact that the membrane structures function only in tension, it means that we are trying to use their positive characteristics to extend their life cycle.

But as every other material even the technical textiles have some negative aspects. They are strongly geometrically non-linear structures which makes it more difficult to design. It means that, in difference to the traditional structures, the membranes change their shape in a process of loading, so we need to separate it step by step. A non-linear analysis with an iteration process has to be carried out to obtain the assessment of the structure.

Another specificity is the fact that they need to be pre-stressed to carry out the external load. Technical textile is just piece of a fabric first and initial conditions requires pre-stress of a membrane surface. And because the membrane has to be pre-stressed, we do not know their initial shape in general. For this purpose, form-finding process is used to find the shape. We can say that it presents equilibrium between stress and displacement (shape).

Form-finding is very specific step in design of the membrane structures without which is not possible to continue in a static or dynamic analysis. It can not be described by simple mathematical methods. We have to use physical models or nowadays more appropriate computational methods.

In general, the membrane structures consist of 3 main parts: technical textile itself as a surface load-bearing element, linear load-bearing elements providing pre-stress and therefore stability (e. g. rope), and primary structure allowing transfer of internal forces from these structures to the foundation.

Structures of light-weighted architecture, which include membranes, we can define as which can transmit the same amount of load at a lower self-weight in terms of internal force transfer. Like any building construction, textile architecture is subject to strict design rules and all tentative requirements for mechanical resistance and stability must be respected, taking into account their specific characteristics listed above.

The usage of the membrane structures is dated thousands of years ago in the form of a temporary dwellings and later used for tents, for example circus tens called *chapiteau*. But light-weighted structures as we know it today began with the dissertation thesis *Das Hägende Dach* from Otto Frei in 1952. In 1967 he participated in the design of the German pavilion (figure 1) at the World Expo in Montreal. Although the original design of the pavilion was a stretched membrane structure, it was ultimately built as a rope net, using the fabric only as a secondary overlay [1].

Figure 1. German pavilion at the World Expo [1].

2. The numerical model

For the numerical analysis, the membrane structure in form of the hyperbolic paraboloid was modeled. Computational software Dlubal RFEM was used for this purpose. 3D hypar structure is made of orthrotropic material with modulus of elasticity in the warp direction (as principal direction) E_x=1057 MPa and modulus of elasticity in the weft direction E_y=612 MPa [2]. It simulates the most important material characteristics of technical textile Serge Précontraint 502.

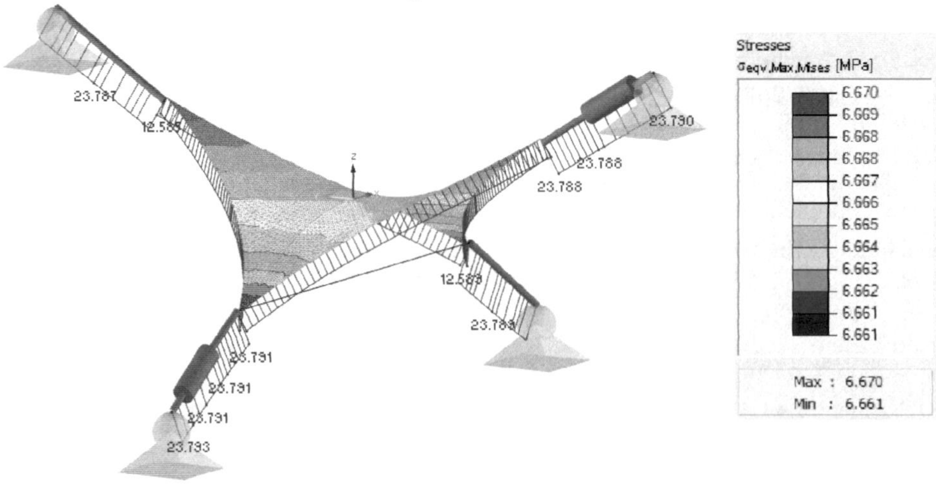

Figure 2. Membrane after form-finding.

Another parameters which characterise the geometrical size of the membrane are floor plan 2×2 m with overall height of 1 m which represent the rise equal to the overhang 0.5 m. On the edges of the surface element are linear elements – ropes with diameter of 8 mm. They are modeled as steinless steel ropes with modulus of elasticity E=130 GPa. They provide the pre-stress of the membrane surface in the form of absolute sag with value of 0.2 m. The pre-stress of the structure is completed with the pre-stress of the surface in both of directions (warp and weft) equal to $n_x=n_y=4$ kN/m [3].

The membrane stabilized by the edge ropes is anchored articulated with usage of 2 anchor rods and 2 actuators. Actuators are elements which can change their length and therfore change the stress distribution of membrane. The structure become adaptive with this process. This part of the model will be use in a future research, for now we consider the structure without change of anochorage.

Anchorages are modeled as a solid circular cross-section (D46 and D145). They represent masses on the experimental device in laboratory on which the whole numerical model was created. Verification with the numerical model will be carry out in a next phase of the research on this laboratory device.

Dlubal RFEM calculate any structure based on Finite Element Method (FEM). Therfore finite elements was used: 1D for edge ropes and anchor rods and 2D elements for membrane surface. The mesh size of 2D elements was 50 mm and we used triangular finite shell element. 1D elements are divided with 10 divisions per member (for cable and anchorage also) [3, 4]. The membrane was modeled as 2 triangulars connected in the middle between lower anchors with third vertices in the upper anchors.

Before the dynamic analysis of natural vibrations, process form-finding was performed to find initial shape according to the established pre-stress. Dlubal RFEM uses Update Reference Strategy (URS) for this aim. It basically loads the membrane structure step by step with pre-stress value and trace the displacement of nodes in mesh [5].

In figure 2 we can see surface stresses and internal forces in achores after form-finding. It is nessesary to point that form-finding is performed without considering dead load.

3. Natural vibration analysis and results

Dlubal RFEM as FEM software has a few add-on modules available. One of them is RF-DYNAM PRO, which serves for dynamic analysis. In this paper, natural vibration analysis was performed to achieve natural frequencies of mode shapes. One mass case was created and defined as permanent. This one represents only self-weight of the structure.

From this mass case one natural vibration case was made of. For the analysis general setting had to be considered. Number of lowest eigenvalues to calculate was established for 10 and scaling of mode shapes was chosen as recommended in manual for $\{u_j\}^T [M] \{u_j\}=1$. This setting is related to non-linear analysis of membrane [6].

For natural vibration case consistent mass matrix and Lanczos method for solving eigenvalue problem were used. It is nessesary remark that because of the membrane structure and form-finding process, stiffness modification was considered from load case of form-finding.

In the table 1 you can see all of 10 mode shapes of mass case with only self-weight, with values of eigenvalues λ, angular frequencies f and natural periods T. In the figure 3 you can see all of the mode shapes for this mass case also with deflection.

Table 1. Mass case with self-weight.

Mode	Angular Frequency v [rad/s]	Natural frequency f [Hz]	Natural period T [s]
1	57.186	9.101	0.110
2	60.415	9.615	0.104
3	66.512	10.586	0.094
4	92.390	14.704	0.068
5	99.473	15.832	0.063
6	180.252	28.688	0.035
7	220.289	35.060	0.029
8	314.905	50.119	0.020
9	343.091	54.605	0.018
10	376.551	59.930	0.017

IOP Conf. Series: Materials Science and Engineering **867** (2020) 012040 doi:10.1088/1757-899X/867/1/012040

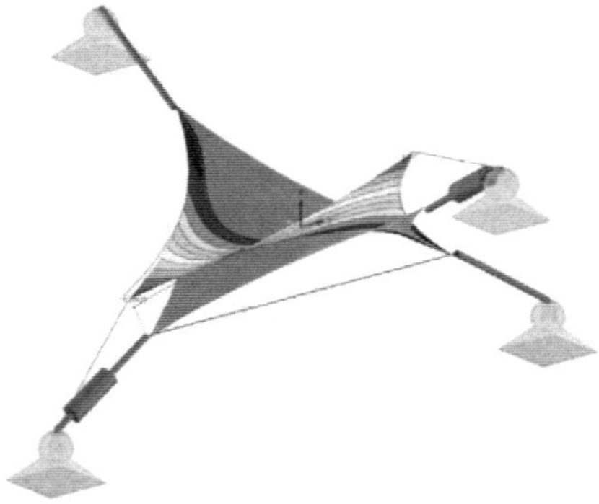

1st mode shape

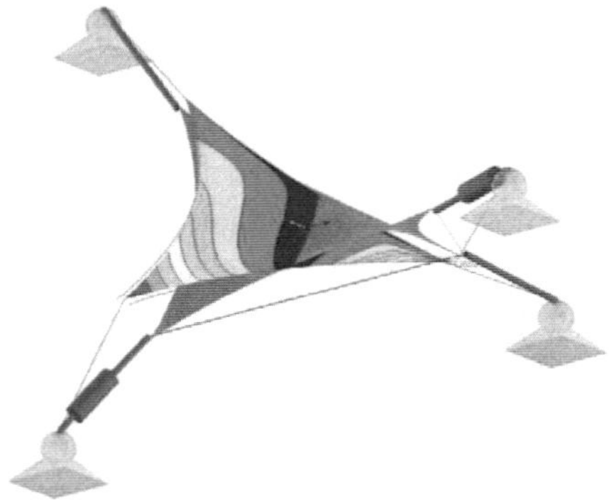

2nd mode shape

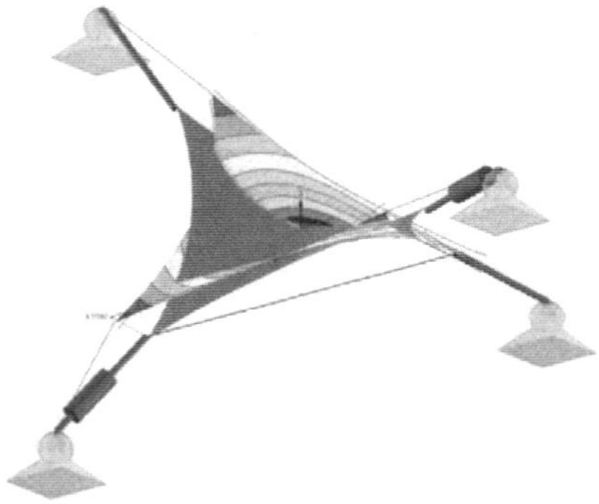

3rd mode shape

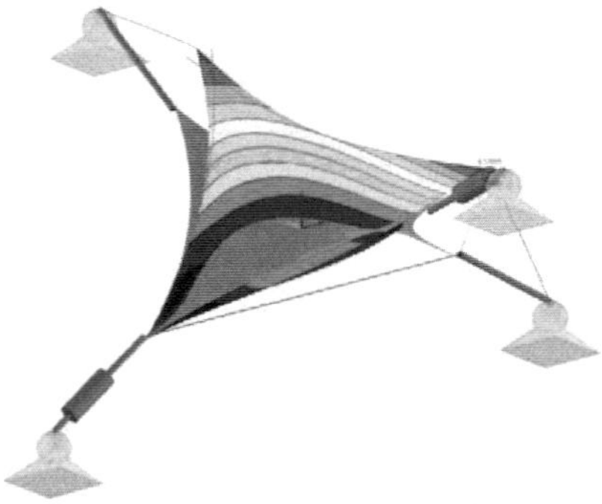

4th mode shape

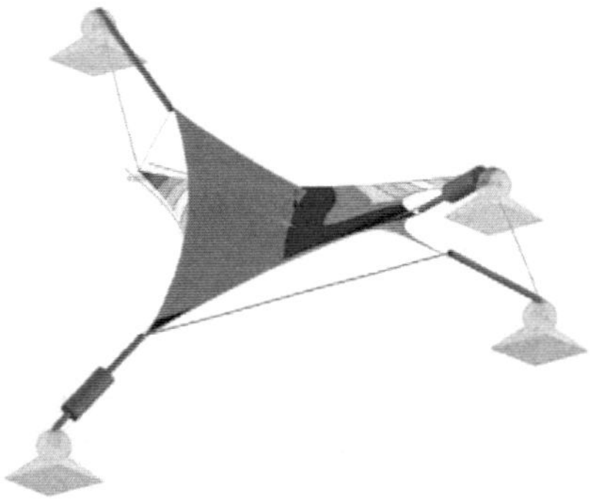

5th mode shape

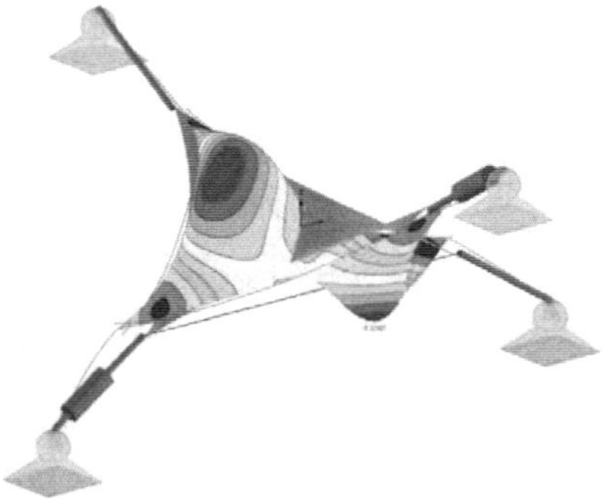

6th mode shape

7th mode shape

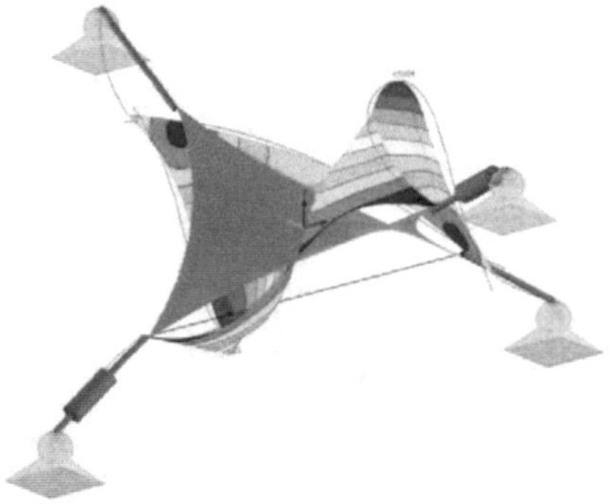

8th mode shape

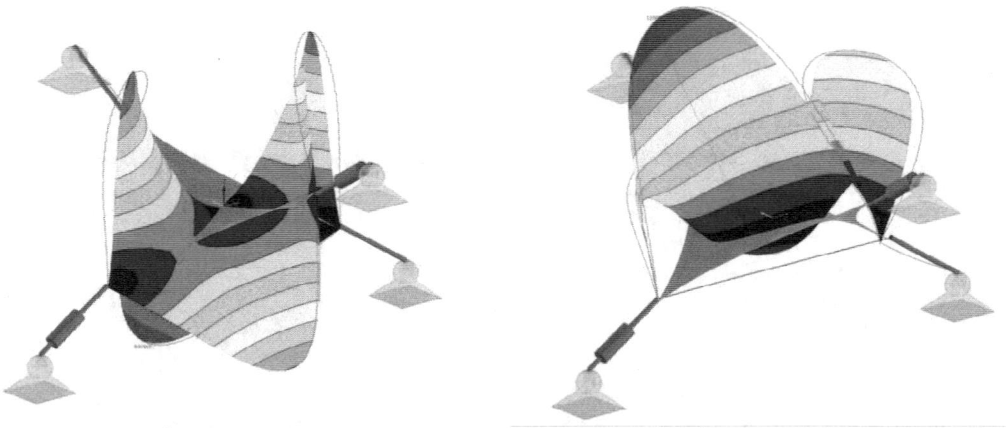

9th mode shape 10th mode shape

Figure 3. Mode shapes under natural vibration.

4. Conclusion

The aim of this contribution was to show modal analysis and natural vibration of the membrane structure. This structure is strongly geometrically non-linear and it requires specific step in the beginning of static or dynamic analysis. This step make membrane structures special in the design stage in compare to the typical materials.

As we can see in the table 1 and figure 3, the lowest natural frequencies begin with value of little bit higher that 9 Hz. We can say that this frequency is lower than 10 Hz which represents dynamic wind load, but it is far more than 5 Hz which can we consider as considerable factor of wind. For this purpose, it is necessary to supplement FEM model also with steel frame, as the experimental device has, so we can suggest that it will have significant influence on these parameters. It will approach real behaviour of the structure and it will make results more precise.

Also in the future research forced vibration will be performed as dynamic load effect and it will extend the results of this paper. We can consider forced vibration caused by harmonic variable or periodic force and every load case can be damped and undamped.

It all can results in comparison and verification of the numerical model with experimental device.

5. References

[1] Foster and Mollaert Marijke 2003 *European Design Guide For Tensile Surface Structures* (Germany: Tensinet) p 354

[2] INOVA Praha s.r.o. 2017 *Dokumentácia k skúšobnému zariadeniu – 17 120 časť 1 – Plachtostroj* (Prague: INOVA Praha s.r.o.)

[3] Kapolka L 2019 Analysis of an experimental tension structure *IOP Conf. Ser.: Mater. Sci. Eng.* **566** 012014

[4] Šoltýs R and Tomko M 2012 Analýza kmitania lana zaťaženého vetrom *Juniorstav 2012: 14. odborná konference doktorského studia: sborník anotací*

[5] Dlubal 2019 RFEM 5 Spatial Models Calculated According to Finite Element Method User Manual (Germany: Dlubal Software) p 650

[6] Dlubal 2019 RF-DYNAM Pro Natural Vibration Analysis, Response Spectra, Time History, Equivalent Static Forces Program Description (Germany: Dlubal Software) p 115

Acknowledgments

This work was supported by The Slovak Research and Development Agency under the contract No.APVV-15-0777 and also by The Slovak Scientific Grant Agency under the contract No.VEGA 1/0129/20.

IOP Conf. Series: Materials Science and Engineering **867** (2020) 012041 doi:10.1088/1757-899X/867/1/012041

Computer modeling of flow on the object of the sewer network

M Šutúš[1], J Hrudka[1], G Rózsa[1] and I Škultétyová[1]
[1] Slovak University of Technology in Bratislava, Faculty of Civil Engineering, Department of Sanitary and Environmental Engineering, Slovakia

Abstract. This work deals with the creation of a hydrodynamic model of the combined sewerage overflow chamber in Banská Bystrica. As an object on the sewer network, we chose a combined sewerage overflow chamber, because its correct functionality directly affects the quality of the environment in the recipient. Simulations were calculated in the program Ansys Workbench 19.2 – Fluid Flow (Fluent). The basis of the work was the creation of a 3D graphic model. The 3D model serves as a base for flow modeling. The core of the work are simulations in the combined sewerage overflow chamber that should correspond to real flow. The goal of our work is to compare the influence of inflow velocity and flow rakes on hydrodynamic flow.

1. Introduction

In the majority of urban areas in Slovakia, wastewater is drained through a single sewer system. At present, CSOs are used to regulate the rain flow and are therefore a typical object for these systems. Their purpose is to reduce the amount of rainwater supplied to the treatment plant and thus reduce the uneven load of the treatment plant by rainwater.

The efficiency of the CSOs depends on appropriate design and regular maintenance. Since not all of the CSOs are properly hydraulically designed, the recipient often becomes polluted and thus the quality of surface and groundwater is reduced.

The assessment of the functioning of the individual CSOs presupposes the use of a relatively economically demanding monitoring system. The use of such a monitoring system is too costly for long-term observation.

Mathematical-computer modeling can be an effective means of simulating possible phenomena and processes on the sewer network. Mathematical simulations will be created using ANSYS Fluent software.

2. Ansys Fluent – CFD

CFD (Computational Fluid Dynamics) stands for the modern calculation method. This method offers us using various software modeling and calculation methods, which the user can then choose according to the problem he solves in the project. In Ansys Fluent, we can model both 2D and 3D fluid flow; liquids and gases. This program has a huge application in terms of hydraulic, water and research [1,2,3,4].
The ANSYS CFD includes the Fluent solver, which provides real-world simulations of both 2D and 3D fluid and gas flow. Materials are available in a comprehensive material library, or you can create your own material and specify their specific properties. The mesh generator offers automatic mesh creation based on the selected simulation with the possibility to set the mesh and refinement parameters in detail. It allows us to solve sophisticated physical phenomena with maximum accuracy. The results show the flow, pressures, and temperatures with the possibility of their animation, maximum and minimum values, specific sectional views, and nozzles [2,4].

ANSYS CFD is integrated into the ANSYS Workbench - a platform designed for efficient and flexible workflows. It has CAD associativity and powerful features for modeling geometry and networking. The integrated parameter manager makes it easy to perform a parametric analysis. ANSYS Workbench Workflow makes it easy to link analyzes and solve multiple tasks [2,3,4].

2.1. Modeling

CSO was modeled in modeling software SpaceClaim according to the drawings. Dimensions of CSO are (h x w x l) 2660 mm, 5450 mm and 14150 mm. The inflow sewer has a diameter of DN 1600, the relief sewer has a diameter of DN 1600 and the bypass to the WWTP has a diameter of DN 600.

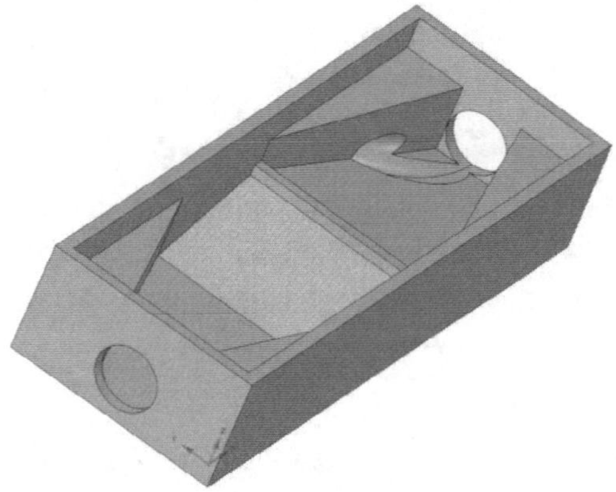

Figure 1. Model of CSO.

2.2. Boundary conditions

Calculation used the SIMPLE method and the k-epsilon model [4,5,6]. During the simulation, the model state of extreme operation with inflow velocity v = 2,0 m.s^{-1} was considered and in normal operation with inflow velocity v = 0,5 m.s^{-1}.

The boundary conditions are defined as follows in table 1.

Table 1. Boundary conditions.

Boundary condition	**Boundary condition**	**Definition**
Velocity Inlet	Inlet	0,5 / 2 m.s^{-1}
Wall	Wall	0,013 mm
Wall0	Water level	0 mm
Outflow	Outlet	-
Solid	Volume	H_2O

3. Simulations

Four scenarios were varied in which the inflow rate and the height of overflow edge changed:

1. 0,5 m.s^{-1}, height of overflow edge (normal) 290 mm
2. 2,0 m.s^{-1}, height of overflow edge (normal) 290 mm
3. 0,5 m.s^{-1}, height of overflow edge (increased) 350 mm
4. 2,0 m.s^{-1}, height of overflow edge (increased) 350 mm

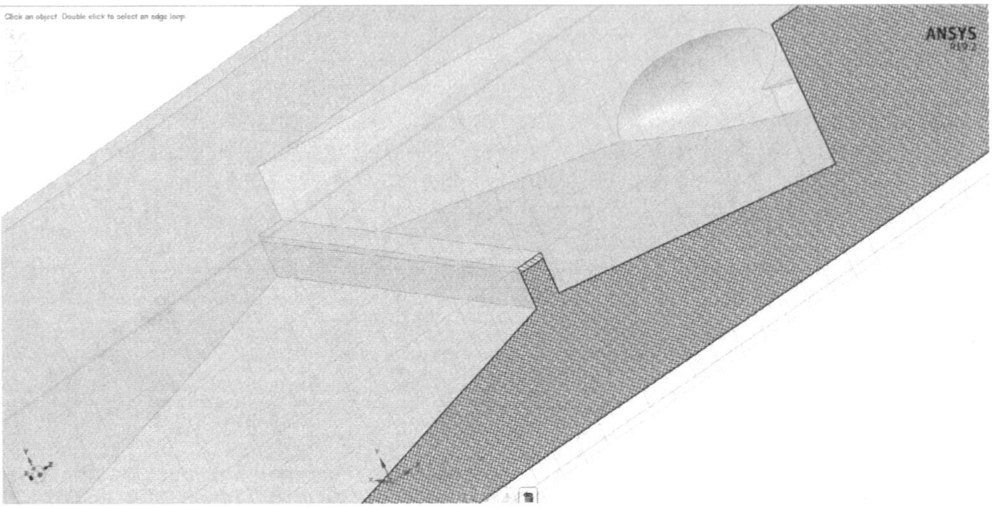

Figure 2. The height difference of edges.

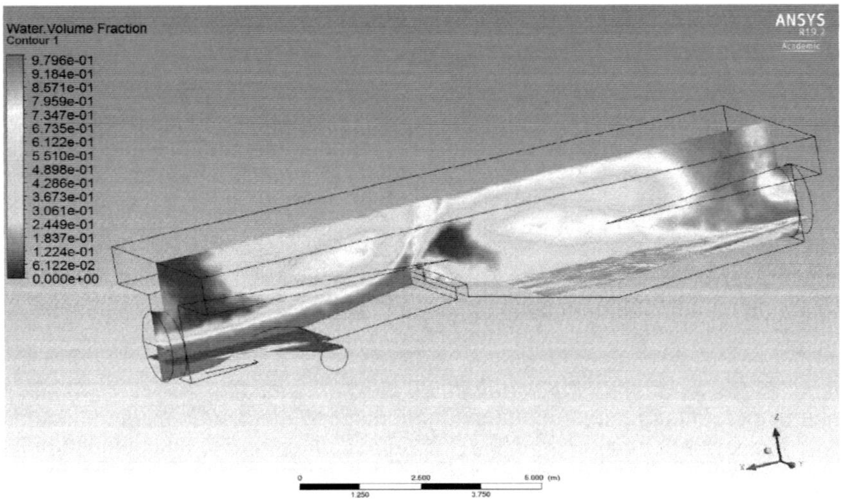

Figure 3. Example of simulations.

4. Results

In simulations, speed velocity was compared. Velocity was measured at the outlet to the WWTP, overflow edge and outlet to a recipient. In each simulation, speed was measured in points with the same coordinates. The results are summarized in table 2.

Table 2. Velocity comparison.

	Height of overflow edge											
	290 mm						350 mm					
	Velocity [m.s^{-1}]											
	2,0			0,5			2,0			0,5		
Time [sek]	**WWTP**	**edge**	**Out let**	**WWTP**	**Edge**	**Out let**	**WWTP**	**Edge**	**Out let**	**WWTP**	**Edge**	**Out let**
5	2,15	-	-	-	-	-	2,18	-	-	-	-	-
15	2,08	1,14	-	0,52	-	-	2,09	1,08	-	0,53	-	-
35	2,15	1,16	2,04	0,53	0,23	-	2,16	1,07	1,98	0,54	0,23	-
50	2,18	1,16	2,05	0,53	0,25	-	2,18	1,09	1,98	0,53	0,26	-

5. Conclusion

This work deals with mathematical modeling and 3D modeling of existing CSO. The core of the work are hydrodynamic simulations of flow in Ansys Fluent. Two load cases were selected that correspond to extreme inflow velocity v = 2,0 m.s^{-1} and a more common situation with inflow velocity v = 0,5 m.s^{-1}. The results show that height change of overflow edge slightly reduced the velocity in the extreme state. In the future, authors want to focus on whether the change of height, change of slope or combination of changes can improve flow in CSO.

6. Acknowledgments

This work was supported by the Scientific Grant Agency of the Ministry of Education, Youth and Sports of the Slovak Republic and the Slovak Academy of Sciences within the project VEGA 1/0574/19, co-funded by the European Regional Development Fund and by the Slovak Research and Development Agency under the contract No. APVV-18-0203.

7. Reference

[1] Molnár V 2011 Computational Fluid Mechanics: Interdisciplinary Approach with CFD Applications *Spektrum STU Bratislava* ISBN 9788081060489
[2] ANSYS FLUENT 12.0 Theory Guide, Canonsburg: ANSYS Inc. . (2009)
[3] Kozúbková M 2008 Modeling of fluid flow FLUENT CFX Ostrava: VSB – Technical University of Ostrava
[4] Bojko M 2010 3D flow – ANSYS Fluent Ostrava: VSB – Technical University of Ostrava
[5] Hrudka J, Stanko Š and Holubec M 2017 Analysis of flow and sedimentation processes in secondary sedimentation tank *Pollack Periodica* vol. 12 no. 2 pp 79 – 89
[6] Gregušová V, Škultétyová I and Holubec M 2017 Measurement of flow velocity in different secondary settling tanks for analysis of the flow *Pollack Periodica* vol. 12 no. 3 pp 15 - 22

IOP Conf. Series: Materials Science and Engineering **867** (2020) 012042 doi:10.1088/1757-899X/867/1/012042

The effect of CO_2 on the strength characteristics of cement composites based on recycled rubber from waste tires

J Svoboda[1], V Vaclavik[1], T Dvorsky[1], J Charvat[1], K Macalova[1] and R Papesch[1]

[1] VSB - Technical University of Ostrava, Faculty of Mining and Geology, Department of Environmental Engineering, Czech Republic

Abstract. This paper presents the results of a research into the use of a combination of recycled rubber from waste tires as a 100% aggregate replacement in the production of cement composites. The aggregate was replaced with rubber recycled material in two ratios. At a ratio of 50/50 and a ratio of 40/60 fraction 0/1 mm and fraction 1/3 mm. The test specimens were further subjected to CO_2 effect in a Lamart laboratory chamber, where CO_2 effect was simulated as equivalent to an aging period of 50 years. The presented results show that the effect of CO_2 reduces the strength characteristics of the composite compared to the comparative samples.

1. Introduction

The boom in the automotive industry has been accompanied by a proportional increase in the production of waste tires, resulting in high environmental burden, because these tires are not properly stored and disposed of [1-3]. Following the example of the use of other waste materials as a secondary raw material [4-7], a potential research can also focus on the incorporation of recycled rubber material in cement composites [8,9] In recent decades, great emphasis has been placed on the specification of the resistance of concrete structures. Along with a better understanding of the individual factors affecting the durability of concrete and reinforced concrete structures, many research teams are dedicated to improving the resistance of concrete [10]. The corrosion of concrete and steel reinforcement is the dominant cause of the deterioration in the properties of concrete structures, which leads to concrete degradation and a reduction in the effective reinforcement cross-sectional area [11-13]. Corrosion occurs as the result of carbonatation of the surrounding concrete or the penetration of chloride ions [14]. Carbonatation is a process during which carbon dioxide (CO_2) from the atmosphere diffuses the porous concrete microstructure and reacts with hydroxides, such as calcium hydroxide released from the hydration process, to form carbonates, thereby neutralizing the alkalinity of the concrete, leading to the degradation effects of the cement matrix [15-18]. At present, most research projects dealing with the effects of CO_2 on cement composites are concerned with the effect of CO_2 as the solidification and hardening accelerator in the early stages of composite formation [19-21] or as a possible CO_2 absorbent used to reduce its concentration in the atmosphere [22-24]. There are also researches focused on the influence of CO_2 on the resistance of cement composites depending on the individual components of the composite [25-27]. This article pays attention to the use of a combination of recycled waste rubber from waste tires as a 100% aggregate replacement in the production of cement composites and their resistance to carbonatation.

2. Materials and Methods

Two formulas with the designations of Z2 and Z3 (table 1) were designed on the basis of the properties already established before. The designed formulas had been exposed to CO_2 effects equal to 50 years and their strength characteristics were examined afterwards.

2.1. Cement

One type of cement was used for the designed formulas, namely CEM III/A 32.5 N Blast Furnace Cement. The amount of the cement used is presented in table 1.

2.2. Mixing water

Water from the water supply system according to ČSN EN 1008 was used as the mixing water [28]. The amount of mixing water used is presented in table 1.

2.3. Rubber granulate

Rubber granulate from waste tires with two fractions of 0/1 mm and 1/3 mm was used as the filler to fully replace standardized sand. The amount of granulate used is shown in table 1. Figure 1 shows the rubber granulate particle size analysis of fr. 0/1 mm and 1/3 mm compared to standard aggregates.

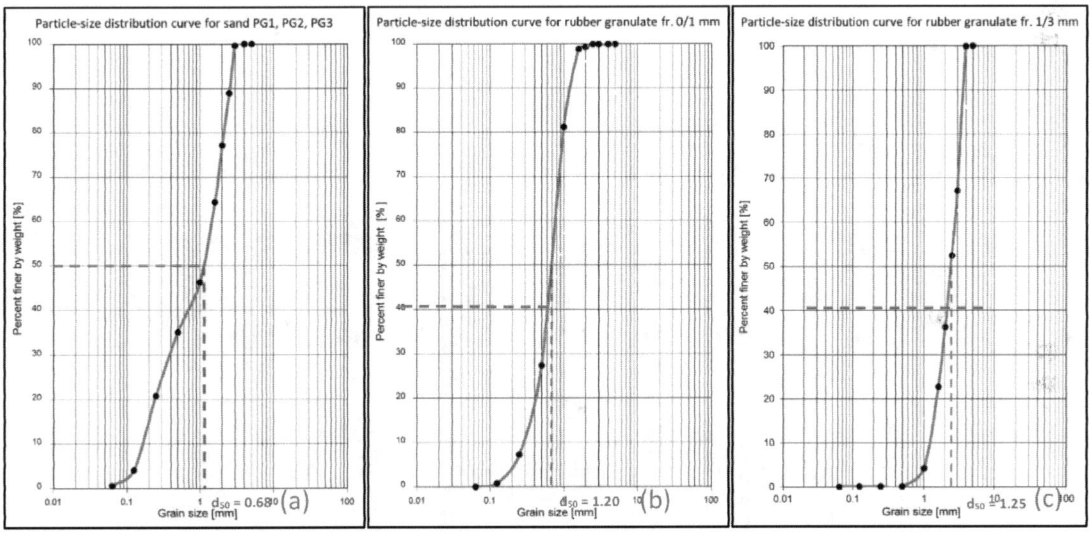

Figure 1. Results of grain-size curve determination: (a) comparative sample of standard aggregates PG1, PG2, PG3; (b) rubber granulate sample with the grain-size of 0/1 mm; (c) rubber granulate sample with the grain-size of 1/3 mm.

2.4. Formula design

Two formulas with the designations of Z2 and Z3 were designed in order to manufacture the test specimens from concrete based on recycled rubber. The compositions of the individual formulas per $1m^3$ of mixture are presented in table 1.

IOP Conf. Series: Materials Science and Engineering **867** (2020) 012042 doi:10.1088/1757-899X/867/1/012042

Table 1. Composition of experimental cement composite formulas containing rubber granulate and pure mixing water per $1m^3$ of mixture.

Formula	Comp.	Z2	Z3
CEM III/A 32,5 N [kg]	502	674	674
Pure mixing water [kg]	251	337	337
Standardized sand [L]	915	-	-
Rubber granulate fr. 0/1 [L]	-	614	491
Rubber granulate fr. 1/3 [L]	-	614	737

2.5. Strength characteristics
Tensile flexural strength and compressive strength tests were carried out in accordance with CSN EN 196-1 [29]. Beams with the dimensions of 40×40×160 mm were used as the test specimens. Form test instrument with a compressive force of 100 kN and 300 kN was used as the test equipment.

3. Results and Discussion

3.1. Surface carbonatation
The comparative samples were cured for 90 days in a humid environment and subsequently exposed to CO_2 imitating a period of 50 years and then they were broken. A phenolphthalein indicator was applied to the fracture surface of the sample, and the area not affected by carbonatation turns pink (figure 2).

Figure 2. Cement composite samples after the application of phenolphthalein of fracture surface: (a) Comp., (b) Z2, (c) Z3.

Table 2. Carbonatation area.

Mixture	Total sample area [mm^2]	Carbonatation area [mm^2]	Area affected by carbonatation [%]
Comp.	1613.1	715.2	44.4
Z2	1648.6	731.6	44.4
Z3	1632.1	709.4	43.5

Table 2 presents the area affected by carbonatation. It shows that the mixtures Z2 and Z3 have an area affected by carbonation of about 44%. The difference from the comparison sample is negligible. It is therefore clear that the use of recycled rubber did not affect the carbonatation properties of the cement composites either in a positive or in a negative way.

3.2. Strength characteristics

Test specimens with dimensions of 40×40×160 mm were used for testing the tensile flexural strength and compressive strength. The determination of tensile flexural strength and compressive strength was performed on samples after 90 days of age and on samples subjected to CO_2 effect imitating a period of 50 years. The measured values of tensile flexural strength and compressive strength are presented in figure 3 and 4.

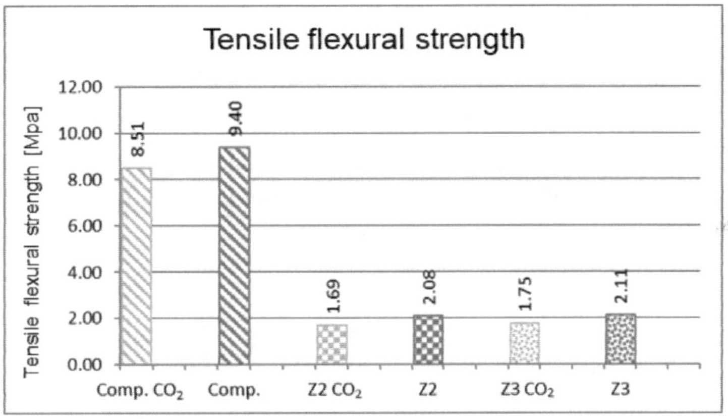

Figure 3. Graphical expression of tensile flexural strength of cement composites exposed to accelerated carbonatation.

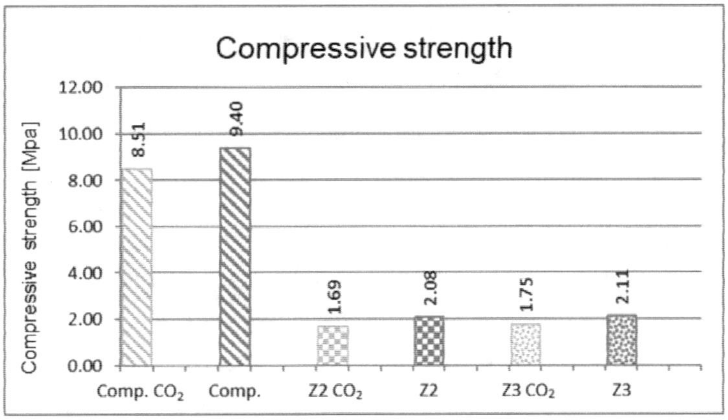

Figure 4. Graphical expression of compressive strength of cement composites exposed to accelerated carbonatation.

It is clear from the results shown in figure 3 and 4 that the samples exposed to CO_2 have reduced tensile flexural strength as well as compressive strength. The comparative sample exposed to accelerated carbonatation showed lower tensile flexural strength by 9.5% and compressive strength by 5.5% compared to the comparative sample not exposed to accelerated carbonatation. The sample of the Z2 formula exposed to accelerated carbonatation showed lower tensile flexural strength by 18.8% and compressive strength by 9.4% compared to the comparative sample not exposed to accelerated carbonatation. The sample of the Z3 formula exposed to accelerated carbonatation showed lower tensile flexural strength by 17.1% and a compressive strength by 8.6% compared to the comparative sample not exposed to accelerated carbonatation.

4. Conclusion

This article presents the results of an experimental research of the use of recycled rubber from waste tires in the production of cement composites. The research was focused on the physical-mechanical properties of cement composites exposed to the effects of CO_2 representing the period of exposure of 50 years.

The results of the tests show that:

- the area of cement composite with recycled rubber affected by carbonatation is comparable to that of the comparative sample, which means that the addition of the recycled rubber did not have any positive or negative effect of the composite to withstand the impact of CO_2.
- the comparative sample exposed to accelerated carbonatation showed lower tensile flexural strength by 9.5% and compressive strength by 5.5% compared to the comparative sample not subjected to accelerated carbonatation.
- both mixtures containing 100% replacement of natural aggregate with recycled rubber showed a decrease in strength compared to composites not exposed to accelerated carbonatation This decrease was almost double compared to the cement composites based on natural aggregate.
- the formula Z3 containing 100% aggregate replacement with recycled rubber in the ratio of 40% fraction 0/1mm and 60% fraction 1/3mm was rated as the most suitable formula containing rubber granulate.

5. References

[1] Rawahi Z A and Waris M B 2017 *MATEC Web Conf.* **120**
[2] Thomas B S, Gupta R C and Panicker V J 2016 *J. Clean. Prod.* **112** 504–13
[3] Xue J and Shinozuka M 2013 R *Constr. Build. Mater.* **42** 196–204
[4] Brožová S and Jursová S 2012 *SGEM 2012* **4** 745–52
[5] Pustějovská P, Brožová S and Jursová S 2014 *SGEM 2014* **1** 11–4
[6] Ondova M, Stevulova N and Estokova A 2012 *Procedia Eng.* **42** 1863–72
[7] Junak J, Stevulova N and Ondova M 2014 *Pollack Period.* **9** 95–104
[8] Si R, Wang J, Guo S, Dai Q and Han S 2018 *J. Clean. Prod.* **180** 823–31
[9] Li G, Stubblefield M A, Garrick G, Eggers J, Abadie C and Huang B 2004 *Cem. Concr. Res.* **34** 2283–9
[10] Ferrer B, Bogas J A and Real S 2016 *Cons. Build. Mat.* **120**, 161-171
[11] Barberon F, Bouny V B, Zanni H, Bresson B, Caillerie J B, Malosse L and Gan Z 2005 *Mag. Reson. Imag.* **23**(2), 267-272
[12] Moreno m, Morrism W, Alvarez M G and Duffo G S 2004 *Corros. Sci.* **42**(11), 2681-2699
[13] Talakokula V, Bhalla S, Ball R J, Bowen C R, Pesce G L, Kurchania R, Bhattacharjee B, Gupta A and Paine K 2016 *Sens. Actuat.: Phys.* **242**, 79-91
[14] Papadakis V G, Fardis M N and Vayenas C G 1992 *ACI Mat. J.* **82**(2), 119-130
[15] Bertolini L, Elsener B, Pedeferri P and Polder R 2004 *Corrosion of Steel in Concrete* (Wiley-VCH, Weinheim, Germany) p 392
[16] Claisse P, Elsayad H and Ganjian E 1997 *ACI Mat. J.* 96(3), 378-381
[17] Marangu J M, Thiong'o J K and Wachira J M 2019 *J. Chem.*
[18] Lazarean H N A, Iliescu M, Ciont N and Abrudan I F 2019 *Con. Build. Mat.* **212**, 745-754
[19] Bogas J A, Real S and Ferrer B 2016 *Cem. Concr. Compos.* **71**, 110–21

[20] Liang C, Pan B, Ma Z, He Z and Duan Z 2020 *Cem. Concr. Compos.* **105**
[21] Lo T Y, Tang W C and Nadeem A 2008 *Constr. Build. Mater.* **22**, 1648–55
[22] Ghayeb H H, Razak H A and Sulong N H R 2020 *J. Clean. Prod.* **242**
[23] Souto-Martinez A, Arehart J H and Srubar W V 2018 *Energy Build.* **167**, 301–11
[24] Kurda R, de Brito J and Silvestre J D 2019 *J. CO2 Util.* **29**, 12–9
[25] Singh S P and Singh N 2016 *J. Mater. Eng. Struct.* **3**, 35–57
[26] Monkman Sean and Shao Yixin 2010 *J. Mater. Civ. Eng.* **22**, 296–304
[27] Xie X, Feng Q, Chen Z, Jiang L and Lu W 2019 *Cons. Build. Mat.* **218**, 119-125
[28] CSN EN 1008. Mixing water for concrete–Specification for sampling, testing and assessing the suitability of water, including water recovered from processes in the concrete industry, as mixing water for concrete. Praha: Czech office for standards, metrology and testing, 2003.
[29] CSN EN 196-1. Methods of testing cement – Part 1: Determination of strength. Praha: Czech office for standards, metrology and testing, 2016.

Acknowledgments
This research was funded by VŠB–TUO, Faculty of Mining and Geology—grants numbers SP2020/30 and SP2019/29; Moravian-Silesian Region—grant number RRC/10/2018.

IOP Conf. Series: Materials Science and Engineering **867** (2020) 012043 doi:10.1088/1757-899X/867/1/012043

LCA analysis of Slovak residential building

Ž Tomková[1]

[1] Technical university of Košice, Faculty of Civil Engineering, Department of architectural engineering

Abstract. Building industry is main contributor to greenhouse gas (GHG) emissions. According to goals of Paris agreement to achieve temperature rise below 2°C a year and then below 1,5°C a year, we need to find where are our biggest possibilities of improvement. Focus of this paper is on residential buildings built in Slovakia. With population growth also demand for apartments rises. With residential building development comes possible place for improvement. That is why main goal of this assessment is to analyse current state of our residential building fond using eToolLCD software and cradle to grave approach. Selected environmental impact of this analyse are, global warming potential (GWP), ozone depletion (OP), acidification potential (AP), eutrophication potential (EP), photochemical ozone creation potential (POCP), abiotic depletion potential of elements (ADPE), abiotic depletion potential of fossil fuels (ADPF) and also use of primary energy resources (PENERT). Assessed residential buildings are first part of larger investigation in this sector.

1. Introduction

The building industry is leading industry in conjunction with negative environmental impacts and consumes large portion of resources. This sector is responsible for consuming up to 40% of the total energy consumption in EU, with a related 39% of carbon dioxide (CO_2) emissions if upstream power generation is included [1]. EU wants to comply Kyoto Protocol to the United Nations Framework Convention on Climate change (UNFCCC), which committed to reduce greenhouse gas emissions by at least 20% below 1990 levels, and by 30% in the event of an international agreement being reached [2]. Furthermore, long term goal is to reduce emission of greenhouse gasses even more till 2050 compare to levels from 1990, with aim to secure high energy efficient, decarbonized interstate building fond. One part to achieve these goals is by refurbish existing buildings and improve their energy efficiency [3]. Second part is built new buildings according to even higher criteria in terms of energy efficiency. But with this requirement rises also demand for focus on whole life cycle. This is supported by findings of Martin Röck at al. which came to conclusion that contribution of embodied GHG emissions increases up to and beyond ratio of 1:1 (embodied : operational) when 50 year life period is considered, for buildings that have adopted new energy performance standards for building operation [4]. Another review acknowledged the changing trend in percentage of caused GHG emission by each life cycle stage. One of conclusions was that, for low energy buildings, the other life cycle stages, than use stage, become more important, representing up to 65% of the building life cycle contribution [5]. With population growth, comes demand for new accommodation. Residential buildings show to be important area which to focus on, in relation to environmental impacts, with upcoming EU plan for new buildings to meet criteria of nearly zero energy buildings from 2020. Aim of this whole research is to examine residential building fond of Slovak republic and find possible areas for improvement. Subject of this paper is evaluation of new residential building located in Eastern Slovakia. This analysis focuses on several environmental impacts such as: global warming potential (GWP), eutrophication potential (EP), acidification potential (AP), photochemical ozone creation potential (POCP), ozone depletion potential

(ODP), abiotic depletion potential – elements (ADPE), abiotic depletion potential - fossil fuels (ADPF), and total use of non-renewable primary energy resources (PENRT). For modelling the life cycle impacts eToolLCD program was used. It uses the available database BRE IMPACT, version 4. To display the resulting data by the amount of environmental impacts for each category, the so-called inventory tables are used. Material information was gathered from project documentation and budget calculation. Energy consumption and water supply were gathered from technical report. Afterwards the collected data from documentation were one by one picked in the database of the program. If there was not the material with perfect match available the closest one was picked instead or parameters were slightly modified in program. If there was a possibility an EPD was added.

2. Methods and Research Process

Residential building located in Eastern Slovakia in Košice was chosen for this study. In time of conducting this analyse the building was still under construction.

2.1. LCA

Environmental life cycle analysis is a systematic tool used for assessing the environmental impacts associated with a specific product or service or building over their whole life cycle, including raw material extraction, manufacturing, stage of use, and end of life, disposal and potential of recycling [6] [6][7]. This type of analysis is helping us to show compact picture about impacts related to the whole building. The requirements and procedure of implementation LCA analysis are defined in ISO14040 and ISO 14044 standard methods [8][8].

2.2. Goal and scope

The main objective of this study was analysis of new multi-story residential building with polyfunction. This building will be then part of the wider study sample consisting of existing buildings, new buildings and planned projects. System boundaries are set to cradle to grave. The functional unit utilized in this study was one square meter of gross floor area and estimated design life of 60 years.

2.3. Case study

Case study building is residential building located in Eastern Slovakia in Košice. In time of conducting this analyse the building was still under construction.

Table 1. Building information.

Built-up area	1286.0 m^2
Usable floor area	4608.1 m^2
- usable area for polyfunction	177.4 m^2
- usable area for apartments	3600.6 m^2
- usable area of apartment storage spaces	249.7 m^2
- usable area of apartment communication spaces	536.4 m^2
- usable area of technical background	44.0 m^2
Terrace area	1162.1 m^2
Number of apartments	50
Number of polyfunctional spaces	2
Number of floors	7

2.3.1. Characterization of material and services used for reconstruction. Foundations of this building are designed as foundation feet mounted on piles. Piles are pyramid shape, made of reinforced concrete and prefabricated. Reinforced concrete foundation beams with a cross-section of height 700 mm and width 400 mm are designed around the perimeter of the building. Underground constructions are insulated with XPS polystyrene and protected with profiled foil. On the 1st above ground floor is designed monolithic reinforced concrete slab with variable thickness. Vertical load-bearing structures are designed as reinforced concrete monolithic multi-storey tree-field frames. Horizontal load-bearing structures are designed as continuous monolithic reinforced concrete slabs which will be mounted on mentioned frames. On them are floors isolated with EPS boards. Vertical communications in building are three-spoke reinforced concrete monolithic slab staircases with double-folded middle plate and edge plates. In building are also elevators, they are placed in monolithic reinforced concrete shafts. Building envelope consists of filler brick made of ceramic POROTHERM blocks. Envelope will be insulated with contact thermal insulation system based on mineral wool and skirting will be insulated with XPS polystyrene boards. Roof is designed as walk on terraces and impassable flat roofs with PVC waterproofing and EPS insulation boards. In living spaces with sanitary installation are made of suspended plasterboard ceiling. Windows frames and glass walls were made of aluminum profiles with isolating triple glazing. Entrance apartment door are designed as security doors class 2 installed in security frame. Interior doors are wooden installed in wooden frame. Exterior wall finish is made of silicone plaster, on terraces is floor made of frost-resistant concrete pavement. Step layer on staircases and hallows is made of glued anti-slip gress tiles. In apartments floor finish is made of ceramic tiles and laminated floor panels in rooms. Walls in interiors are lined with ceramic tiles in wet operation rooms and other walls will have gypsum plaster coating. Technical rooms and storage areas will be coated with lime-cement plaster.

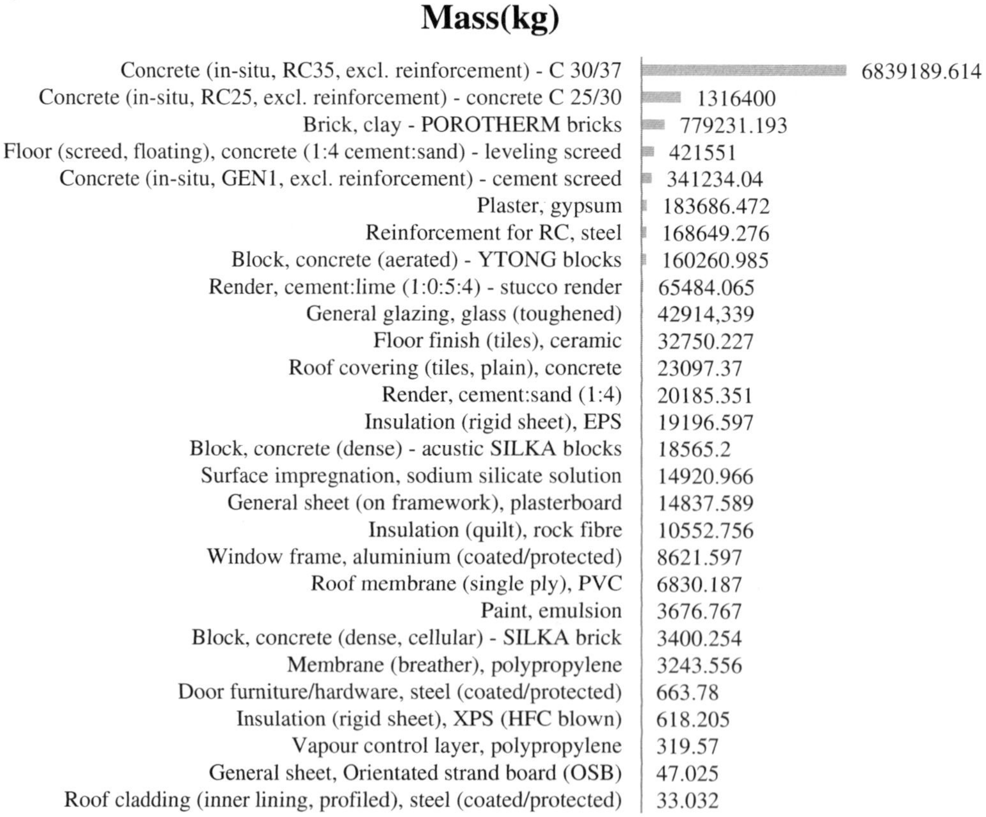

Figure 1. Material composition sorted by mass (kg).

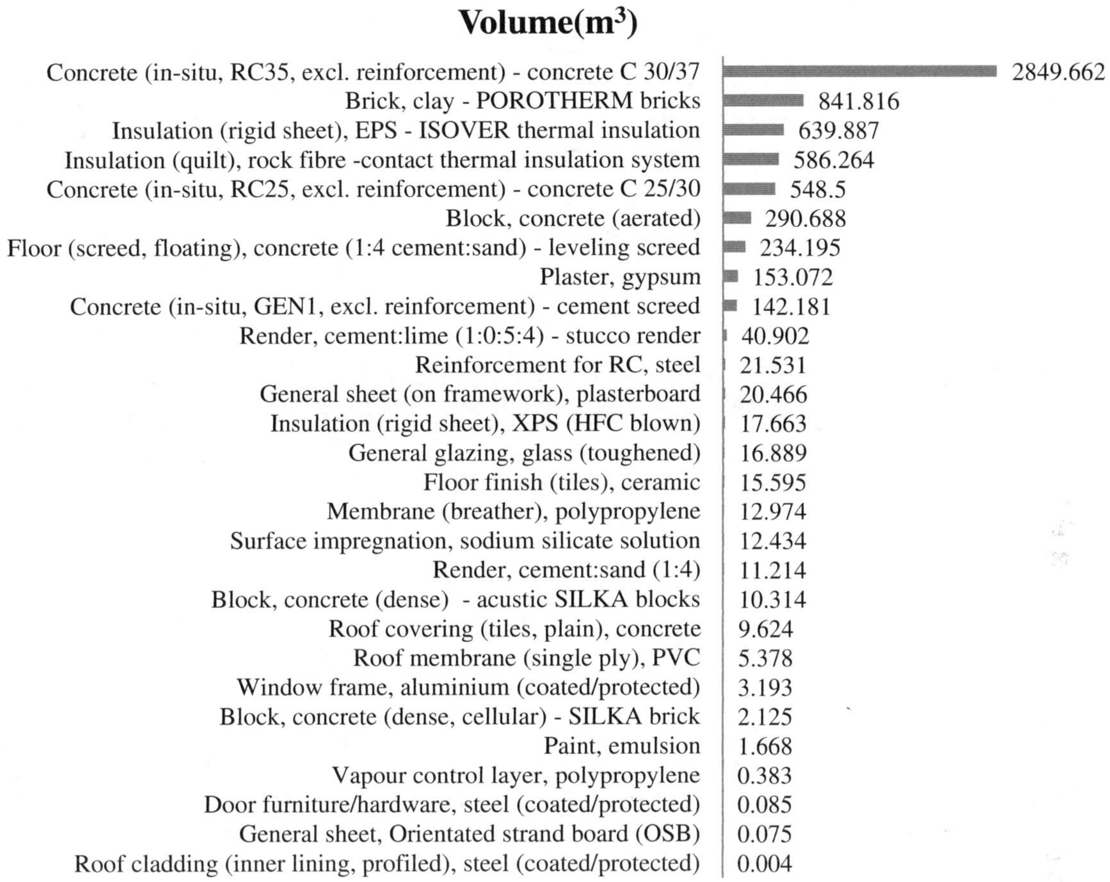

Figure 2. Material composition sorted by volume (m^3).

2.3.2. Building technical equipment. Water for drinking, sanitary facilities and for the building fire safety system is provided by the public water supply. The heat source will be a hot water connection from district heating and a heat exchanger station in the building on the 1st floor. In the common corridors are designed electric radiant heating panels on the ceiling. Floor heating will be realized by Uponor - Tacker wet system. Exhaust radial fans with built-in non-return valve are designed for ventilation of sanitary facilities of toilets and bathroom. Ventilation of the ATS spaces is provided by a fan into the circular duct. The fan is used for air exhaust. ATS ventilation is vacuum based. Air supply to the room is solved at the floor through a fire grille. Same applies for OST space, distribution room. Engine room has in one case vacuum based ventilation in other overpressure ventilation. CO bunker has individual ventilation system which is not use during normal service. Electricity will be provided from commercial grid.

2.3.3. Operation of building. Since building is still under construction, information about operation energy is obtained from the Project evaluation of energy performance of buildings. This part of building documentation has calculated values of energy demand for heating according to valid technical standards for climatic conditions – Košice. Results of calculation confirms that building meet the criteria for energy class A0.

3. Results and discussion

Results of the LCA analysis of selected residential building, are presented in table 2. and also table 3. To be able to compare the results with different studies, two types of data tables were carried out. Table 2. shows environmental impact per m2 of usable floor area in estimated life span of 60 years. Table 3. shows another common measurement unit, environmental impact per m^2 of usable floor area impact per year. These two sets of data will also make later use of data easier.

Table 2. Indicators characterizing environmental impacts for individual phases of building life per m^2 of usable floor area.

Characterized impacts per m2 usable floor area		GWP kg CO_{2eq}	ODP kg CFC-11$_{eq}$	AP kg SO_{2eq}	EP kg $(PO_4)^{3-}_{eq}$	POCP kg $C_2H_{4\,eq}$	ADP-E kg Sb_{eq}	ADP-FF MJ
Materials and construction	A1-A3	530	1.60E-05	2	0.48	0.32	9.30E-04	4800
	A4	43	2.60E-06	0.097	0.021	0.017	2.30E-07	580
	A5	0.16	1.10E-10	0.0019	0.0028	0.016	4.80E-10	0.014
Use stage	B1	-7.9	0	0	0	0	0	0
	B2	0	0	0	0	0	0	0
	B5	99	6.90E-06	0.83	0.31	0.094	1.20E-03	1500
	B6	1700	2.70E-06	0.48	0.11	0.063	5.30E-06	840
	B7	41	2.10E-06	0.2	0.047	0.012	9.60E-06	630
End of life	C4	3.1	2.5	0.019	0.041	0.00086	6.10E-04	0.37
Total		2400	2.80E-04	3.6	1	1.8	2.60E-03	8300

Impact keys: over 40% (red); 31%-40% (blue); 21%-30% (orange); 10%-20% (yellow)
Product stage: A1 = Raw Material supply, A2 = Transport, A3 = Manufacturing; Construction stage: A4 = Transport, A5 = Construction; Use stage: B1 = Use, B2 = Maintenance, B5 = Refurbishment, B6 = Operational energy use, B7 = Operational water use; EoL stage: C4 = Disposal

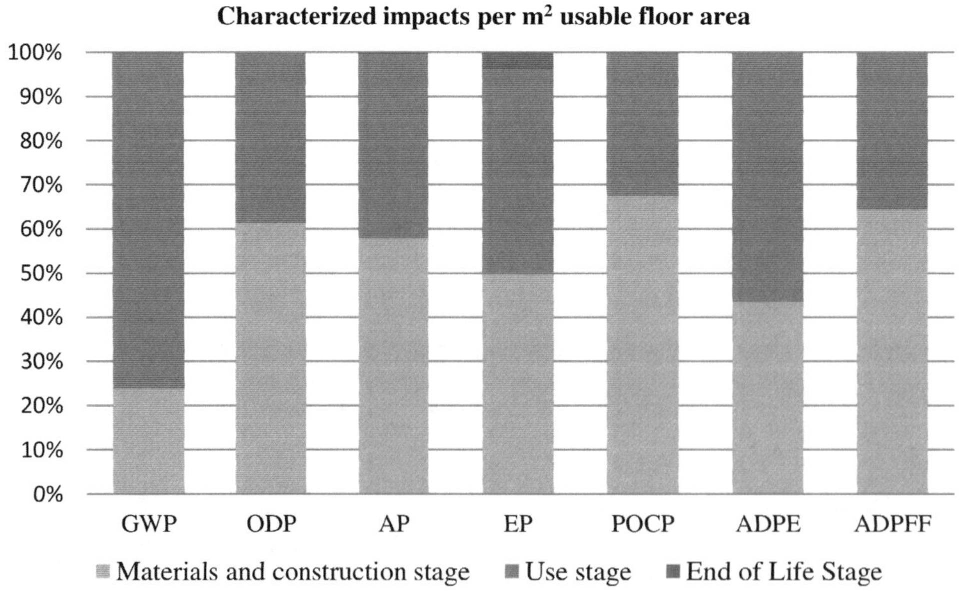

Figure 3. Characterised relative impacts of three stages per square meter for fully enclosed floor area.

From figure 3 we can see that impact on GWP is greater in use stage. Almost 25% compared to almost 75%. This corresponds to findings from review of Röck M at al. Building categorized as new standard, which corresponds to analysed building, show similar percentages for each of these stages [4]. Impacts causing ODP and AP are greater in material and construction stage. Impacts causing EP is almost similar for material and construction stage (49.9%) and use stage (46.2%). In this impact category is more visible also impact in end of life stage around 4%. For impact category POCP is more significant material and construction stage. Use stage has second largest impact in ADPE impact category with 56.6% impact. Material and construction stages are more significant for last assessed impact category ADPFF.

Table 3. Indicators characterizing environmental impacts for individual phases of building life per m^2 of usable floor area per year.

Characterized impacts per m2 usable floor area per year		GWP	ODP	AP	EP	POCP	ADP-E	ADP-FF
		kg CO_{2eq}	kg CFC-11_{eq}	kg SO_{2eq}	kg $(PO_4)^{3-}_{eq}$	kg $C_2H_{4\,eq}$	kg Sb_{eq}	MJ
Materials and construction	A1-A3	8.8	2.60E-07	0.033	0.008	0.053	1.60E-05	79
	A4	0.71	4.40E-08	0.0016	0.00035	0.00029	3.80E-09	9.6
	A5	0.0027	1.80E-12	0.000032	0.000047	0.00026	8.00E-12	0.00024
Use stage	B1	-0.13	0	0	0	0	0	0
	B2	0	0	0	0	0	0	0
	B5	1.6	1.20E-07	0.014	0.0052	0.0016	2.00E-05	25
	B6	28	4.50E-08	0.0079	0.0019	0.001	8.90E-08	14
	B7	0.68	3.50E-08	0.0034	0.000079	0.0002	1.60E-07	11
End of life	C4	0.03	2.50E-11	0.00032	0.00068	0.000014	1.00E-10	0.0061
Total		39	5.00E-07	0.06	0.017	0.0086	3.60E-05	140

Impact keys: over 40% (red); 31%-40% (blue); 21%-30% (orange); 10%-20% (yellow)

Product stage: A1 = Raw Material supply, A2 = Transport, A3 = Manufacturing; Construction stage: A4 = Transport, A5 = Construction; Use stage: B1 = Use, B2 = Maintenance, B5 = Refurbishment, B6 = Operational energy use, B7 = Operational water use; EoL stage: C4 = Disposal

The residential building totally produces 2400 kg $CO_{2\,eq}/m^2$. Study of Monica Lavagna et al. [9][9] created sample of 24 representative EU dwellings with aim of creating a benchmark sample. Results for GWP impact variated from 34.4 - 99.9 kg $CO_{2\,eq}/m^2/y$. Result of this study is 39 kg $CO_{2\,eq}/m^2/y$. Study of Szalay Z on different types of building based on Hungarian building stock shows results from nearly 40 kg $CO_{2\,eq}/m^2$ for single family house to around 17 kg $CO_{2\,eq}/m^2$ for medium high multi-family building [10]. The ODP value of analysed building is 0.0000005 kg CFC-11 $_{eq}/\ m^2/y$. This value is even lower than values mentioned in comparing study range 0.00000385 – 0.0000202kg CFC-$11_{eq}/\ m^2/y$ [9]. Value in study by Szalay are approx. 0.000006 CFC-11 $_{eq}/\ m^2/y$ for single family house and approx. 0.0000027 CFC-11 $_{eq}/\ m^2/y$ for medium high multi-family building.[10] Value for AP is 0.06 kg $SO_{2\,eq}/\ m^2/y$. Values from Lavagna´s study are not comparable. Values from Szalay´s study are almost 0,10 kg $SO_{2\,eq}/\ m^2/y$ for single family house and approx. 0.04 kg $SO_{2\,eq}/\ m^2/y$ for medium high multi-family building.[10] Total value for EP is 0.017 kg $(PO_4)^{3-}_{eq}/\ m^2/y$ this values is also not comparable with proposed benchmark sample. Values in study done by Szalay are approx.0.009 $(PO_4)^{3-}_{eq}/\ m^2/y$ for single family house and approx. 0.0037$(PO_4)^{3-}_{eq}/\ m^2/y$ for medium high multi-family building [10]. Value for POCP is 0.0086 kg $C_2H_{4\,eq}/\ m^2/y$ and it also is not comparable with results from study by Lavagna et al. Stdy by Szalay shows results approx. 0.009 kg $C_2H_{4\,eq}/\ m^2/y$ for single family house and approx. 0.003 kg $C_2H_{4\,eq}/\ m^2/y$ for medium high multi-family building [10]. Value for ADP-E is 0.000036 kg Sb $_{eq}/\ m^2/y$ reported values in study of Monica Lavagna et al. range from 0.00186 – 0.00942 kg Sb $_{eq}/\ m^2/y$ [9]. Presented values are again much lower, more then 51 – times. Values from study of Szalay are missing for this impact. Value for ADP-FF which is 140 MJ/ m^2/y is also not comparable and missing in other study.

4. Conclusions

In conclusion aim of this study was evaluation of residential building as one of the samples for further research. Analysed object belongs to category of new buildings, results shows that it doesn't have larger impact then sample of buildings from EU. This agrees with the EU's objective of reducing the environmental impact. Other conclusions are however that there is a need for standardization of methodologies. Even same impact categories don't use the same impact or the impacts are similar but still not enough to be comparable.

5. References

[1] Abergel T, Dean B, Dulac J, Hamilton I, Wheeler T 2018 Global Status Report - Towards a zero-emission, efficient and resilient buildings and construction sector; ISBN 978-92-807-3729-5

[2] 2010, *Directive 2010/31/EU of the European parliament and of the council of 19 may 2010 on the energy performance of buildings* (Official Journal of the European Union), OJ L 153, p 13

[3] 2018, *Directive (EU) 2018/844 of the European Parliament and of the Council of 30 May 2018 Amending Directive 2010/31/EU on the Energy Performance of Buildings and Directive 2012/27/EU on Energy Efficiency.* https://eur-lex.europa.eu/legal-content/EN/TXT/?uri=uriserv:OJ.L_.2018.156.01.0075.01.ENG&toc=OJ:L:2018:156:FULL

[4] Röck M, Saade M R M, Balouktsi M, Rasmussen F N, Birgisdottir H, Frischknecht R, Habert G, Lützkendorf T and Passer A 2020 Embodied GHG emissions of buildings – The hidden challenge for effective climate change mitigation, *Applied Energy* **258** 114107

[5] Mirabella N, Röck M, SaadeM R M, Spirinckx C, Bosmans M, Allacker K and Passer A 2018 Strategies to improve the energy performance of buildings: A review of their life cycle impact, *Buildings* **8** 105

[6] Ciambrone DF 1997 *Environmental life cycle analysis* (New York: Lewia, Boca Ration)

[7] Cabeza L F, Rincón L, Vilariño V, Pérez G, Castell A 2014 Life cycle assessment (LCA) and life cycle energy analysis (LCEA) of buildings and the building sector: A review, *Renewable and Sustainable Energy Reviews* **29** 394-416

[8] ISO. ISO 14040. *In Environmental Management-Life Cycle Assessment-Principles and Framework*; ISO: Geneva, Switzerland, 2006

[9] Lavagna M, Baldassarri C,Campioli A, Giorgi S, Valle A D, Castellani V, Sala S 2018 Benchmarks for environmental impact of housing in Europe: Definition of archetypes and LCA of the residential building stock *Building and Environment* **145 260-275**

[10] Szalay Z 2007 *Life cycle environmental impacts of residential buildings* (Budapest University of Technology and Economics, Hungary) p 106

Acknowledgments

This paper was created with the support of grant projects of Slovak Republic to support projects No. VEGA 1/0512/20 and 1/0674/18.

IOP Conf. Series: Materials Science and Engineering **867** (2020) 012044 doi:10.1088/1757-899X/867/1/012044

The design of the technical flood protection measures on Svitava River, Czech Republic

H Uhmannova[1], V Ondrejka Harbulakova[2] and L Cetkovska[1]

[1] University of Technology Brno, Department of Water Structure, Faculty of Civil Engineering, Zizkova 17, 60200 Brno, Czech Republic
[2] Technical university of Kosice, Faculty of Civil Engineering, Institute of Environmental Engineering, Vysokoskolska 4, 042 00 Kosice, Slovakia

Abstract. The paper present the design of the technical flood protection measures on Svitava River in Březová above Svitava town. The current state of the study area was evaluated and determined based on related documents available for public as well as on the personal survey of the landscape. Technical solution as proposal of adjustment of the Svitava River (by composite trapezoid), extracting sediments and designing, stabilization of the bank and construction of the protective floodwall is shown in the paper. The riverbed slopes were fortified with a stone riprap with a sieve of grass mixture.

1. Introduction

Natural floods are caused by natural phenomena, where the water from the riverbed occurs outside the channel. These natural phenomena might be [1]:
- reaching the water level or maximum discharge of the stream,
- intensive and long-term rainfall,
- snow melting
- ice barriers on the flow and ice passage, etc.

However, other factors also contribute to the origin of the flood, for example [1]: size, shape and morphological characteristics of the river basin; shape of the channel; type of surface; density and ordering of the river network; geological, hydrogeological and pedological conditions; occurrence of objects and reservoirs on the stream [2,3].

The pre-flood environment, such as soil saturation, vegetation cover or retention capacity of the river network, etc. [1,4], also affects flood threats.

In order for flood protection measures to work well, their interdependence is important. It is a combination of [4,5]:

- flood prevention (e.g. adaptation of buildings to floods risk, appropriate land use, etc.),
- protection (e.g. technical and non-technical/natural flood control measures flows to reduce the probability and impact of floods),
- preparedness (e.g. informing the population about the situation and how to behave when flood situation occur, in case of emergency),
- rescue system (e.g. rescue plans).

By non-technical measures are meant delimitation of flood zones, forecasting and warning systems and education of the public how to behave in case of emergency before the flood occurred and during the flood. In flood zones the restrictions where construction, permitting or construction works is not permitted is applied. The exception is water structures, which have a positive effect on flood protection. [1,4-6]

Technical measures on watercourses are mainly riverbed adjustment in order to ensure sufficient capacity and stability of the bottom and banks for design flow [7]. These include the establishment of the retention areas in the reservoirs and dry polders, the construction of the weirs, temporary dams, dams, protective walls, natural mounds, embankments, increasing of the capacity of the riverbed and its maintenance, sediment extraction from the river, controlled flooding, bypass of the streams, reduction of depth and side erosion, and more [1,5,8].

In the paper is presented the flood protection on the Svitava River by adjusting the shape of the riverbed, extracting sediments and designing a protective floodwall. Controlled flooding is considered in places where it is not necessary to protect private property. Outside the area of interest, a theoretical flood control measure is proposed, which present how is necessary to limit the large flood into the urban area of the city. It was considered to use a protective wall, a natural embankment and a mobile barrier.

2. Material and Methods

2.1. Study area

The Svitava River belongs to the sub-basin of the Dyje, which is part of the International Danube River Basin District. The sub-basin of the Dyje River is the second largest basin in the Czech Republic, which is adjacent to the sub-basin of the Morava, to the tributaries of the Váh, Upper and Middle Labe, Lower Vltava and Upper Vltava. The area is located in the south-eastern part of the Czech Republic. It is managed by Povodí Moravy enterprise. The water from the basin of Svitava River flows into the Black Sea.

Svitava is 98.39 km long and its catchment area is 1,149.4 km^2. The river springs on northwest of Svitavy town near the Javorník village at an altitude of 471 m above sea level. Along its entire length, the river is densely populated. It flows southwards through the towns Březová above Svitava, Letovice, Rájec-Jestřabí, Blansko and Adamov, where it has a predominantly natural and meandering shape flowing in a deep valley (nearby Svitavy town, figure 1). The industrial zones began to be formed around the stream and its meanders were regulated in late 19th century in the southern part of Brno (at an altitude 191 m above sea level). The river was straightened gradually in Brno [5,8,9].

Figure 1. Sub-basin of Dyje River with the study area of Svitavy River.

The study of area of Svitavy River is located in the urban area of the town of Březová above Svitavou, which lies on the Moravian-Czech border in the Pardubice Region. Nowadays, around 1690 inhabitants live in the town. The built-up area is mostly located in close proximity to the flow. There is a railway line and road running along the whole section, which creates a barrier to possible flooding into the floodplain. The site is known for its important resources of groundwater, which are pumped and used for the treatment and distribution of drinking water, especially in Brno and its surroundings [9].

The original riverbed in the study area was non-capacity in some places even for flows smaller than $Q_1 = 5.9\ m^3/s$. On the basis of the calculations of the course of water levels using program HEC-RAS, a flood control measure was proposed for the design flow rate $Q_{20} = 24\ m^3/s$. The proposed adjustment was to extend the channel in five sections and a design of the protective wall on four sections of the stream. The design consists of adjusting the shape of the riverbed to a 2-stage trapezoid with a deepen bottom. The flow will develop and change freely within the designated range at normal flow rates. The example of this riverbed is depicted in figure 2. The proposal took into account the ownership of the land, so there was an effort to minimize land acquisition.

Figure 2. The example of composite trapezoid channel with the movable bottom with the protective wall.

The adjustment does not change the original riverbed in any way. The shape of the riverbed is adapted from a simple trapezoid to a composite trapezoid. Ordinary daily flow rates are concentrated into a smaller part of trapezoid of wide 2m flow in the bottom. The shape of this part is unfortified and therefore it can evolve spontaneously in the defined area. The bottom part is excavated approximately 0.5 m below the original flow bottom. Its capacity is about $Q_{30d} = 1.35\ m^3/s$ with the slopes of 1: 1.5. The shape of the designed channel is shown in the figure 3.

The width of the channel varies according to the possibility of utilize the space along the banks. The shape of a composite trapezoid with a slope of 1: 2 is designed almost along the entire length of the study area in the river km 74.479 to 76.744. In the last part of the section, the protective walls on both banks are designed. The shape of the channel is in the last section is rectangle with a movable bottom in r km from 76.744 to 76.924. The height of the banks is almost the same as that of the original channel what helps with a smooth connection to the original terrain along the banks.

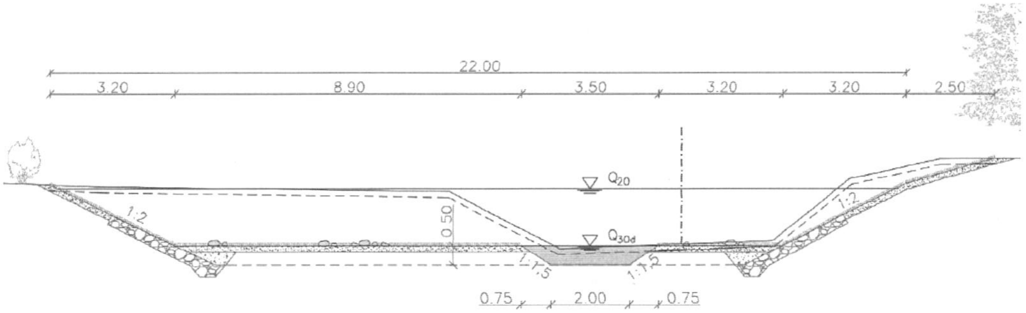

Figure 3. The composite trapezoid channel with the movable bottom.

From the geological data of the Svitava River in the sections of interest, it was clear that the riverbed was massively clogged. Therefore, the excavation of the bottom that leads to removing of these deposits and bottom levelling was designed.

3. Results and Discussion

The removal of the bed sill of the former sluice gate will be removed and the channel will be smoothly connecting the original bottom to the adjusted one in r km 76.479. The removal of this bed sill should not in any way impair the hydrological regime of the flow.

There is also a bed sill, which will be removed in r km 76.403. It is located in in a part of the channel that will be extended, and therefore its construction would be an obstruction in the adjustment. A given section of the flow is completely fortified with a concrete, which will also be removed. It can therefore be assumed that after removal of the concrete fortification the step would have to be repaired or possibly extended. Both constructions have a height of approx. 0.5 m. These bed sills present a migration barrier; therefore, their removal will improve the conditions for the animal migration.

By adjusting the bottom with sediment, deposits and bed sills removing, the average longitudinal slope changes to 3.8‰.

In four parts of the flow section, a protective wall is designed as a technical flood control measure due to insufficient channel capacity for the design flow (figure 4). The wall is designed in places where the wall was already in the original channel or where there was no space to channel expanding or implementation of the natural water protection measures.

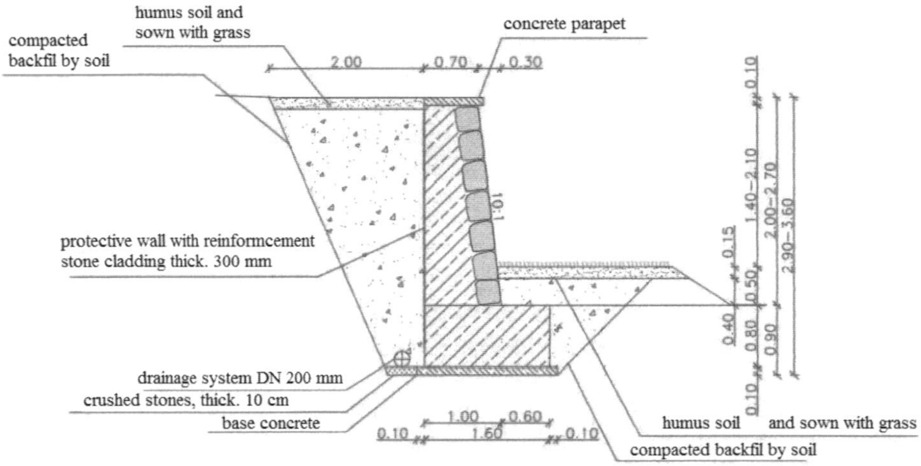

Figure 4. Protective wall.

Based on geological survey it was found out that in some parts of the river there is subsoil with low-bearing capacity. Respecting this, the walls are designed to be founded at least 1.5 m below the ground level. The walls are made of reinforced concrete with stone cladding on the inside. The wall is 2.0 - 2.7 m above the level of the excavated bottom and 0.9 m below it. During the construction, a line drainage system will be installed along the wall, which serves to drain the water and so protect the substructure. A more detailed description with dimensions is shown in the figure 4.

Due to the absence of the granularity curve, the roughness of the channel and the effective grain were estimated based on a personal inspection and from the tables as d_{ef} = 40 mm. The design of the new slopes´ fortification was based on the previous fortification of the original bank, what was evident from the personal inspection. It is necessary to stabilize the slopes in the riverbed up to the theoretical bottom level so that the spontaneous changes in the excavated bottom do not cause the banks erosion and leads to the collapse of the fortification.

The fortification is made of stone riprap up to a height of 0.8 m from a quarry stone sown with the grass mixture in a thickness of 250 mm with a gradual extension up to 300 mm. The aggregate weighs

80 – 200 kg. The slope fortification is designed along the entire length of the channel of the interest, with the exception of banks with protective walls. A more detailed description is shown in the figure 5.

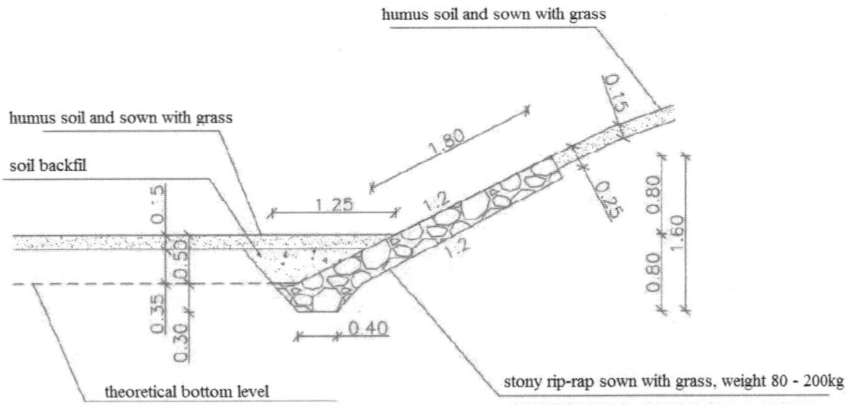

Figure 5. The stabilization of the bank.

4. Conclusion

Flood protection along the Svitava River was designed with respect to ownership. The proposal consists of the stream adjustment of into the shape of a composite trapezoid with a migratory ridge. The stream can spontaneously meander in the defined space. The riverbed slopes were fortified with a stone riprap with a sieve of grass mixture. Original trees are designed along the river as accompanying green respecting the system of its planting described in [10]. Protective floodwalls were designed in the least-capacity sections of the stream. Increasing of the high of the bridge structures so that they were capacity for the design flow was design.

5. References

[1] Drab A and Riha J 2010 *Flood Protection Measures: Modul M01 University of Technology Brno,* Faculty of Civil Engineering, Brno

[2] Ondrejka Harbulakova, V Zelenakova and M Sugarekova M 2019 Flood Management in Municipalities in the Prešov Region *People, Buildings and Environment: International scientific conference* Bristol Publishing House **222** pp 1-8

[3] Junáková N and Bálintová M 2014 The study of bottom sediment characteristics as a material for beneficial reuse *Chemical Engineering Transactions* **39** pp 637-642

[4] Ondrejka Harbulakova, V Lechman and M Zelanakova M 2017 Flood protection measures in Slovakia - polders *Public recreation and landscape protection - with nature hand in hand!* Brno: Mendel University pp 127-132

[5] Possibilities of dealing with flood in Czechoslovakian border area: Flood protection measures [online] [ref 2019-01-01] Available at: http://cs-povodne.eu/Protipovodnova-ochrana-apovodne/ /in Czech/

[6] Vokurka A and Zuna J 2017 Basis for updating the methodology for designing natural flood retention measures (NFRM), ČSKI prepared for Ministry of Environment of Czech Republic /in Czech/

[7] Slezingr M and Fialova J 2012 An examination of proposals for bank stabilization: the case of the Brno water reservoir (Czech Republic) *Moravian Geographical reports* **20** (2) pp 47-57

[8] Pelikan P, Slezingr M and Markova J 2018 The Efficiency of a Simple Stabilization Structure in Water Reservoir *Polish Journal of Environmental Studies* **27**(2) pp 793-800

[9] Nabrezisvitavy.cz About Svitava River [online] Copyright 2016 [ref 2019-01-01]. Available at: http://nabrezisvitavy.cz/o-rece-svitave/ /in Czech/

[10] Jedlička L and Slezingr M 2010 Bankside trees and shrubs – basic information, *Colloquium on Landscape Management* Location Mendel Univestiry, Brno, Czrch Republic pp14-17

IOP Conf. Series: Materials Science and Engineering **867** (2020) 012044 doi:10.1088/1757-899X/867/1/012044

Acknowledgments
Thanks to The National Scholarship Programme of the Slovak Republic there was an opportunity to prepare this paper with the colleagues from the University of Technology Brno and also I would like to thank the Project "Specifický výzkum" VUT FAST Brno 2018 – 2020 BD12800026.

IOP Conf. Series: Materials Science and Engineering **867** (2020) 012045 doi:10.1088/1757-899X/867/1/012045

The treehouses

V Vacenovská[1]

[1] Vysoké učení technické v Brně, Fakulta Stavební, Veveří 331/95, Brno 602 00

Abstract. The goal of this research is to evaluate and find new solution in the area of wooden constructions placed in the treetops, which are interesting mainly due to their independence and amazing view among the treetops, and enable the utilization of the land below the treehouse itself. Treehouses can be constructed as a detached object without having to interfere with the trunk, or anchored into the trunk using special anchoring elements, where compact root system of surrounding trees is essential for healthy grow of the tree. Wood is also a specific natural material, which offers a possibility to create architecturally interesting and modern construction. This article is focused on evaluation of anchoring suitability of these constructions in treetops, evaluation of using these houses for recreational purposes as well as permanent living, effect of climate conditions on suitably chosen wood type and construction of these objects.

1. Introduction

Modern wooden constructions use wood as a specific material, from which highly functional and architectonically interesting buildings can be created. Such buildings use industrially manufactured building system, where primary load-bearing structure consists of glued, composite or stacked blocks. Those components are used for vertical, horizontal and roof constructions. Such flat structural systems are set with insulation on the exterior side. With this type of buildings, we can achieve huge effective load transfer and high dimensional stability, while preserving interesting visual properties of wood. Besides large solid constructional components, we can see also small format structural members can be used to assemble a flat construction at building site.

Besides using timber when handling the load-bearing structure, we can see it in other architectonically-constructional elements as a part of structural composition, interior or exterior. Wood as a quality material has many ways of usage when designing wooden constructions and it is desirable that the carpentry is articulated. When designing wooden constructions like treehouses, the tree trunk is mostly used as structural member into which the load-bearing structure is anchored. The significant element of the whole construction is then ladder, staircase or ramp, which becomes main and the most important part of the construction.

During residential building planning it is customary to supply the carpentry with wooden façade. Less common is to supply this construction with more solid wood elements, either in the interior or exterior. Nevertheless, wood is very variable structural material with great potential for use. It is also possible to suitably complement such building with metal structures, while considering convenience of the solution as well as climate conditions where the building is placed.

The treehouses are the embodiment of freedom and pleasantly comfortable safety. Their independence is captivating and brings a stunning view among the treetops. One of the original purposes of a treehouse, to protect us from predators, is now rare and its function is fulfilled only in areas with wildlife, which is dangerous to humans. Each house has its own structure and a ladder as a way into the house itself, which can be pulled up any time, enabling us to disappear among the treetops and hide from the chaotic world on the ground.

The interior of such houses creates individual distinctive and spiritual space, which depicts mainly our inner path. With such space we can experiment boldly by letting this mysterious area to ascend to the surface. It is not entirely necessary to own a tree for building a "treehouse" –freestanding slab construction, which doesn't interfere its structure into the tree itself (or surrounding trees), can perfectly assimilate its qualities. A treehouse does not have to be available only for individuals with the luxury of owning a robust and structurally binding tree. Urban treehouse has as much authenticity as the one spread in the treetop with trunk carrying its weight. Every modern building which draws inspiration from treehouses, but without founding and anchoring into the tree itself, deserves our attention as well. In many ways structural engineering has offered satisfying environmental answers that respect the tree itself.

2. Characteristics of the wooden construction

Solid-wood buildings and architectonically-constructional elements of solid wood are entirely specific in their characteristics. Therefore, it is necessary to know physical and mechanical properties of wood and regularities of building physics precisely, as well as to keep basic principles of designing using wood. Considering these properties and principles helps to reduce negative effects and material degradation, while achieving desired outcomes and utilizing natural properties and benefits of solid wood adequately. Buildings with wood in their structure are very specific from the building physics perspective; therefore we needed to take their special characteristics into consideration during the designing process, whether it is for carpentry, space frame or skeletal structure. Based on characteristics of wood, such as moisture regain, change in dimensions and susceptibility to damage by various factors, it is necessary to design both construction and all wooden elements based on structural principles which can help to increase lifetime of a wooden product. Wood as a natural material has special characteristics also from environmental point of view. Besides all that, its usage in structure and interior positively influences microclimate of the building and living comfort.

For using wood in building industry its life span is crucial. Many factors affect deterioration of wood and its life span. These can be influenced to a large extent, resulting in prolonging the longevity of wooden elements to its maximum.

3. Specifics for designing wooden structures

Since wood is a natural material, which significantly reacts to a direct contact with water and its presence in the air, weather conditions and other factors, it is necessary to follow certain principles when designing wooden buildings or architectonically-constructional elements to achieve as high durability and resistance as possible. Wood as a biological material is naturally inhomogeneous on all levels from submicroscopic to macroscopic one. Moreover, other factors influence the characteristics of wood during its growth. Various kinds of wood differ in resistance against degradation factors; therefore, the initial phase of the design consists in choosing the suitable material based on the purpose and surroundings in which the building will be constructed. European wood species with high durability include larch, oak, ash, pine or elm [4].

Of course, natural durability of wood can be improved with impregnation or other artificial proofing. Nevertheless, constructional protection of wood is still primarily preferred as it eliminates exposure of wood to degradation effects [2].

Indeed, we choose timber without defects. The most serious degradation factors are dry rot and insects that significantly lower strength of the wood. Their occurrence requires temperature around 20°C, increased relative atmospheric humidity and moisture content in the wood above 20% [1].

For all this it is crucial to prevent the wood from being exposed to water already during storage and mounting. Wooden elements should be stored directly at construction site for a minimum period of time. Also, during mounting wooden elements and buildings should be sufficiently protected when weather conditions are bad. With finished building, the most dangerous is warm and humid environment, where moisture easily penetrates into the wood and cumulates there. It is similarly dangerous if walls are not vapour permeable. Especially during winter period, when air pressure on the inside of the building is higher than outside, it is crucial that diffusion resistance towards outer wall layers is decreasing and structure contains vented air layer for vapour outtake. With diffusion-closed outer wall structures, we use vapour control layers placed as close to the interior as possible, to minimize vapour penetration into the structure. Condensed water drainage needs to be placed in the structure wherever we assume higher occurrence of vapour condensation. Wood should be incorporated into the structure with dampness corresponding approximately with the future operational dampness. This is usually around 8% in the interior [1]. This should prevent both formation of dry-out cracks, which could enable insects or fungus to contaminate the wood, and bulking in case of overdrying. It is also essential to prevent carpentry or wooden elements to come in contact with ground dampness, splashing rainwater or snow. That is why wooden buildings and all exterior elements are set to be 300 to 400 mm above surrounding ground [3].

Due to its flammability we need to keep certain fire safety precautions when planning to use wood at certain situations. Wooden elements should be sized based on safety factor, which is secured by smaller wood surface in proportion to its volume (therefore also higher fire resistance), and finished with smooth surface or inflammable top. Of course, wooden construction has to consider fire risk, comply with applicable legislation and should be equipped with fire safety precautions [1].

4. Lifespan and wear of wooden structure

When designing modern wooden structures timber is used as a specific material to create architectonically interesting and highly functional buildings. This material is very shapeable and has a great potential use in construction and architecture. In wooden structures it's standard to supply load-bearing structure with wooden enclosing shell [4].

All structures wear out physically and morally. Physical wear-out indicates useful value decline owing to defect rise. This wear-out can be influenced by structure protection and taking care of it. Moral wear-out rises owing to development and increasing demands on the construction. Wear-out value is connected with lifespan of structure, which includes physical, moral and economic life. Moral life is connected especially with aesthetics and requirements for construction and elements. Economic life is based on economy review from point of view of comparing running cost, abatement and maintenance cost to structure usability. Length of these lifespans differ and is also different for individual structural parts. Parts with longer physical life are load-bearing, beam or roof structure. On the other hand, floor, roofing or enclosing shell can be among the elements with shorter life. Nevertheless, all of these always depend on particular conditions, placing of the elements and their execution and on level of protection and maintenance. Among most common factors negatively influencing life span of wooden structure are insufficiencies and mistakes in design, its execution or mounting or during usage [1], [2].

When designing structures and buildings it is important to choose suitable type of wood for the enclosing shell so that it would perfectly resist weather conditions in which the building is placed and sub-sequentially to choose suitable surfacing. From the architectural point of view it's preferable to focus on lightfastness and color changes of wooden element so that it would comply with architectural brief. So when designing enclosing shell from particular type of wood we have to consider color change in advance, or assume such changes and choose the pattern according to expected color changes. Wood surfacing significantly influences lifespan and lightfastness of the wooden element.

Significant way to apply solid wood on the exterior of modern wood construction is wooden façade. Such façade solution consists of exfiltrated load-bearing grid, which is covered with layer of solid wood elements. These elements can be planed planks, unplanned boards or split shingles. For correct functionality and achieving maximum lifespan of wooden façade it is necessary to keep the same principles of structural protection as with external elements. Façade should be protected by the roof overlapping against direct sunlight and rainwater as much as possible. Individual elements of joinery facing have to be protected against withholding water in joints and its capillary action in them. Details design therefore has to enable draining of water; weatherboarding boards have to overlap one another in direction of rainfall. For fluent drainage the façade should be offset to the face of substructure at least 30 mm and should have water nib milled [3].

Besides all this, air flow around wooden elements should be ensured via keeping sufficient spacing between them, so that they can dry fast and completely. Development in area of wooden facades focuses mainly on prolonging their lifespan and on reducing and simplifying the care of it. The question being discussed is, if treated or untreated wood should be used. Some types of surfacing and impregnation have negative effects on ecological aspect of the material. Additionally, surfacing needs to be repeated regularly. Untreated façades naturally degrade by sunlight and greys unevenly due to the sunshine. More significant greying happens on surfaces more exposed to the sunlight, that is, on southern side of the façade and on sunspaces. Color changes do not influence durability or quality of the wood.

Untreated wooden façades are commonly used in Austria, as opposed to Scandinavian countries. The reason for this can be the fact that in Central Europe quality and resistive larch wood is accessible, whereas in Scandinavia façades are constructed of spruce or fir wood.

Suitable, but very costly alternative to larch could be oak wood. To execute wooden façade we can use thermowood as well. Wooden façade can be either vertical or horizontal. The most common type of wooden elements on façade in our country is planed plank. Somewhat more durable version would be unplaned board, which is used mainly in northern Europe. Probably the longest lifespan has split-shingle façade, where splitting does not intact wood fibers [5].

5. Modern constructions

Modern wood constructions are based on principles of using wood as a specific building material, which can be used to create architecturally interesting and highly functional buildings. Aesthetical legacy of traditional wood constructions prevails in modern wood constructions in the form of visually admitted wood. Modern wood constructions use prefabricated building system, where the load-bearing structure is made up of flat solid blocks formed by gluing, lamination or folding. These flat components can be used in horizontal, vertical or roof structure. This is a flat-acting load-bearing structure with insulation on the exterior side, where at least half of the load-bearing layer is made of solid wood. With this type of constructions, it is possible to achieve very efficient transmission of large loads and to reach high stability of dimension while keeping interesting visual characteristics of the solid wood. Besides using large-size solid-wood structural components we can use small-size load-bearing elements, which are assembled into flat construction at the construction site [6].

6. Possibilities of treehouses and their types

We can divide treehouses based on composition type of the perimeter structure – per thermal insulation thickness used in such composition and way of use of these engineering structures. Such structures are specific in particular by their load-bearing structure, which lifts the whole structure and has to be adjusted to designed composition of the perimeter structure.

First type is recreational housing – in Czech Republic we can see such houses in Lelekovice near Brno – which serves mainly for amusement purposes during summer months. In our country such small house is the most common and its structural composition brings to mind classic summer houses. Conception of perimeter structure is simple and do not apply load to load-bearing structure or trunk, which the structure is anchored to. Interior of such housing is equipped and furnished simply and functionally, and its internal space does not even enable more complicated internal layout. No utility networks are installed in such buildings.

The other type is housing for yearlong occupancy with more complicated conception of perimeter structure. These houses require more stable and stronger main load-bearing structure or anchoring into the trunk. Such engineering structures are not very common in Czech Republic and can be found more often abroad where climate is more favorable and temperature swings are smaller. Interior of such building is more articulated and more complex than in the first type, also internal layout more resembles common family house and installation of utility networks is customary there.

7. Anchoring elements of treehouses

Based on the load-bearing structure, treehouses can be divided into two categories. Either freestanding treehouses, which interfere with the treetops and do not need or require anchoring of the load-bearing structure into the trunk; or houses, which require anchoring into the trunk of the tree (or trees), or to be hung on ropes with trees serving only as anchoring element, trunk is not mechanically disrupted and tree is not damaged. With every such type of construction stress and loading of trees need to be evaluated [7].

Treehouses are interesting architectural works; however, it is important to evaluate if intervention in the trunk is inevitable and consider possible impacts on lifespan of such tree. It is not only about the mechanical intervention – that is anchoring with bolts directly into the trunk, but also any suspension of load-bearing structure on trees. It is also necessary to review the quality of the root system and generally it is a fact, that the more trees are surrounding the damaged tree, the longer its lifespan. Therefore, it is inappropriate to choose solitary trees (or tree) with sparse root system for such types of constructions.

Figure 1. Mechanical anchoring to the tree.

8. Composition of enclosing shells – treehouses

Coefficient of thermal transmittance and thermal resistance (basically it is so-called inverse value of U) are basic quantities characterizing thermal insulating properties of any building structure. When choosing particular house, or better said, particular building system, we need to observe the composition of the enclosing shell, its thermal insulating properties and its ability of heat accumulation. Materials used for thermal insulation have more or less the same thermal conductivity, which defines how well (or poorly) the insulant conducts the heat. Heat capacity, on the other hand, is very different – it influences thermal inertia and so-called phase shift.

When the house is designed correctly, we can use solar gain to compensate heat loss and to support improvement of energy balance of the building and to lower heating expenses. To prevent overheating, windows need to be equipped with shading for the summer. Windows with efficient thermal insulation are becoming a necessity instead of needless comfort.

Wood is an organic matter and as such it has to breathe. Therefore, edge venting and venting of roof cladding is essential part of wood construction. Those are done as diffusely-open, so that natural convection of air outtakes all the dampness from the structure and secure its dry state. For outer walls this issue is solved by exfiltrated façade systems.

From the point of thermo-insulation properties, it is important to mention, that suitable composition of structure should be chosen based on the building type and its usage. For recreational housing simple composition would be sufficient, eventually with higher standard it is possible to choose a composition with thicker thermal insulation layer. In case of building for year-round housing we need to deal with coefficient of thermal transmittance for enclosing shell and suggest suitable composition to comply with ČSN (Czech Technical Standards).

9. Conclusion

The purpose of the research is to find and evaluate new solution in the area of treehouses that are interesting mainly due to their independence and amazing view among treetops, and which enable us to utilize the land below the house itself. Treehouses can be constructed as freestanding object without necessity to interfere with the tree trunk or anchoring it there using special anchoring elements, where surrounding roots system is crucial for healthy growth of the tree. Wood is a specific natural material and offers possibilities to create architectonically interesting modern structures. This work focuses on evaluating suitable anchoring elements for these structures among treetops, also usability of these objects for recreational activities and permanent housing, and last but not least, evaluating the effects of climate conditions on suitably chosen type of wood and on the structure of these buildings.

10. References

[1] Havířová Z 2006 *Dům ze dřeva* (Brno: ERA. Stavíme) ISBN 80-7366-060-1
[2] Svatoň J 2000 *Ochrana dřeva* (Brno: Mendelovazemědělská a lesnická univerzita) ISBN 80-7157-435-X
[3] Ševců O, Vinař J and Pacáková M *Metodika ochrany dřeva* 2000 (Praha: Státní ústav památkové péče v Praze) ISBN 80-86234-14-2
[4] Stará J *Jakou zvolit fasádu, aby nevyžadovala údržbu* 2013 [online] [cit. 2014-10-20]
[5] *This gorgeous modern treehouse hides a surprising interior – Insight found* Insight found [online].
[6] Novatop *Nosná konstrukce zároveň finálním vnitřním povrchem* 2012 [online] 2012 [cit. 2013-10-20].
[7] *Is this Scottish eco treehouse the most romantic in the world?* | Daily Mail Online. [online]. Copyright © Associated Newspapers Ltd [cit. 03.05.2018]

Acknowledgments
The research originated with the support of specific research FAST-J-19-5999 "Study of effects that UV radiation and surface finish have on lightfastness and photodegradation of different kinds of wood" and within the scope of TACR (Technology Agency of the Czech Republic) project – The programme ZETA for the support of applied research TJ02000072 "Technology development of surface carbonization of constructional wooden elements: elimination of moisture and biodegradation effects".

IOP Conf. Series: Materials Science and Engineering **867** (2020) 012046 doi:10.1088/1757-899X/867/1/012046

Testing of continuous shear connector

P Vaňová[1] and V Kvočák[1]

[1] Technical university of Kosice, Faculty of Civil Engineering, Institute of Structural Engineering, Vysokoskolska 4, 042 00 Kosice, Slovakia

Abstract. Composite steel-concrete bridges are one of the most recent additions to the bridge science. By joining the two of the most commonly used materials it is possible to use their strengths combined. This was firstly presented for composite ceilings. It was assembled by encasing the whole I-beams into the concrete deck. Nowadays, it can also be used for short span railway bridges. Ongoing research at Technical University of Košice (TUKE), Faculty of Civil Engineering, is currently trying to prove the unnecessity of usage of the entire I-beam as well as the entire block of concrete by encasing only part of the beam, the so called perforated steel strip, into the concrete.

1. Introduction

There have been several researches investigating this topic. Oguejiofor and Hosain were one of the first researchers to publish studies of 'perfobond rib shear connectors' [1]. They performed several tests and compered the results with common headed shear stud connectors, proving the capabilities of the strip connectors. Kim and Jeong considered perfobond ribs welded onto decks, experimentally verifying their effectiveness [2]. In Slovakia, Rovňák *et al* have done extensive research comparing several geometric possibilities of connectors and adjusting their resistance formulas via regression analysis [3].

2. Geometry

The composite bridges consist of two key elements – steel and concrete. In this paper, the steel part in form of a strip fulfils a function of rigid longitudinal reinforcement.

2.1. Steel Connector

The shear connection is usually secured by friction between the surfaces of steel strip and concrete, as well as by concrete shear studs, which are formed in the holes of the strip. In the new design, developed at Faculty of Civil Engineering (TUKE), there was present an extra connection – via curved reinforcement welded onto the strips wall from both sides.

The continuous shear connector was made out of IPE220 cutted in the middle longitudinally, with 20 mm holes, 300 mm apart from each other. The reinforcement used was of 12 mm diameter and it was welded at the edge of the holes. Every 100mm by length it was kinked in 90° angle, with height of kinked parts equal 50 mm, as visible in figure 1 and 2.

2.2. Concrete beam

Steel-concrete composite bridges containing continuous shear connector are usually entirely or partially prefabricated. For this reason any changes in basic rectangle cross-section do not complicate the bridge construction, which makes it easy to remove the unnecessary parts of the concrete – the ones, where tension is present. The new, lightweight, cross-section is portrayed in figure 3. The beam prepared for experiments (figure 2) was 6 m long.

Figure 1. The shear connector in a push-out test layout.

Figure 2. The shear connector before poured into concrete beam.

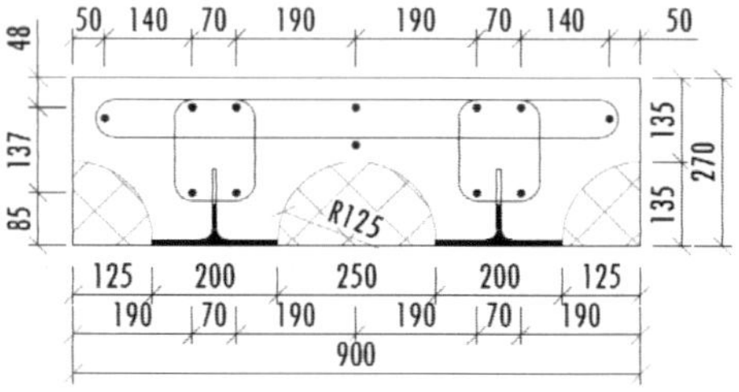

Figure 3. Cross-section of the lightened beam.

3. Material

Yield point of the steel was found out to be 315.3 MPa, via tensile tests performed by Faculty of Metallurgy, TUKE. Concrete was tested for Young's Modulus, tensile and compressive strength at Laboratory of Excellent Research, TUKE. The results were 33 GPa, 3.7 MPa and 39.7 MPa, respectively. The reinforcement used was label as B500 B of which normative yield strength is 500 MPa [4].

4. Experiments

In order to determine the resistance of the composite bridge containing this type of shear connector, several experiments had to be performed. Each test was executed onto three specimens, to exclude the statistical error.

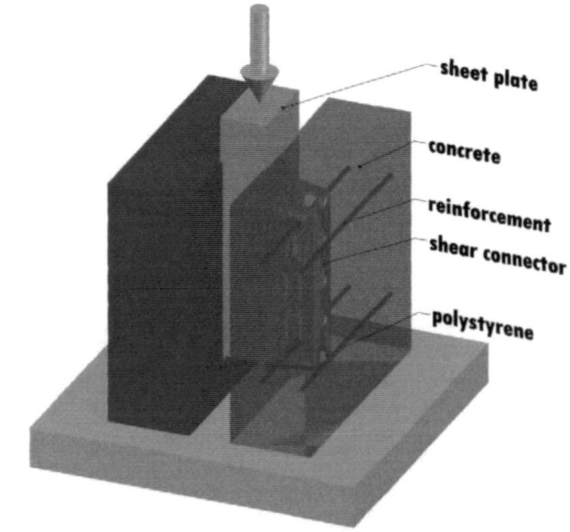

Figure 4. General layout of push-out tests.

Figure 5. Four-point flexural test.

4.1. Push-out test

To find out the shear resistance of the shear connector, push-out tests (POT) had to be carried out. For the POT, the strip had to be turned into vertical position and poured into concrete block with 10mm layer of polystyrene between the two materials, as depicted in figure 4. This way the extrusion of steel into concrete was prevented and only shear force was taken into account. The searched values were measured by hydraulic press software (as input data) and inductive displacement sensors.

4.2. Four-point flexural test

One of the most important internal forces in bridges is bending force. To prove the suitability of the connector for bridge usage, the four-point flexural test had to be executed. Its basic layout are two supporting points at the both ends of the beam and two points of loading, distance of which is ½ of beam length. This way, the beam had to face to maximum bending stress. To measure the searched values the same instruments were used as for POT, plus strain gauges attached to the steel strip, as visible in figure 2.

5. Results

5.1. Push-out test

Four specimens were made for the testing. First tests were manually stopped after reaching the searched value even though the specimens were not broken due to the first intention of the research being only the resistance value (see figure 6).

The estimated shear resistance was established as 889.67 kN [5]. The real shear resistance – determined at the end of the first, linear, section of the diagram (see figure 7), after which the increase of the slip was higher than the increase of the load and diagram passed to its horizontal part – was 1125 kN. The ratio between the estimated and real resistance was 1.26, which proved the safe usage of the formula.

The failure of the specimens occurred in concrete parts, where visible cracks were observed. They demarcated the concrete failure area, which is highlighted in the figure 9.

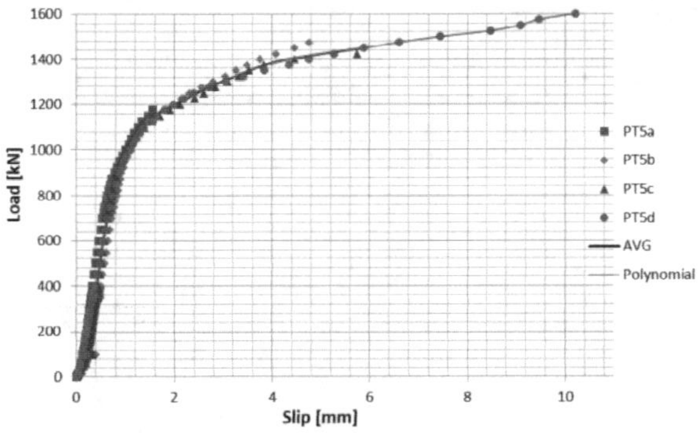

Figure 6. Push-out tests results.

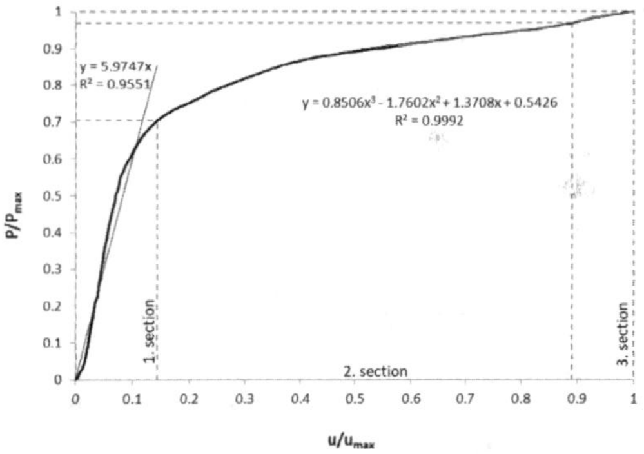

Figure 7. Diagram dependence of load onto the slip.

5.2. Four-point flexural test

The four-point flexural test resulted with bending resistance equal 370 kN. At the course of the first part of the loading, two cycles were executed in which the specimens were unloaded from 210 to 50 kN and 350 to 150 kN and back (see figure 8). The cracks developed during loading at the sides of the beam closed almost entirely and the displacement of the specimen lowered nearly to its original value, which demonstrated the elastic behaviour of the specimen.

The test was stopped at 440 kN, after which the specimen was unloaded in three loading conditions. The deformation of the beam dropped about approximately 50 mm which suggests that the specimen was still in its plastic-elastic part and the plasticization did not occur. Although after the testing the specimen was not completely damaged and it had some elastic behaviour, there were still several cracks remaining, with the maximum width of 0.45 mm (see figure 10).

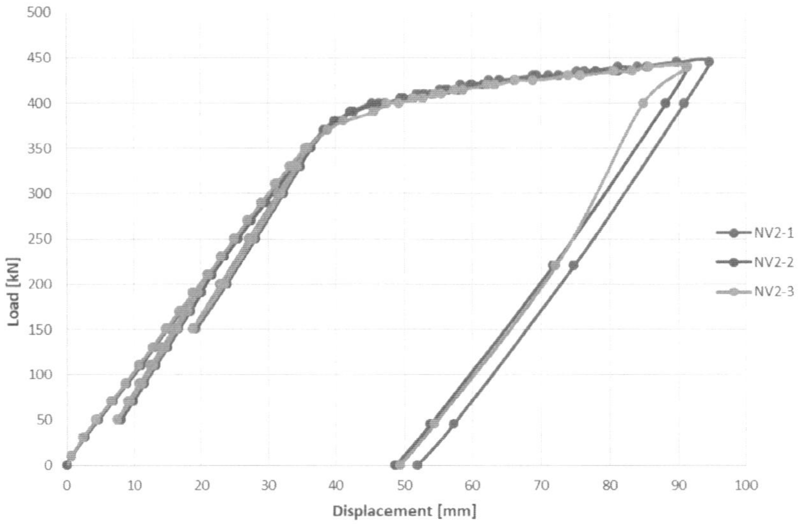

Figure 8. Four-point flexural tests results.

Figure 9. Concrete failure area after push-out test.

Figure 10. Cracks originated from four-point flexural test.

6. Conclusion

The resistance of the continuous shear connector with welded curved reinforcement was determined. For application in practice, the fatigue tests as well as long-term tests have to be provided.

7. References

[1] Oguejiofor E C and Hosain M U 1991 Behaviour of perfobond rib shear connectors in composite beams: full-size tests *Can. J. Civ. Eng.* **19** 224-235

[2] Kim H Y and Jeong Y J 2005 Experimental investigation of steel-concrete composite bridge decks with perfobond ribs *J. Construct. Res.* **62** 463-471

[3] Rovňák M, Ďuricová A, Kundrát K and Naď Ľ 2006 *Spriahnuté oceľovo-betónové mosty* (Košice: elfa s.r.o.)

[4] Vaňová P, Kvočák V, Kožlejová V, Dubecký D and Kanishchev R 2019 FEM analysis of steel-concrete composite bridges with encased perforated steel strips *Preprint*

[5] Kvočák V, Kožlejová V, Dubecký D, Kanishchev R and Vaňová P 2019 *Experimental and Software Analysis of Composite Action in Steel-Concrete Composite Bridges with Continuous Shear Connectors* (Ernst & Sohn Verlag für Architektur und technische Wissenschaften GmbH & Co. KG, Berlin)

Acknowledgments

The paper has been supported by The projects: VEGA 1/0172/20 ”Stress and deformation analysis of load bearing components made out of steel, glass and composite materials“ of the Scientific Grant Agency of the Ministry of Education, science, research and sport of the Slovak Republic and the Slovak Academy of Sciences and by the project Slovak Research and Development Agency under the contract No. APVV-15-0486 “Analysis of Shear Connection Influence in Bridges with Encased Beams”.

Building materials for 3D print

I Wawrek[1]

[1] Technical University of Košice, Slovakia, Faculty of Civil Engineering, Institute of Architectural Engineering, e-mail: igor.wawrek@tuke.sk

Abstract. The paper maps the current level of knowledge in 3D printing of construction objects. It introduces different approaches to creating wall structures of two world companies dealing with the problem of 3D print. It confronts their solution with the normative requirements for the territory of the Slovak Republic valid after 2021. It also presents proposals for possible improvement of the thermal-technical properties of the perimeter wall, modification of the design of the internal structure and properties of the printed material towards the normative requirements.

1. Introduction

The beginning of the use of concrete in construction practice was a major breakthrough in construction. It allowed to build higher, lighter structures and span longer distances. It has given freedom to architectural design and in many cases concrete become a material that defines architectural styles or ages. In combination with steel reinforcement, it became a representative of a progressive way of construction. The development of concrete as a building material, in its various transformations, has essentially never ended. It even appears to be one of the best usable materials for 3D printing in buildings. This is evidenced by its wide application in many world research centres dealing with the implementation of printing into the construction process.

2. Current way of construction

One of the main disadvantages of 3D printed structures is their low thermal insulation property. This weakness of the structure is caused by the design of the shape of the inner wall structure, which in most realized examples is adapted to create ideal paths for the print head and to help shorten the construction time of the object. It focuses only on one of the advantages of 3D printing technology, namely the speed of construction. However, 3D printing technology has other advantages, such as accuracy and the ability to shape the layered object freely. That are precisely these advantages that underpin the proposal for adjustment, which are presented in this research. In order to start thinking about possible solutions to a problem, we need to define the requirements that the solution must meet.

3. Requirements

The design of the modification has to meet the requirements for perimeter structures according to the standard STN 73 0540-2: 2012 [4].

According to energy efficiency, buildings are divided into energy classes. By the end of 2015, it was sufficient to reach the upper limit of Class B (global primary energy indicator) for new buildings for their successful certification. From January 1, 2016 to the end of 2020 it is necessary to get to class A1 and after 2020 to class A0. Table 1 shows the heat transfer coefficient U requirements for building envelope structures set by the standard.

Table 1. U values for building envelope structures according to STN 73 0540-1: 2012.

BUILDING STRUCTURES CREATING PACKAGING CONSTRUCTION OF THE BUILDING	**U (W/(m².K))**		
	Do 31. 12. 2015 NED ...B	Od 1. 1. 2016 UNB ...A1	Od 1. 1. 2021 TNB (PD) ...A0
Exterior wall, pitched roof with slope above 45°	0,32	0,22	**0,15**
Flat and sloping roof (up to 45°) ceiling above exterior	0,20	0,10	0,10
Ceiling over unheated space	0,25	0,15	0,15
Windows and glazed walls, skylights	1,40	1,00	0,60
Door with / without vestibule	3,00/4,00	2,50 / 3,00	2,00
Average U building (for shape factor 0.3 to 1.0)	0,39 - 0,58	0,27 - 0,38	0,20 - 0,25

4. Increasing effectivity of building materials

If we look at traditionally used building materials such as. brick can be seen as its size, shape and internal structure gradually evolved figure 1. This development was motivated by the ever-increasing requirements for streamlining the building's perimeter structures. The peripheral structures created by 3D printing are currently at the beginning of their development. By knowing how to modify traditional materials to improve their thermal insulation properties, we are able to propose possible solutions for improving the properties of 3D printed structures.

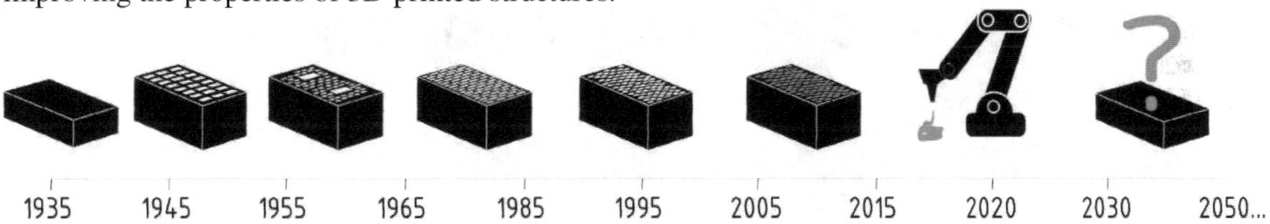

Figure 1. Historical development of building materials – bricks.

Modification of the building elements to improve their thermal insulation properties is realized in traditional materials mainly in three ways.

Adjusting the internal structure of the preform

By keeping the original material by adjusting the inner structure of the preform we can increase its insulating property. The increase in thermal resistance creates a reduction in heat transfer through the enclosed air cavity. Such higher the density of these cavities is in the fitting as the higher its resulting thermal resistance R is.

Increasing the thickness of the structure while maintaining the original material and the spatial arrangement of the internal structure of the preform. By increasing the parameter d (thickness of the structure), the thermal resistance R of the given cross-section resulting from the relation

$R = d / \Delta$ (W / (m².K)) is directly proportional to it.

Insulation of the fitting with another material with a higher thermal resistance parameter R (the material is applied to the outer wall of the cross-section or directly inserted into the cavities of the internal structure, thus increasing the total sum of resistances $R = R1 + R2 + R3...$ of the perimeter wall construction.

By adjusting the properties of the material of which the block is made to improve the thermal insulation properties of the product. By reducing the thermal conductivity coefficient Δ, we increase the ability of the material to withstand the temperature changes induced by heat conduction.

IOP Conf. Series: Materials Science and Engineering **867** (2020) 012047 doi:10.1088/1757-899X/867/1/012047

Design optimization of wall construction for printing combining above mentioned methods. Comes from knowing way of increasing insolation values properties of traditional materials.

5. Material requirements

For optimal material design it was necessary to determine the input requirements to suit the way of construction by 3D printing.

Requirements:

-flow through nozzle d = 10mm
-set speed of solidification
-consistency
-strength
-use of recycled materials as concrete admixture
-reducing the proportion of cement in concrete

Figure 2. The final form of the concrete mix.

The mixing ratio of the resulting concrete mixture, which meets predefined material requirements, came from experimental verification figure 3. At the moment when the mixture met the properties of the 3D printing machine layering, it began to improve its thermal insulation properties. The polystyrene granulate was mixed into the concrete mixture in various percentages to the total weight of the sample of figure 3.

The samples were then measured by the pulse method at different time intervals during their drying time. Comparison of the measured values of λ can be seen in the graphical representation of figure 4 for concrete mortars and figure 5 for concrete pastes. The measured values show that the use of polystyrene granulate as a filler indeed increases the thermal resistance R for the material. By comparing the results of measurement of thermal conductivity and compressive and tensile strength of given samples, we are able to focus more precisely the area of development of concrete material for printing in terms of its more effective action in the peripheral bearing structure.

6. Impulse measure method

Measurement is based on analysis of the temperature response of the analysed material to heat flow impulses. Heat flow is excited by electrical heating of resistor heater inserted into the probe which is in direct heat contact with the tested specimen. Evaluation of thermal conductivity and volume heat capacity is based on periodically sampled temperature records as function of time, provided that heat propagation occurs in unlimited medium.

Figure 3. Prepared samples for pulse measurement.

Table 2. Measured properties of tested materials.

	cca 28 days					20.1.
	Bend f_c MPa	std MPa	**pressure** f_{cf} MPa	std MPa	**Volume weight** D kg.m^{-3}	**lambda**
Paste						
0	2,80	1,30	106,6	4,9	2150	0,231
20	3,00	0,35	35,9	3,5	1660	0,358
40	2,60	0,20	21,8	1,9	1250	0,209
60	2,15	0,55	6,2	0,8	1025	0,149
80	na	na	na	na	325	0,131
Mortar						
0	7,85	0,70	53,8	2,1	2090	0,208
20	4,60	0,30	23,1	1,5	1635	0,169
40	2,50	0,20	8,0	1,1	1205	0,138
60	0,75	0,05	na	Na	710	0,103

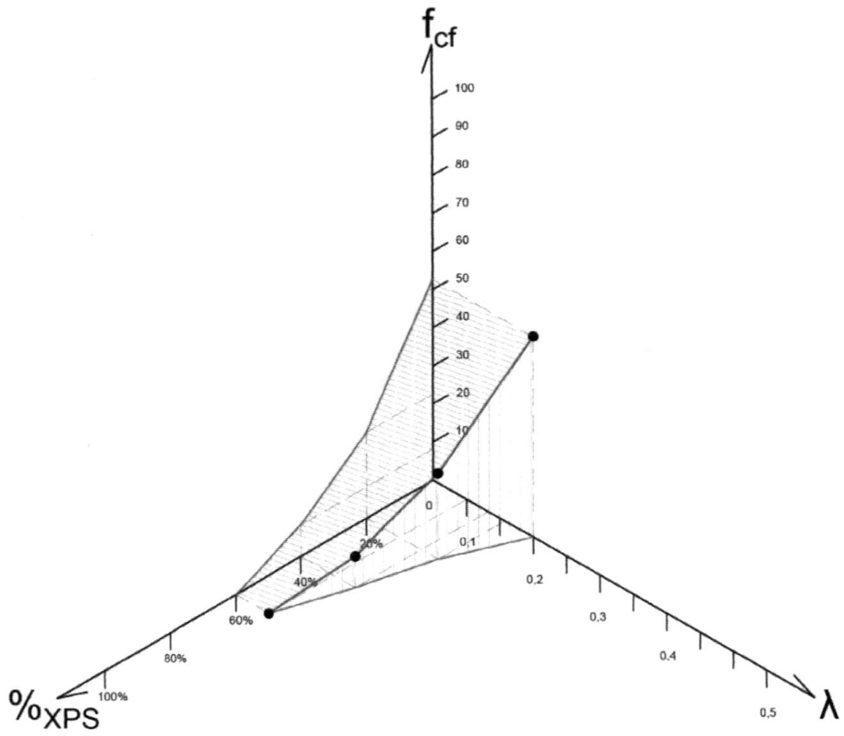

Figure 4. Graph of measured values for mortar.

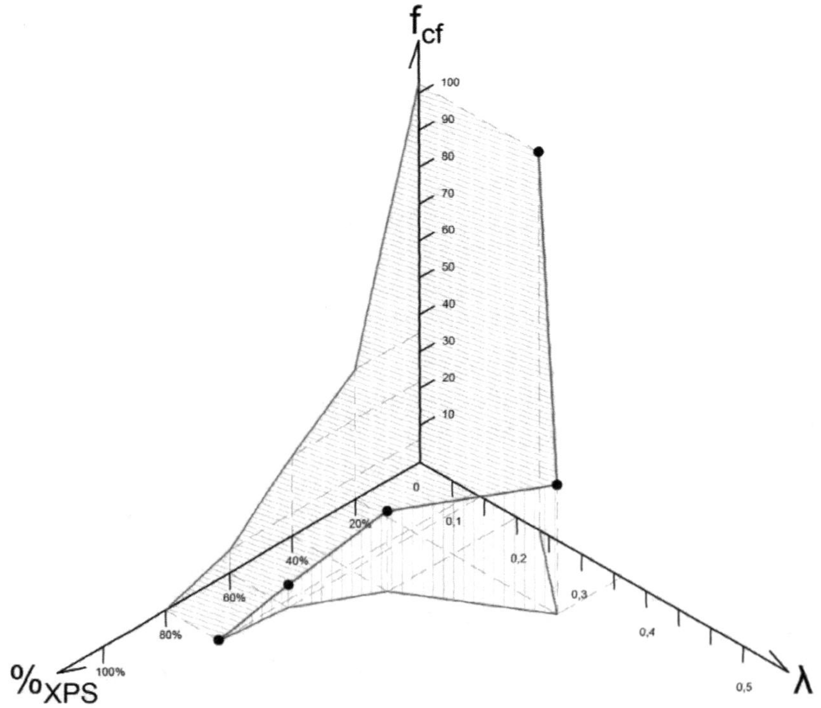

Figure 5. Graph of measured values for pastes.

IOP Conf. Series: Materials Science and Engineering **867** (2020) 012047 doi:10.1088/1757-899X/867/1/012047

7. Conclusion

The measured values of the thermal conductivity coefficient λ in combination with the measured values of strength are the starting points for a more precise orientation of the future research of building material for 3D printing. The measured values confirmed the assumption of an increase in the thermal resistance R of the material using polystyrene granulate. The measurement results will be used as input values in the future simulation of the perimeter wall for printed buildings.

8. References

[1] Sobotka A and Pacewicz K 2016 Building Site Organization with3D Technology in Use *World Multidisciplinary Civil Engineering-Architecture-Urban Planning Symposium* (Prague)
[2] Pifko H 2017 NEED *Navrhovanie energeticky efektívnych domov* Eurostav, spol s.r.o. Bratislava
[3] Winsun 2019 *winsin3d* http://www.winsun3d.com/En/About/
[4] STN 73 0540-2:2012/71: 216
[5] Pacewicz K, Sobotka A and Gołek1 Ł 2018 *Characteristic of materials for the 3D printed* MATEC Web of Conferences **3**
[6] A L Mohd 2017 Cost viability of 3D printed house in *UK IOP Conference: Materials Science and Engineering* Yogyakarta Indonézia 773
[7] Pablo C H 2016 Hove to print architecture *XX Congreso de la Sociedad Ibero-americana de Gráfica Digital* Buenos Aires Argentina
[8] Irina Mayatskaya I 2017 Building constructions: architecture and nature *International Science Conference* SPbWOSCE-2016 “SMART City” **106**
[9] Shiva Khoshtinat S and Shirvani A 2016 Inspirations from nature as a Method for Sustainable Architectural Design *4th. international congress on civil engineering architecture and urban development* At. Iran
[10] ElGhazi Y 2017 Responsive Plant-inspired skins *Conference: plea2017* (Edinburgh) **3**
[11] Mariappan P and Akhilesh K 2018 Old buildings are green buildings *International Journal of Academic Research and Development* **3**
[12] Wu Q and Gaoc W 2016) Research on the design of ecological energy-saving building based on the climate condition of Hangzhou *Urban Planning Architecture Design for Sustainable Development*
[13] Nagy E 2015) Manuál ekologickej výstavby Alter-Nativa o.z. Bratislava
[14] Petráš D 2001 Nízkoteplotné vykurovanie a obnoviteľné zdroje energie Jaga group Bratislava
[15] Keppl J 2013 *Rukoväť udržateľnej architektúry* SKA Bratislava
[16] Apis-cor 2019 Apis-cor. https://www.apis-cor.com/

IOP Conf. Series: Materials Science and Engineering **867** (2020) 012048 doi:10.1088/1757-899X/867/1/012048

Analysing the cumulative energy demand of external bearing walls

M Wolfova[1]

[1] Technical University of Kosice, Faculty of Civil Engineering, Institute of Environmental Engineering, Vysokoskolska 4, Kosice, 042 00, Slovak Republic

Abstract. The assessment of environmental properties of buildings is now commonly using a life cycle analysis (LCA). Due to the rapidly evolving construction industry, LCA research is carried out in many areas, from building materials and components to overall building system. Life Cycle Assessment (LCA) is a tool used to compare environmental impacts along the life cycle of products or services. This paper presents an evaluation and comparing of two external bearing walls of different material compositions regarding the materials´ total energy consumption within the cradle to gate boundaries. Impact categories were calculated by the software SimaPro using Cumulative Energy Demand (CED) method and applying the Ecoinvent database. A renewable (biomass, wind, solar, geothermal, water) and non-renewable (fossil, nuclear, biomass) energies were under consideration. The composition with brick as a core wall material achieved worse results in all categories than the wall composition with aerated concrete block.

1. Introduction

The construction industry and the built environment is important contributor to economic and also social development. These parameters represent a large proportion of capital in most countries and it is a primary source of environmental impacts too. In addition, existing buildings require continuous investments in repairs and renovations [1].

Due to the price crisis, energy analysis of products as well as processes involving the supply of energy itself began to be carried out by scientists at the start of the 70 years. In a publication called Handbook of Industrial Energy Analysis has reached the peak. This handbook consists of energy requirements tables for about 400 industrial processes [2]. Nevertheless, energy consumption as well as pressure on the environment continued to increase. In the late 1960s and early 1970s was stylized the idea of a comprehensive LCA at the American Midwest Research Institute (MRI), while nearly identical ideas were developed in Europe at approximately the same time [3]. Resource and environmental profile analysis: a life cycle environmental assessment for products and procedures [3, 4].

Today, this assessment method is considered an important methodology to quantify the environmental impacts of the life cycle of products and services [5]. Cumulative energy demand (CED) or also primary energy consumption is one of the key indicators that have been solved since the first studies LCA. This impact assessment method is now quantified in studies, which is based on life cycle assessment [6].

The International Organization for Standardization (ISO) 2006a, b) [7] do not require or address the energy consumption indicator. However, the European Standard EN 15643-2 and EN 15978 address the sustainability of construction works (EN 15643-2 and EN 15978) [8, 9] and also the European Standard on environmental product declarations (EPD) deal with energy consumption with 9 indicators (EN 15804) [10]. Although this method is often used, there is no (does not exist) harmonized approach. The cumulative energy in the above standards and guidelines is defined in a different way [11]. In addition, it is still under discussion whether the CED as an indicator of impacts belongs in the life cycle assessment or only in the life cycle inventory [5].

The paper presents the results of an evaluation of two external bearing walls of different material compositions regarding the materials total energy consumption (CED) within the cradle to gate boundaries.

2. Material and methods

This study focused on evaluation of two material alternatives of external bearing walls (table 1). The first analysed wall (A) consisted of 300 mm thick brick and the expanded-polystyrene thermal insulation with graphite (EPSg), (thickness of this insulation was 120 mm and bulk density 28 kg/m^3). The second analysed wall (B) consisted of 400 mm thick aerated concrete block (AC) with the same thermal insulation of thickness of 70 mm and bulk density 28 kg/m^3. The other components were the same for both variants. The scheme of the material composition of the walls is illustrated in figure 1.

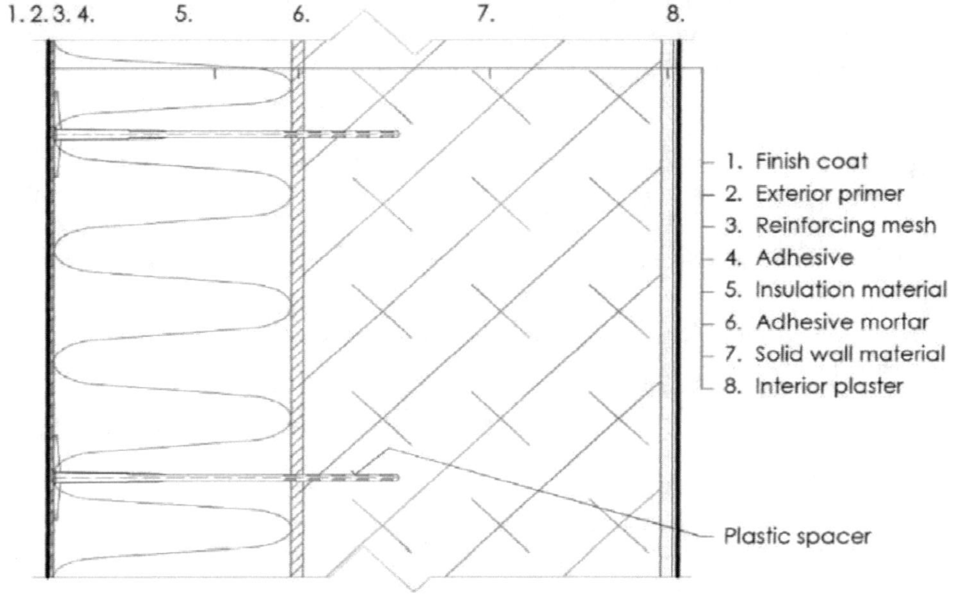

Figure 1. Scheme of external bearing wall.

Material composition of the analysed walls are described in table 1.

Table 1. Wall alternatives and composition.

Wall alternative	Density [kg/m^3]	Thickness [mm]	Weight [kg]	Interior Layer	Core	Thermal Insulation	Exterior Layer
Wall A	2000	15	30	Lime–cement plaster	Brick 300 mm	EPSg	Silicate plaster
Wall B	2000	10	20	Lime–cement plaster	AC 400 mm	EPSg	Silicate plaster

The impact assessment method Cumulative energy demand (CED), which is available in Software SimaPro was used for the quantification the environmental impacts. This method includes both energies (direct and indirect) that are associated in all life cycle processes of material, i.e. the energy needed to extraction of the material, manufacturing, use phase and disposal. [12, 6].

The functional unit was established of 1m^2 of the structure. The objective of the LCA was a cradle-to- gate assessment. The idea of sustainable design of buildings, their construction and operation in the management and construction sectors needs improvements adapted to the current demands and requirements at all stages of the building's life. Building life is a set of interrelated processes. The first step is an initial architectural and construction solution that leads to real construction and subsequent maintenance, inspection and, last but not least, renovation or demolition.

In accordance with building regulations on the basis of lifespan, requirements have been generated in terms of the basic spheres of society (social, economic and environmental) issues related to high-efficient energy-saving building systems [13]. In this study, the lifespan was considered of 50 years for the core materials and thermal insulation, and 25 years for both the exterior and interior plasters. In practice, however, the lifetime of construction products is shorter than had been designed for [14].
The calculations were performed using software SimaPro based on LCA analysis and Ecoinvent database [15].

2.1. Cumulative energy demand (CED)

The method was applied to calculate the energy impacts of the constructions. This method has a long tradition. It has been developed in the early seventies after the first oil price crisis [12].

The CED methodology can be understood as a screening indicator for environmental [16]. It is also possible to compare the results of the CEDs with the results of a detailed LCA study that only covers primary energy demand. The plausibility check can be performed based on the final results of the CED. In this case, it is easy to determine if there were any serious errors or not [17]. In the Ecoinvent database, the CED method is divided into eight categories (see table 2), because there are different concepts of perception of this method and also an unclear basis for characterizing the different carriers of primary energy [18]. The table does not contain any aggregate values.

Table 2. Impact assessment method cumulative energy demand (CED) implemented in Ecoinvent.

	subcategory	includes
non- renewable resources	fossil	hard coal, lignite, crude oil, natural gas, coal mining off-gas, peat
	nuclear	uranium
	primary forest	wood and biomass from primary forests
renewable resources	biomass	wood, food products, biomass from agriculture, e.g. straw
	wind	wind energy
	solar	solar energy (used for heat & electricity),
	geothermal	geothermal energy (shallow: 100-300m)
	water	run-of-river hydro power, reservoir hydro power

Common to all categories is the thesis that all energy carriers have an intrinsic value. This intrinsic value is determined by the amount of energy taken from nature. The unit of all categories was MJ.

3. Results and discussion

LCA quantification of all impacts of the entire energy supply chain, represented by the cumulative energy demand (CED) for the production of buildings materials in the present study, which were used in the external bearing wall, is presented in figure 2.

IOP Conf. Series: Materials Science and Engineering **867** (2020) 012048 doi:10.1088/1757-899X/867/1/012048

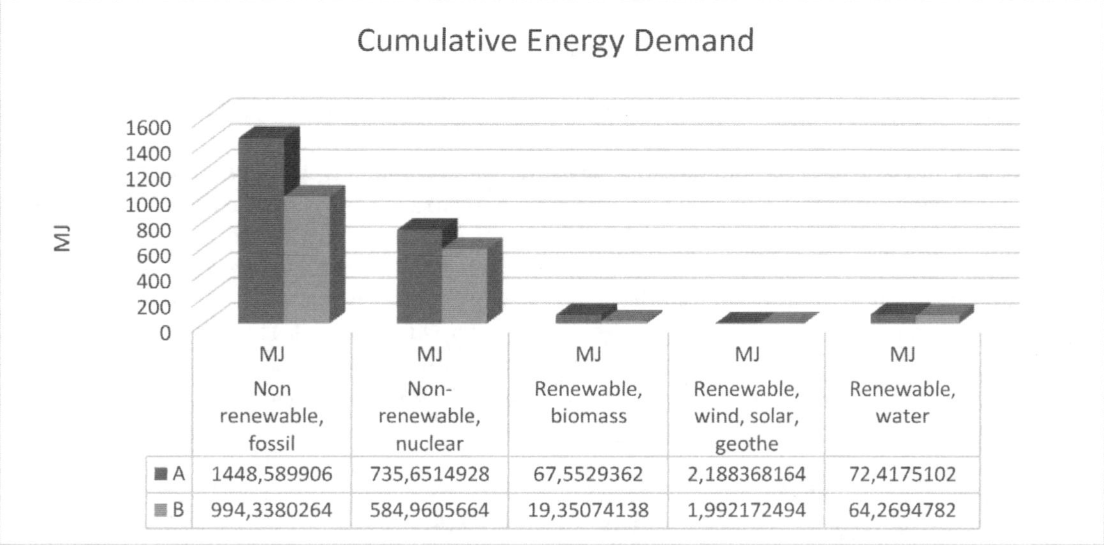

	MJ Non renewable, fossil	MJ Non-renewable, nuclear	MJ Renewable, biomass	MJ Renewable, wind, solar, geothe	MJ Renewable, water
A	1448,589906	735,6514928	67,5529362	2,188368164	72,4175102
B	994,3380264	584,9605664	19,35074138	1,992172494	64,2694782

Figure 2. Comparison of the energy demand of the wall structures.

The CED method involves the use of both energy sources (renewable and non-renewable). Figure 2 presented the results of the environmental impact of 1m^2 of the external bearing wall materials, with brick and aerated concrete as a core material. The figure shown, that the wall composition with brick was in all categories worse, then with using the aerated concrete. This is due to the manufacturing process, because the brick needed more energy and raw materials during extraction and production phase. In the category non- renewable fossil and renewable biomass show the results bigger differences as in other categories. It means, that these two categories have the greatest impact on overall results. The smallest negative impacts on the environment were recorded in the category non-renewable biomass for both alternatives. The lowest values and therefore the smallest negative impact for environment achieved the category non- renewable biomass.

The comparison of technical parameters (weight and thickness of the analysed walls) and calculated energy indicators of the material alternatives are presented in figure 3. The accessary values are expressed as percentage shares of the overall values.

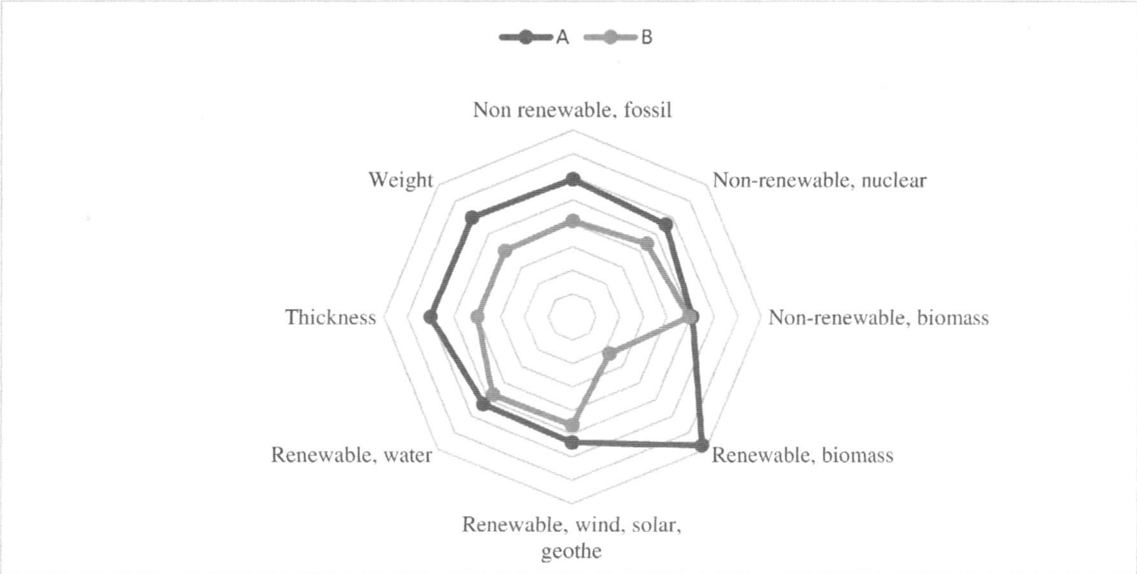

Figure 3. Comparison of the percentage share of the material.

As can be seen in figure 3, option B was presented as a better option than A, even considering the technical parameters such as weight and thickness.

4. Conclusion

Life Cycle Assessment (LCA) is a tool to better understand the various aspects of product development and its potential impact throughout the product lifecycle, which means from cradle to grave. The cradle-to-grave boundary encompasses all processes from raw material acquisition, processing, production, use and final disposal of materials. The lifetime of a building is influenced by several factors. This factors includes market demands and lifetime of building materials be it reconstruction or demolition, which results in a reduction in the life of the building. The lifespan in this paper was considered of 50 years for the core materials and thermal insulation, and 25 years for the plasters. A LCA study of the environmental impacts of the construction of external bearing wall with brick (A) and aerated concrete block (B) were compared in this paper. Based on the results, the option B was shown as a better alternative than A.

5. References

[1] Hovde P J and Moser K 2004 (Rotterdam: Copyright) pp 294
[2] Boustead I and Hancock G F 1979 Handbook of industrial energy analysis (United Kingdom: N. p.) pp 422
[3] Hunt R G, Sellers J D and Franklin W E 1992 *Environ. Impact Assess.* **12**(3) 245-269
[4] Hunt R G, Franklin W E, & Hunt 1996 *Int. J. Life Cycle Assess.* **1**(1) 4-7
[5] Frischknecht Rolf et al. 2015 *Int. J. Life Cycle Assess.* **20**(7) 957-969
[6] Huijbregts M A et al. 2006 *Environ. Sci. Technol.* **40**(3) 641-648
[7] International Organization for Standardization 2006 Environmental Management: Life Cycle Assessment; Principles and Framework No. 2006
[8] European Committee for Standardization (CEN) 2011 EN 15643–2: 2011–Sustainability of construction works–Sustainability assessment of buildings–Part 2: Framework for the assessment of environmental performance.
[9] STN EN 15978 2012 Sustainability of construction. Assessment of the environmental performance of buildings. Calculation methods 56
[10] EN, CEN et al. 2012 Sustainability of Construction Works-Environmental Product Declarations-Core Rules for the Product Category of Construction Products 15804
[11] VDI Richtlinien 1997 Cumulative energy demand–terms, definitions, methods of calculation
[12] Frischknecht R et al. 2007 Implementation of life cycle impact assessment methods. Data v2. 0. Ecoinvent report No. 3. Ecoinvent Centre
[13] Chen Z et al. 2006 *Energy Build.* **38**(5) 393-409
[14] Ashworth A 1996 *Structural Survey* **14**(2) 4-8
[15] Frischknecht R et al. 2005 *Int. J. Life Cycle Assess.* **10**(3) 3-9
[16] Wagner H J & Pick E 2004 *Energy* **29**(12-15) 2289-2295
[17] Frischknocht R & Jungbluth N 2007 Ecoinvent v. 2 report: overview and methodology. PRé Consultants, the Netherlands
[18] Gürzenich D et al. 1999 *Int. J. Life Cycle Assess.* **4**(3) 143-149

Acknowledgments
This paper has been prepared within the Grants No. 1/0648/17 of the Slovak Grant Agency for Science.

The Effect of the changing cover thickness on the stability of cellars

J Zenah[1] and P Görög[1]

[1] Budapest University of Technology and Economics, Faculty of Civil Engineering, Department of Engineering Geology and Geotechnics, Budapest H-1111 Műegyetem rkp. 3, Hungary

Abstract. The stability of natural and man-made cavities is a very important topic nowadays especially if it is located under new construction areas. There are a lot of residential areas in Budapest and other cities of Europe such as Rome, Paris, and Cracow with stability problems because of cavities. For stability calculations, the geometry of the cavity, strength properties of the rock mass and joints are needed as input parameters, which can be obtained by field and laboratory tests. The FEM software is effective for stability calculations and with them; it is not so difficult to perform parametric analysis. This paper is focusing on the effect of the thickness of the cellar cover on their stability. The investigation site is in Budapest, Hungary, where there are several cellars cut into porous limestone. These cellars are more than 100 years old and nowadays cause difficulties for the building improvements. The geometrical parameters of the cellars and strength properties were measured. The cover of the studied cellar is around 6.0 m, but other cellars in the area have various cover thicknesses from 2.0 to 9.0 m. Therefore, the paper studied the effect of the cover on the stability to find the relation between the depth and the cellar stability. The modelling has been done by Rocscience software (RS2), using a real cross-section that crosses six branches of the cellar system.

1. Introduction

Investigation and stability analysis of cellars and other underground cavities are an important topic, especially when these cavities are under urban areas. Several cities around the world have areas which are undercut by cellars and suffering from stability problems caused by these openings. Well known examples are found in France [1][2], in Spain [3], in Italy [4][5], in China [6], and in Hungary as well in Miskolc [7][8] and in Budapest [9][10]. The cellars and cavities often have historical background and are mainly results of previous quarrying activities. Cellar systems and caverns were established under towns in Hungary form the Middle Age. These were built at a low depth and with thin cover [11]. Nowadays, they are usually abandoned or have different uses, such as wine cellar, mushroom production or storage facility. Most of the above-listed cities, cellars are cut into soft rocks like porous limestone [9], rhyolite tuff [12] or andesite tuff [7]. Because of the need for a new construction area to cover the expansion of the big cities such as Budapest, sometimes the new construction area located above these kinds of cavities; therefore, the stability of these cavities became a serious engineering problem.

There are several different parameters which affect the stability of such cavities such as geometrical parameters including dimensions of the cavity, cover of the cellar, pillar width between branches; strength parameters of the rock mass including joint system, UCS and water resistance of the rock material, long term behaviour. In this paper, the effect of the cavities cover on the stability is investigated.

2. Description of the studied area

The studied cellar located in the southern part of Budapest – Buda side (Central Europe), it had been cut into Miocene limestone. Above the cellar system, a new construction area is planning to be built, the ground surface is almost flat, the depth of the cellar system around 6 m. The investigated cellar system is consisted of a main corridor and from this corridor; seven cellar branches opened with an average length of 30 metres. The cross-sectional dimensions of the branches are the following: the average width of it is 5 m and the average height of it is 3.5 m. Masonry arches were built in the cellars against local failures.

3. Methodology

Specimens obtained from the cellar, and tested under laboratory conditions. Several rock physical tests were done: the dry and saturated density was calculated according to EN 1936:2007. The uniaxial compressive strength (UCS) test and Brazilian tests for both dry and saturated conditions were performed according to ASTM D7012-14e1 and ASTM D 3967, respectively. Water absorption tests were performed based on EN 13755:2008. The ultra-sonic wave propagation velocity test made following the guidelines given by EN 14579:2005. According to the test result, the following parameters were determined as input parameters for the stability calculations:

γ_{dry} = 1600 kg/m^3, γ_{sat} = 1790 kg/m^3, UCS_{dry} = 7.47 MPa, UCS_{sat} = 2.85 MPa, E_{dry} = 3.93 GPa, E_{sat} = 1.81 GPa.

Terrestrial Laser Scanner (TLS) was used to get the geometry of the cellar system. TLS is a useful and effective method for measuring the geometry of buildings [13], rock slopes [14] [15] [16] and 3D geometry of the cellars [16]. The output from this technology is a point cloud in 3D dimension and 2D sections, in a lot of formats like dxf and obj format. The studied cross section is located 21 m distance from the axis of the main corridor; this section is crossing six branches of the seven existing branches of the cellar 'figure 1'.

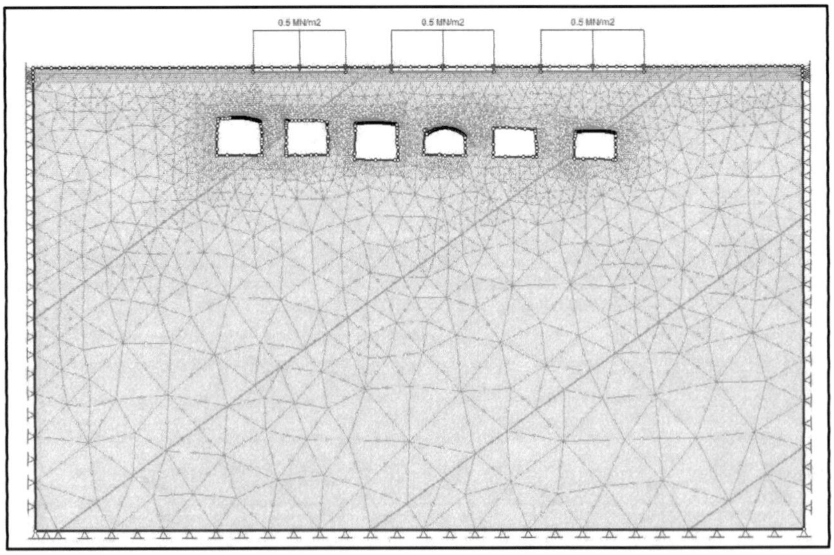

Figure 1. The studied cross-section.

For the upper 1.5 m, the UCS reduced by 20% because the rock at this depth is weathered. The generalised Hock-Brown failure criterion was used for the material model. According to the in-situ measurements, there was only one main joint direction with dip direction = 90 deg and dip angle = 74 deg and the joint spacing is about 20 m. The cover above the cellar branches is different therefore the minimum cover 'figure 2' modified in the different models later. The cover above the first branch is 5.5 m 'figure 2'. During the modelling, this cover reduced by 0.5 m in every step until it reaches 2 m and increased by 0.5 m until it reached 7 m, then the covers 8m and 9m also calculated. The modelling has done for surface loads 0.3, 0.4 and 0.5 MN/m^2, which applied normally to the surface in air dry and water saturated conditions which means the model run 156 times. The modelling built with the RS2 FEM software of the Rocscience software package.

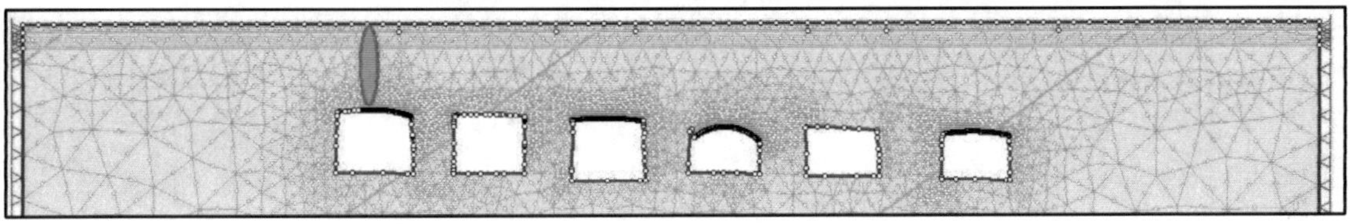

Figure 2. Cross section A21, the minimum cover.

4. The results

After running the 156 models, the results for maximum displacement and factor of safety in both air dry and water saturated conditions were recorded for different loading conditions table 1, table 2 and table 3.

Table 1. The result of modelling for load 0.3 MN/m^2.

Cover		9	8	7	6.5	6	5.5	5	4.5	4	3.5	3	2.5	2
Dry	Fos	2.4	2.38	2.41	2.26	2.23	2.32	2.24	2.39	2.34	2.34	2.21	1.69	1.16
	Max. Dis	3.2	3.15	3.13	3.19	3.54	3.21	3.38	3.24	3.28	3.43	3.52	3.77	5.08
Sat	Fos	1.36	1.36	1.39	1.35	1.33	1.32	1.37	1.37	1.34	1.32	1.26	0.81	0.70
	Max. Dis	6.21	6.1	6.02	5.99	7.12	6.89	6.72	6.26	6.27	6.64	7.22	446.00	531.00

Table 2. The result of modelling for load 0.4 MN/m^2.

Cover		9	8	7	6.5	6	5.5	5	4.5	4	3.5	3	2.5	2
Dry	Fos	2.18	2.13	2.15	1.99	1.97	2.04	1.96	2.10	2.06	2.05	1.96	1.42	1.08
	Max. Dis	4.07	4.02	4	4.08	4.52	4.15	4.42	4.21	4.24	4.45	4.59	4.99	14.70
Sat	Fos	1.26	1.25	1.27	1.23	1.2	1.24	1.24	1.23	1.18	0.97	0.89	0.73	0.61
	Max. Dis	8.07	7.97	7.86	8.11	9.12	9.11	9.13	8.70	8.92	43.70	64.10	705.00	528.00

Table 3. The result of modelling for load 0.5 MN/m^2.

Cover		9	8	7	6.5	6	5.5	5	4.5	4	3.5	3	2.5	2
Dry	Fos	2	1.94	1.96	1.81	1.78	1.85	1.86	1.88	1.83	1.82	1.73	1.06	1.00
	Max. Dis	4.94	4.89	4.88	4.98	5.53	5.12	5.39	5.23	5.22	5.50	5.69	6.36	60.50
Sat	Fos	1.18	1.19	1.18	1.13	1.12	1.12	1.15	1.16	1.10	1.01	1.00	0.67	0.57
	Max. Dis	10	9.92	10.1	10.9	14.5	19.20	21.90	18.50	41.40	55.90	118.00	727.00	1280.00

The safety factor is shown in graphs of 'figure 3'. It shows that the safety factor (strength reduction factor) increases with the increasing cover from 2 m until 4.5 m in both air dray and water saturated conditions, from 7 m cover the safety factor started to be constant. From previous results, it's clear that the cellar system will not affect the ground surface settlement when the cover is equal or greater than 7 m.

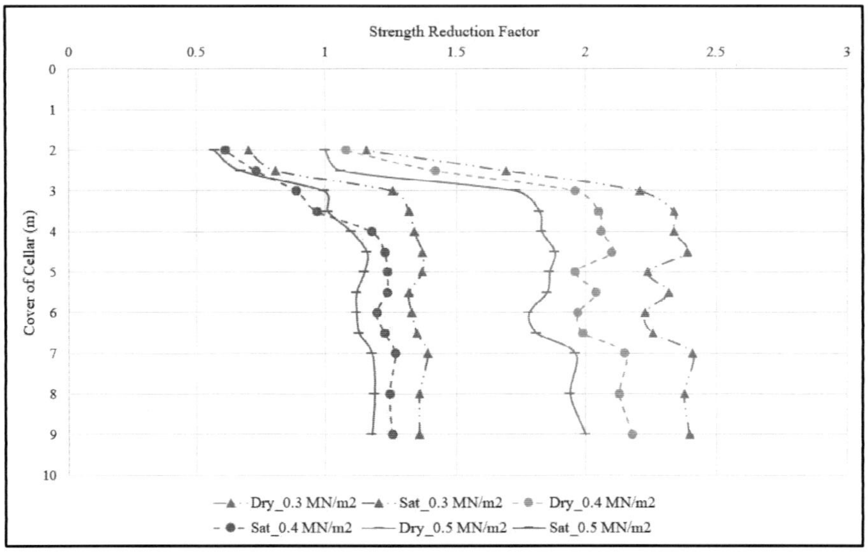

Figure 3. Strength Reduction Factor.

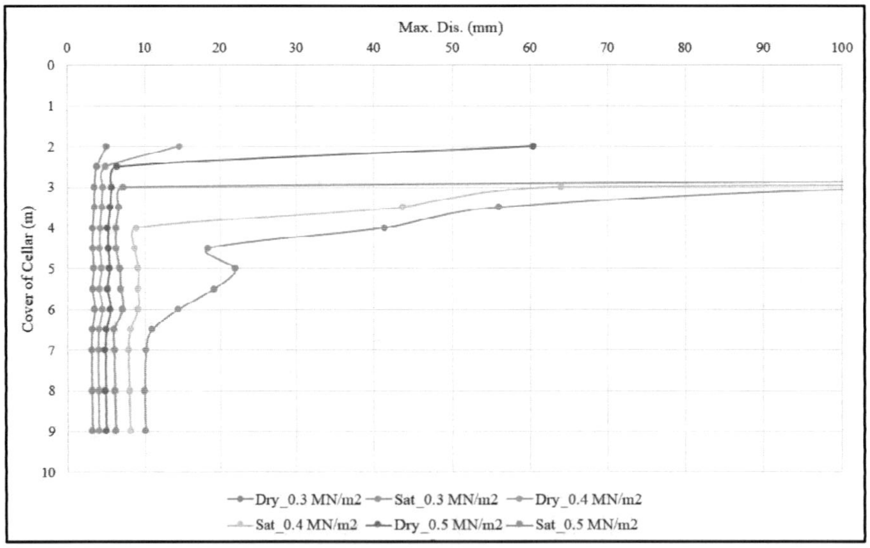

Figure 4. Max. Dis.

The maximum displacement for every cover thickness in every load step is also recorded and drawn in 'figure 4'. It shows that the maximum displacement didn't change significantly when the depth is equal or bigger than 2.5 m in dry condition for the three studied loads. While in saturated condition this depth is 3 m for 0.3 MN/m^2, 4 m for 0.4 MN/m^2 and 7 m for 0.5 MN/m^2.

From 4.5 m until 7 m cover, some uncertainty can be seen in safety factor, where the safety factors are increasing and decreasing. The reason for that is the location of the joints is in a high shear strain zone, which is in the pillar between branches. 'Figure 5-A' shows this area of shear strain for cover 6 m and load 0.5 MN/m^2 in dry condition. It shows very clearly that the joint is located exactly in this shear strain area, which leads to the decreasing safety factor and increasing maximum displacement. After slight modification of the studied models, by changing the location of the joint to be out of the high shear strain area, the modified example cross section can be seen in 'figure 5-B'. By moving the joint, the resulted new safety factors are increased and make the graph smooth and fit to the expected value as in 'figure 5-C'.

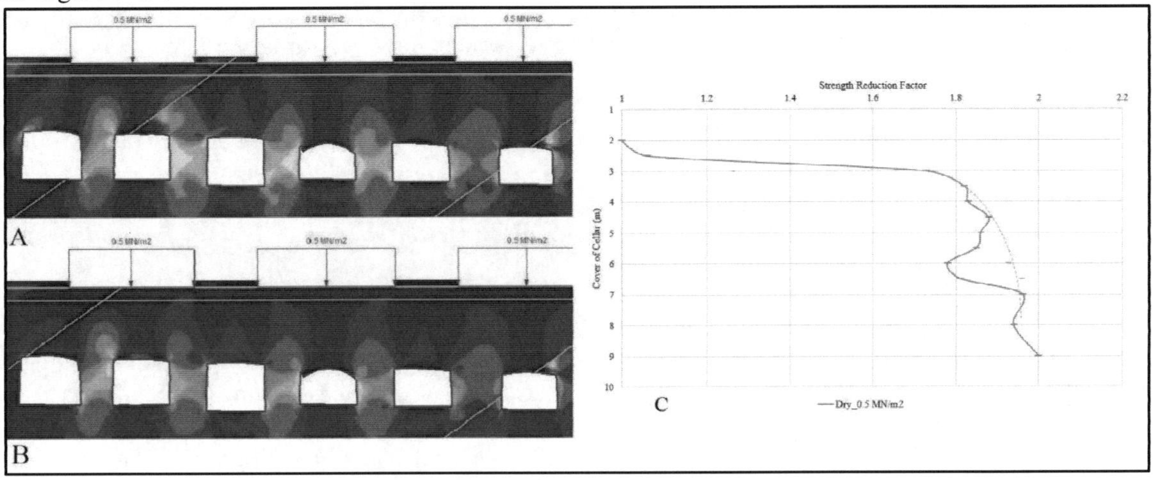

Figure 5. The maximum shear strain area with the joint - 0.5 MN/m^2
(a) The original location, (b) Modified location, (c) Original and modified graphs.

5. Conclusion

Around the investigation area, there are thousands of square metres area, which is undercut by similar cellars. The cover of these cellars usually between 3 and 8 metres, it was the reason for this investigation. The modelling results of this study showed obviously that the safety factor increasing and the maximum displacement decreasing with increasing the thickness of cellar's cover. But we can determine the critical depth of the cellars which has no further effect neither the stability nor the displacement. This critical depth is for safety factor is 3 meters for dry conditions and 4 meters for saturated conditions and for displacements is 3 meters for dry and 6.5 meters for saturated conditions. The surface load cannot be higher than 0.4 MN/m^2 above the cellars in saturated conditions. Comparing the air dried and water saturated conditions, the reduction in the safety factor was around 50% for the three determined loads. The location of the exciting joints is important in stability point of view if the location of them is not favourable; it reduces the stability of the cellar system.

6. References

[1] Smeray J, Mandin D, and Chaumont J 2000 Annual variations of airborne fungal propagules in two wine cellars in French Jura *Cryptogamie Mycologie* **vol. 21** no. 3 pp 163–169
[2] Al Heib M, Duval C, Watelet J M and Gombert P 2015 Back-analysis of the collapse of the Clamart chalk underground quarry - Paris (France) *Geotechnical engineering for infrastructure and development: proceedings of the XVI European Conference on Soil Mechanics and Geotechnical Engineering* (London) pp 1525–1530
[3] Fuentes J M, Gallego E, García A I and Ayuga F 2010 New uses for old traditional farm buildings: The case of the underground wine cellars in Spain *Land Use Policy* **vol. 27** no. 3 pp 738–748
[4] Tinti F, Barbaresi A, Benni S, Torreggiani D, Bruno R and Tassinari P 2015 Experimental analysis of thermal interaction between wine cellar and underground *Energy and Buildings* **vol. 104** pp 275–286

[5] Ciotoli G, Ferri G, Nisio S and Succhiarelli C 2015 The underground cavities in the territory of Rome: typologies, distribution and sinkhole susceptibility *International Congress of Speleology in Artificial Cavities* (Italy Rome) pp 433–439
[6] Liu M, Tang Y, Zhao K, Liu Y, Guo X, Ren D, Yao W, Tian X, Gu Y, Yi B and Zhang X 2017 Determination of the fungal community of pit mud in fermentation cellars for Chinese strong-flavor liquor, using DGGE and Illumina MiSeq sequencing *Food Research International*, **vol. 91** pp 80–87
[7] Mocsár-Vámos M, Görög P and Török A 2015 Engineering geological characterization of the host rocks of underground cellars in Avas hill, Northern Hungary *Geotechnical Engineering for Infrastructure and Development* (London)
[8] Mocsár-Vámos M, Görög P, Borostyáni M, Vásárhelyi B, and Török A 2015 Stability analysis of wine cellars cut into volcanic tuffs in Northern Hungary *Engineering Geology for Society and Territory* vol. 8 pp 153–158
[9] Görög P, Hangodi Á, and Török Á 2013 Stability analyses of underground structures cut into porous limestone *18th International Conference on Soil Mechanics and Geotechnical Engineering* (Paris) pp 1707–1710
[10] Hajnal G 2006 *Hydrogeological Study of the Castle Hill in Budapest* (Hungary: Szinergia Ház Közhasznú Egyesület)
[11] Gálos M, Kertész P, and Kürti I 1981 Engineering geological problems of cellars and caverns under historical centres of towns *Subsurface Space* (Elsevier) pp 119–126
[12] Vincze Á and Görög P 2016 Engineering geological investigation of a cave spa cut into rhyolite tuff *25th European Young Geotechnical Engineers Conference* Romania) pp 119–122
[13] Rehany N and Lovas T 2017 Supporting Historic Building Reconstruction by Laser Scanning *AIS 2017 - 12th International Symposium on Applied Informatics and Related Areas organized in the frame of Hungarian Science Festival 2017* (Óbuda University: Budapest Hungary) pp 157–162
[14] Kordić B, Lužar-Oberiter B, Pikelj K, Matoš B and Vlastelica G 2019 Integration of Terrestrial Laser Scanning and UAS Photogrammetry in Geological Studies: Examples from Croatia *Periodica Polytechnica Civil Engineering*
[15] Vlastelica G, Miščević P and Fukuoka H 2014 Rockfall Monitoring by Terrestrial Laser Scanning - Case Study of the Rock Cliff at Duće, Croatia *Landslide and Flood Hazard Assessment, Proceedings of the 1st RSLABR with the 3rd Workshop of the Japanese - Croatian Project* (Zagreb) pp 51–55
[16] Török Á, Bögölz G, Cyinder B, Görög P, Kleb B, Vásárhelyi B, Lovas T, Barsi Á, Molnár B, Koppányi Z, Somogzi J Á 2016 Terrestrial laser scanner aided survey and stability analyses of rhyolite tuff cliff faces with potential rock-fall hazards, an example from Hungary *Ulusay, Reşat; Aydan, Omer; Gerçek, Hasan; Hindistan, Mehmet Ali; Tuncay, Ergün (ed.) Rock Mechanics and Rock Engineering: From the Past to the Future: Eurock 2016* (London, UK: Taylor and Francis) pp 877–881

AUTHOR INDEX

Institute of Physics
Dirac House, Temple Back
Bristol BS1 6BE UK

ISSN: 1757-8981
ISBN 978-1-7138-3619-3

CURRAN ASSOCIATES INC.
proceedings.com